# Lecture Notes in Physics

Founding Editors: W. Beiglböck, J. Ehlers, K. Hepp, H. Weidenmüller

T0156016

## The Lecture Notes in Physics

The series Lecture Notes in Physics (LNP), founded in 1969, reports new developments in physics research and teaching – quickly and informally, but with a high quality and the explicit aim to summarize and communicate current knowledge in an accessible way. Books published in this series are conceived as bridging material between advanced graduate textbooks and the forefront of research and to serve three purposes:

- to be a compact and modern up-to-date source of reference on a well-defined topic
- to serve as an accessible introduction to the field to postgraduate students and nonspecialist researchers from related areas
- to be a source of advanced teaching material for specialized seminars, courses and schools

Both monographs and multi-author volumes will be considered for publication. Edited volumes should, however, consist of a very limited number of contributions only. Proceedings will not be considered for LNP.

Volumes published in LNP are disseminated both in print and in electronic formats, the electronic archive being available at springerlink.com. The series content is indexed, abstracted and referenced by many abstracting and information services, bibliographic networks, subscription agencies, library networks, and consortia.

Proposals should be sent to a member of the Editorial Board, or directly to the managing editor at Springer:

Christian Caron
Springer Heidelberg
Physics Editorial Department I
Tiergartenstrasse 17
69121 Heidelberg / Germany
christian.caron@springer.com

C. Gattringer
C.B. Lang

# Quantum Chromodynamics on the Lattice

An Introductory Presentation

 Springer

Christof Gattringer
Universität Graz
Institut für Physik
Universitätsplatz 5
8010 Graz
Austria
christof.gattringer@uni-graz.at

Christian B. Lang
Universität Graz
Institut für Physik
Universitätsplatz 5
8010 Graz
Austria
christian.lang@uni-graz.at

Gattringer C., Lang C.B., *Quantum Chromodynamics on the Lattice: An Introductory Presentation*, Lect. Notes Phys. 788 (Springer, Berlin Heidelberg 2010),
DOI 10.1007/978-3-642-01850-3

Lecture Notes in Physics ISSN  0075-8450          e-ISSN  1616-6361
ISBN  978-3-642-26095-7          e-ISBN  978-3-642-01850-3
DOI 10.1007/978-3-642-01850-3
Springer Heidelberg Dordrecht London New York

*Cover design:* Integra Software Services Pvt. Ltd., Puducherry

Printed on acid-free paper

Springer is part of Springer Science+Business Media (www.springer.com)

To our wifes,
who endured
absent-minded husbands
working on weird texts.

# Preface

Quantum chromodynamics (QCD) is the fundamental quantum field theory of quarks and gluons. In order to discuss it in a mathematically well-defined way, the theory has to be regularized. Replacing space–time by a Euclidean lattice has proven to be an efficient approach which allows for both theoretical understanding and computational analysis. Lattice QCD has become a standard tool in elementary particle physics.

As the title already says: this book is introductory! The text is intended for newcomers to the field, serving as a starting point. We simply wanted to have a book which we can put into the hands of an advanced student for a first reading on lattice QCD. This imaginary student brings as a prerequisite knowledge of higher quantum mechanics, some continuum quantum field theory, and basic facts of elementary particle physics phenomenology.

In view of the wealth of applications in current research the topics presented here are limited and we had to make some painful choices. We discuss QCD but omit most other lattice field theory applications like scalar theories, gauge–Higgs models, or electroweak theory. Although we try to lead the reader up to present day understanding, we cannot possibly address all ongoing activities, in particular concerning the role of QCD in electroweak theory. Subjects like glueballs, topological excitations, and approaches like chiral perturbation theory are mentioned only briefly. This allows us to cover the other topics quite explicitly, including detailed derivations of key equations. The field is rapidly developing. The proceedings of the annual lattice conferences provide information on newer directions and up-to-date results.

As usual, completing the book took longer than originally planned and we thank our editor Claus Ascheron for his patience. We are very grateful to many of our colleagues, who offered to read one or the other piece. In particular we want to thank Vladimir Braun, Dirk Brömmel, Tommy Burch, Stefano Capitani, Tom DeGrand, Stephan Dürr, Georg Engel, Christian Hagen, Leonid Glozman, Meinulf Göckeler, Peter Hasenfratz, Jochen Heitger, Verena Hermann, Edwin Laermann, Markus Limmer, Pushan Majumdar, Daniel Mohler, Wolfgang Ortner, Bernd-Jochen Schaefer, Stefan Schaefer,

Andreas Schäfer, Erhard Seiler, Stefan Sint, Stefan Solbrig, and Pierre van Baal.

It would be surprising if there were not mistakes in this text. We therefore set up a web companion to this book: `http://physik.uni-graz.at/qcdlatt/` On that page we document errata and provide further links and information.

Graz,                                                    *Christof Gattringer*
March 2009                                              *Christian B. Lang*

# Contents

# 1

# The path integral on the lattice

The basic tool for quantizing fields on the lattice is the Euclidean path integral. Our first chapter is dedicated to the introduction of the path integral formalism and to its interpretation. In order to develop the idea without getting lost in technicalities we introduce the path integral for the simplest case, a scalar field theory. We derive and discuss the two key equations of lattice field theory. The first key equation is

$$\lim_{T \to \infty} \frac{1}{Z_T} \operatorname{tr} \left[ e^{-(T-t)\hat{H}} \hat{O}_2 \, e^{-t\hat{H}} \hat{O}_1 \right] = \sum_n \langle 0|\hat{O}_2|n\rangle \langle n|\hat{O}_1|0\rangle \, e^{-t\,E_n} \,, \quad (1.1)$$

where $Z_T$ is a normalization factor given by $Z_T = \operatorname{tr}[e^{-T\hat{H}}]$. The left-hand side of (1.1) is the Euclidean correlation function of two operators $\hat{O}_1, \hat{O}_2$ and $\hat{H}$ is the Hamiltonian of the system. On the right-hand side the Euclidean correlator is expressed as a sum over eigenstates of the Hamiltonian operator labeled by $n$. The terms in the sum contain matrix elements of the operators $\hat{O}_i$ taken between the vacuum $|0\rangle$ and the physical states $|n\rangle$. These matrix elements are weighted with exponentials containing the energy eigenvalues $E_n$ of the system. The right-hand side of (1.1) can thus be used to extract matrix elements of operators as well as the energy spectrum of the theory.

In the second key equation

$$\frac{1}{Z_T} \operatorname{tr} \left[ e^{-(T-t)\hat{H}} \hat{O}_2 \, e^{-t\hat{H}} \hat{O}_1 \right] = \frac{1}{Z_T} \int \mathcal{D}[\Phi] \, e^{-S_E[\Phi]} \, O_2[\Phi(.\,,t)] \, O_1[\Phi(.\,,0)] \,,$$

$$(1.2)$$

the Euclidean correlator on the left-hand side is expressed as a path integral, which is an integral over all possible configurations of the field $\Phi$. This is a crucial point. The left-hand side is formulated in the operator language of quantum field theory. The right-hand side knows nothing about field operators. In the integrand the two operators $\hat{O}_i$ are translated to functionals $O_i$ of the fields and then weighted with the Boltzmann factor containing the classical Euclidean action $S_E[\Phi]$. The right-hand side of (1.2) can be evaluated numerically on the lattice.

Gattringer, C., Lang, C.B.: *The Path Integral on the Lattice*. Lect. Notes Phys. **788**, 1–23 (2010)
DOI 10.1007/978-3-642-01850-3_1        © Springer-Verlag Berlin Heidelberg 2010

## 1.1 Hilbert space and propagation in Euclidean time

This first section is dedicated to a detailed discussion of the Euclidean correlators (1.1). Before we actually introduce the Euclidean correlators we prepare the ground with a brief summary of the definition and properties of Hilbert spaces (see, e.g., [1–4] for introductory texts).

### 1.1.1 Hilbert spaces

A Hilbert space $\mathcal{H}$ is an infinite dimensional vector space. Its elements can be added and multiplied with scalars and $\mathcal{H}$ is closed under these operations, i.e., for vectors $|u\rangle, |v\rangle \in \mathcal{H}$ and complex numbers $\alpha, \beta$ we find

$$\alpha |u\rangle + \beta |v\rangle \in \mathcal{H} . \tag{1.3}$$

We will often refer to vectors in Hilbert space as states. In addition, a Hilbert space is equipped with a scalar product, i.e., a sesquilinear functional $\langle u|v\rangle$ which maps a vector $|v\rangle$ and a dual vector $\langle u|$ into the complex numbers. The scalar product obeys the properties (the $^*$ denotes complex conjugation)

$$\langle u|v\rangle = \langle v|u\rangle^* , \tag{1.4}$$

$$\langle w|\alpha\, u + \beta\, v\rangle = \alpha \langle w|u\rangle + \beta \langle w|v\rangle . \tag{1.5}$$

The Hilbert spaces we are interested in here have a complete, countable basis, i.e., a set of linearly independent vectors $|e_n\rangle$ such that every vector $|u\rangle \in \mathcal{H}$ can be written as a linear combination of the basis elements,

$$|u\rangle = \sum_n \alpha_n |e_n\rangle , \tag{1.6}$$

where the coefficients $\alpha_n$ are complex numbers. If the basis vectors $|e_n\rangle$ obey $\langle e_m|e_n\rangle = \delta_{mn}$ the basis is called orthonormal.

An operator $\widehat{O}$ acting on the Hilbert space $\mathcal{H}$ maps vectors onto other vectors in the Hilbert space,

$$|u\rangle \in \mathcal{H} \longrightarrow \widehat{O}|u\rangle \in \mathcal{H} . \tag{1.7}$$

The corresponding adjoint operator $\widehat{O}^\dagger$ is defined by

$$\langle u|\widehat{O}|v\rangle = \langle v|\widehat{O}^\dagger|u\rangle^* . \tag{1.8}$$

If the operator obeys $\widehat{O}^\dagger = \widehat{O}$ for all $\mathcal{H}$ it is called self-adjoint or hermitian.

An eigenvector or eigenstate $|u\rangle$ of an operator $\widehat{O}$ is a vector obeying the equation

$$\widehat{O}|u\rangle = \lambda |u\rangle , \tag{1.9}$$

for some complex number $\lambda$, the so-called eigenvalue. The eigenvalues of self-adjoint operators are real and eigenvectors with different eigenvalues are orthogonal to each other. Eigenvectors with equal eigenvalues can be made orthogonal and after normalization the eigenvectors of a self-adjoint operator form an orthonormal set. For a large class of operators the corresponding set of eigenvectors provides an orthonormal basis for the Hilbert space.[1]

The unit operator $\mathbb{1}$ can be written in terms of the vectors of a complete orthonormal basis as

$$\mathbb{1} = \sum_n |e_n\rangle\langle e_n| \,. \tag{1.10}$$

This is also called the completeness relation. Finally, we define the trace $\mathrm{tr}[\widehat{O}]$ of an operator as

$$\mathrm{tr}[\widehat{O}] = \sum_n \langle e_n|\widehat{O}|e_n\rangle \,, \tag{1.11}$$

where the sum runs over the vectors of an orthonormal basis. Operators in which the trace exists are called *trace class*. We stress that the trace is invariant under a change of the basis used for its evaluation since different orthonormal bases are related by unitary transformations.

In some of the calculations in this book we will use non-countable complete sets of states, such as momentum eigenstates. For these sets the sums in (1.6), (1.10), and (1.11) are replaced by integrals.

### 1.1.2 Remarks on Hilbert spaces in particle physics

Let us add a few remarks about the Hilbert spaces relevant to particle physics. We do not attempt a systematic construction of these Hilbert spaces (see, e.g., [4]) but concentrate on discussing some properties needed here.

An important feature of Hilbert spaces in particle physics is the fact that their vectors can be multiparticle states. The underlying physical reason is that, when relativistic field theories are used to describe a physical system, particle creation and annihilation processes take place. In order to incorporate this feature, the Hilbert space is a direct sum of 0-particle, 1-particle, 2-particle, etc., Hilbert spaces. Such a Hilbert space is known as a *Fock space*.

A typical example of a vector in a Fock space is a state with a pion at some position $x$ and a proton at some other position $y$ (a 2-particle state). An example of a 1-particle state is a state with a single $\Delta$-baryon at the origin. States of particles may be described by their quantum numbers and their position, but also the particle momentum can be used to characterize the state. An example of such a case would be a state consisting of two pions with momenta $p$ and $-p$.

---

[1]We remark that our short summary of Hilbert spaces focuses on the algebraic properties we actually need in subsequent calculations. For more subtle aspects such as the completeness of an eigensystem or the discussion of eigenvectors of the continuous spectrum we refer the reader to the more mathematical literature [5, 6].

The 0-particle state is referred to as the vacuum and is denoted by $|0\rangle$. We remark that in some situations, such as spontaneous breaking of symmetries, the vacuum state is not necessarily unique.

An important class of operators are field operators creating and annihilating particles. In many cases the adjoint of a field operator is the field operator for the corresponding antiparticle. An example which illustrates this feature is the pion system: Let $\widehat{O}^\dagger_{\pi^+}$ be the operator that creates a state with the quantum numbers of the $\pi^+$ meson (actually this operator is constructed as a product of quark and antiquark field operators). The adjoint $\widehat{O}_{\pi^+}$ of this operator either annihilates that state or creates a state with the quantum numbers of its antiparticle, the $\pi^-$ meson. We will discuss the construction of hadron operators and their properties in more detail in Chap. 6.

We remark that all the states and operators we have discussed so far are time independent. Their evolution in Euclidean time will be discussed now.

### 1.1.3 Euclidean correlators

Having prepared the ground we can now start to discuss Euclidean correlators. Let us begin with defining the Euclidean correlator $\langle O_2(t)\, O_1(0)\rangle_T$ by

$$\langle O_2(t)\, O_1(0)\rangle_T = \frac{1}{Z_T}\, \mathrm{tr}\left[\mathrm{e}^{-(T-t)\widehat{H}}\, \widehat{O}_2\, \mathrm{e}^{-t\widehat{H}}\, \widehat{O}_1\right], \qquad (1.12)$$

where the normalization factor $Z_T$ is given by

$$Z_T = \mathrm{tr}\left[\mathrm{e}^{-T\widehat{H}}\right]. \qquad (1.13)$$

Later we will show that in the Euclidean path integral formalism the normalization factor $Z_T$ is a partition function of a system of statistical mechanics. Here $\widehat{O}_1$ and $\widehat{O}_2$ can be operators that create or annihilate states, or operators that measure observables or combinations of all of these. The self-adjoint operator $\widehat{H}$ is the Hamiltonian operator of the system which measures the energy in the system and also governs the time evolution. The parameters $T$ and $t$ are real, non-negative numbers denoting Euclidean time distances of propagation. Whereas $t$ is the actual distance of interest to us, $T$ is a formal maximal distance, which will eventually be taken to infinity. The exponentials of the Hamiltonian operator can in most cases (bounded operators) be defined through their power series expansion

$$\mathrm{e}^{-t\widehat{H}} = \sum_{j=0}^{\infty} \frac{(-t)^j}{j!}\, \widehat{H}^j. \qquad (1.14)$$

Let us evaluate the partition function $Z_T$. According to the definition (1.11) of the trace we need to sandwich the operator between vectors of an orthonormal basis and then sum over all basis vectors. The result is independent of which particular orthonormal basis one uses. For the trace in (1.13) a natural choice is the basis of the eigenstates of $\widehat{H}$. Denoting the eigenstates of $\widehat{H}$ as $|n\rangle$, the corresponding eigenvalue equation reads

$$\widehat{H}\,|n\rangle = E_n\,|n\rangle\,. \tag{1.15}$$

The energy eigenvalues $E_n$ are real numbers and, assuming discrete spectrum, we order the states such that

$$E_0 \leq E_1 \leq E_2 \leq E_3\ldots \tag{1.16}$$

The index $n$ labels all the combinations of quantum numbers describing the states such that their energies are ordered according to (1.16).

When using the basis $|n\rangle$ for evaluating the trace in (1.13) one finds

$$Z_T = \sum_n \langle n|e^{-T\,\widehat{H}}|n\rangle = \sum_n e^{-T\,E_n}\,, \tag{1.17}$$

where in the second step we used (1.15) and the normalization of the eigenstates, $\langle n|n\rangle = 1$. The partition function $Z_T$ is simply a sum over the exponentials of all energy eigenvalues $E_n$.

The Euclidean correlator $\langle O_2(t)\,O_1(0)\rangle_T$ can be evaluated in a similar way. When computing the trace as before and inserting the unit operator in the form (1.10) to the right of $\widehat{O}_2$ we obtain

$$\langle O_2(t)\,O_1(0)\rangle_T = \frac{1}{Z_T}\sum_{m,n}\langle m|e^{-(T-t)\,\widehat{H}}\widehat{O}_2|n\rangle\langle n|e^{-t\,\widehat{H}}\,\widehat{O}_1|m\rangle$$

$$= \frac{1}{Z_T}\sum_{m,n} e^{-(T-t)\,E_m}\,\langle m|\widehat{O}_2|n\rangle\,e^{-t\,E_n}\,\langle n|\widehat{O}_1|m\rangle\,. \tag{1.18}$$

The next step is to insert the expression (1.17) for $Z_T$ into the last equation and to pull out a factor of $e^{-T\,E_0}$ both in the numerator and in the denominator. This leads to

$$\langle O_2(t)\,O_1(0)\rangle_T = \frac{\sum_{m,n}\langle m|\widehat{O}_2|n\rangle\,\langle n|\widehat{O}_1|m\rangle\,e^{-t\,\Delta E_n}\,e^{-(T-t)\,\Delta E_m}}{1 + e^{-T\,\Delta E_1} + e^{-T\,\Delta E_2} + \ldots}\,, \tag{1.19}$$

where we defined

$$\Delta E_n = E_n - E_0\,. \tag{1.20}$$

Thus, the Euclidean correlator $\langle O_1(0)\,O_2(t)\rangle_T$ depends only on the energies normalized relative to the energy $E_0$ of the vacuum. It is exactly these energy differences that can be measured in an experiment. For convenience from now on we use $E_n$ to denote the energy differences relative to the vacuum, i.e., we use $E_n$ instead of $\Delta E_n$. This implies that the energy $E_0$ of the vacuum $|0\rangle$ is normalized to zero.

We now analyze (1.19) in the limit $T \to \infty$. In this limit (assuming that the vacuum is unique and $E_1 > 0$) the denominator is equal to 1. Similarly, in the numerator only those terms where $E_m = 0$ survive, i.e., only the contributions where $|m\rangle = |0\rangle$. We thus obtain

$$\lim_{T \to \infty} \langle O_2(t)\, O_1(0) \rangle_T = \sum_n \langle 0|\widehat{O}_2|n\rangle \langle n|\widehat{O}_1|0\rangle\, \mathrm{e}^{-t\, E_n} \,. \qquad (1.21)$$

The expression which we have derived here will be central for the interpretation of lattice field theory. It is a sum of exponentials and each exponent corresponds to an energy level. These energies can be calculated as follows: Let us assume that one wants to compute the energy of a proton $p$. For $\widehat{O}_1$ one chooses an operator $\widehat{O}_p^\dagger$ which creates from the vacuum a state with the quantum numbers of the proton and for $\widehat{O}_2$ the adjoint of that operator, the operator $\widehat{O}_p$ that annihilates the proton. Then the matrix element $\langle n|\widehat{O}_p^\dagger|0\rangle$ vanishes for all states $\langle n|$ that do not have the quantum numbers of the proton. The first state $\langle n|$ where we find a contribution is the state describing a proton, i.e., $\langle n| = \langle p|$. Further up in the ladder of states one finds excited proton states $\langle p'|$ , $\langle p''|$, ..., which also have nonvanishing overlap with $\widehat{O}_p^\dagger|0\rangle$. We thus obtain

$$\lim_{T \to \infty} \langle O_p(t)\, O_p(0)^\dagger \rangle_T = |\langle p|\widehat{O}_p^\dagger|0\rangle|^2\, \mathrm{e}^{-t\, E_p} + |\langle p'|\widehat{O}_p^\dagger|0\rangle|^2\, \mathrm{e}^{-t\, E_{p'}} + \dots \,, \;\; (1.22)$$

where we made use of (1.8) to simplify the coefficients in front of the exponentials. For sufficiently large $t$ the sub-leading terms are strongly suppressed since $E_{p'} > E_p$. Therefore, at sufficiently large $t$ we obtain $E_p$ from the exponential decay of the Euclidean correlator. When using other operators $\widehat{O}_1, \widehat{O}_2$ that create other states from the vacuum, we can extract the ground state energies for all particles in the spectrum of the theory in a way similar to what we have just demonstrated for the proton. Furthermore, one can also study the matrix elements $\langle n|\widehat{O}|0\rangle$. Since $\widehat{O}$ can be a product of several operators, a wide range of matrix elements is accessible. Such techniques will be discussed in great detail in Chaps. 6 and 11.

This completes our discussion of the properties of Euclidean correlators. In the next section we introduce the path integral as a technique for calculating the correlators (1.12).

Let us conclude this section with a short remark on the relation of our Euclidean correlators to real time, Minkowski quantum mechanics. In the Heisenberg picture, where the operators are time dependent, an operator at time $\tau$ is given by (note that we use $\hbar \equiv 1$)

$$\widehat{O}(\tau) = \mathrm{e}^{\mathrm{i}\tau\widehat{H}}\, \widehat{O}\, \mathrm{e}^{-\mathrm{i}\tau\widehat{H}} \,. \qquad (1.23)$$

Thus, the Euclidean correlator (1.12) can be seen as the correlator of operator $\widehat{O}_1$ at time 0 with operator $\widehat{O}_2$ at imaginary time $t = \mathrm{i}\tau$ times an extra time transporter $\exp(-T\widehat{H})$ which projects to the vacuum for $T \to \infty$. Four-vectors with real-time $\tau$ correspond to Minkowski metric. When one switches to imaginary time $t = i\tau$, the relative minus sign between time- and space-components vanishes and the metric becomes Euclidean; thus, the name Euclidean correlators for the objects defined in (1.12).

In the next section, where we discuss the path integral representation of the Euclidean correlator, we will encounter a similar transformation of the metric. The change from real to imaginary time is often referred to as *Wick rotation*. We stress, however, that it is not essential to interpret $t$ as time at all. The expression (1.12) can be simply viewed as a convenient mathematical tool which allows one to extract energy levels and matrix elements of operators.

## 1.2 The path integral for a quantum mechanical system

The Euclidean correlators (1.12) which we have introduced and analyzed in Hilbert space in the last section are now rewritten as path integrals – a concept introduced by Feynman. In Sect. 1.3 we will derive the second key equation of lattice field theory, (1.2). However, as a preparatory step, in this section we first derive the path integral for a simple quantum mechanical system. This allows us to focus on the essential steps of the construction without too much technicalities. In Sect. 1.3 we will repeat the derivation of the path integral for a scalar field theory.

The quantum mechanical system we use in our initial presentation of the path integral is a single particle in a potential $U$. To simplify things further we restrict the motion of the particle to the $x$-axis only. The system is described by the Hamiltonian

$$\widehat{H} = \frac{\widehat{p}^2}{2m} + \widehat{U} \ . \tag{1.24}$$

The canonical quantization condition is given by the commutator

$$[\widehat{x}, \widehat{p}] = i \ , \tag{1.25}$$

(we use $\hbar \equiv 1$) and implies that the momentum operator is given by

$$\widehat{p} = -i\frac{d}{dx} \ . \tag{1.26}$$

For this system we want to compute the partition function according to (1.13)

$$Z_T = \text{tr}\left[e^{-T\widehat{H}}\right] = \int dx \, \langle x | e^{-T\widehat{H}} | x \rangle \ . \tag{1.27}$$

The second step in this equation already expresses our intent to use position states $|x\rangle$ to compute the trace.

Let us begin with the evaluation of the necessary matrix elements for the free case:

$$\langle x|e^{-t\,\widehat{p}^2/2m}|y\rangle = \int \mathrm{d}p\,\langle x|e^{-t\,\widehat{p}^2/2m}|p\rangle\langle p|y\rangle = \int \mathrm{d}p\,\langle x|p\rangle\langle p|y\rangle\,e^{-t\,p^2/2m}\,.$$

(1.28)

In the first step we have inserted the unit operator according to (1.10) using the momentum eigenstates $|p\rangle$. In the second step we have used $\widehat{p}|p\rangle = p|p\rangle$. Note that in the exponent on the right-hand side of (1.28), $p$ is no longer an operator but a number – the momentum.

Using the real space representation of the momentum eigenstates, i.e., the plane waves

$$\langle x|p\rangle = \frac{1}{\sqrt{2\pi}}\,e^{ipx}\,,$$

(1.29)

we find for our matrix element

$$\langle x|e^{-t\,\widehat{p}^2/2m}|y\rangle = \frac{1}{2\pi}\int \mathrm{d}p\,e^{ip(x-y)}\,e^{-t\,p^2/2m} = \sqrt{\frac{m}{2\pi t}}\,e^{-(x-y)^2\,m/2t}\,.$$

(1.30)

In the last step we used the well-known Gaussian integral ($c$, $b$ real, $c > 0$)

$$\int_{-\infty}^{\infty}\mathrm{d}x\,e^{-c\,x^2}\,e^{\pm i\,b\,x} = \sqrt{\frac{\pi}{c}}\,e^{-b^2/4c}\,.$$

(1.31)

Let us now look at the general case with nonvanishing potential $\widehat{U}$. The problem here is that the two terms in the Hamiltonian (1.24) do not commute with each other, making the evaluation of the exponential of $\widehat{H}$ somewhat involved. We split the potential $\widehat{U}$ symmetrically, and for an infinitesimal Euclidean timestep $\varepsilon$ we write

$$e^{-\varepsilon\widehat{H}} = e^{-\varepsilon\widehat{U}/2}\,e^{-\varepsilon\,\widehat{p}^2/2m}\,e^{-\varepsilon\widehat{U}/2}\,(1 + \mathcal{O}(\varepsilon))\,.$$

(1.32)

This formula can be obtained from expanding both sides in $\varepsilon$. We abbreviate the leading term as

$$\widehat{W}_\varepsilon = e^{-\varepsilon\widehat{U}/2}\,e^{-\varepsilon\,\widehat{p}^2/2m}\,e^{-\varepsilon\widehat{U}/2}\,,$$

(1.33)

and find

$$\langle x|\widehat{W}_\varepsilon|y\rangle = e^{-\varepsilon U(x)/2}\,\langle x|e^{-\varepsilon\,\widehat{p}^2/2m}|y\rangle\,e^{-\varepsilon U(y)/2}$$
$$= \sqrt{\frac{m}{2\pi\varepsilon}}\,e^{-\varepsilon U(x)/2}\,e^{-\varepsilon U(y)/2}\,e^{-(x-y)^2\,m/2\varepsilon}\,.$$

(1.34)

In the first step we have used $\widehat{U}|x\rangle = U(x)|x\rangle$, where $U(x)$ is a real number – the value of the potential at $x$. In the second step we inserted (1.30).

We can build up a finite Euclidean timestep $T$ from infinitesimal steps $\varepsilon$ using the Trotter formula (see, e.g., [7] for a proof),

$$e^{-T\widehat{H}} = \lim_{N_T\to\infty}\widehat{W}_\varepsilon^{N_T}\quad\text{with}\quad \varepsilon = \frac{T}{N_T}\,.$$

(1.35)

We obtain

$$Z_T = \int dx_0 \langle x_0 | e^{-T\hat{H}} | x_0 \rangle = \lim_{N_T \to \infty} \int dx_0 \langle x_0 | \widehat{W}_\varepsilon^{N_T} | x_0 \rangle \tag{1.36}$$

$$= \lim_{N_T \to \infty} \int dx_0 ... dx_{N_T-1} \langle x_0 | \widehat{W}_\varepsilon | x_1 \rangle ... \langle x_{N_T-2} | \widehat{W}_\varepsilon | x_{N_T-1} \rangle \langle x_{N_T-1} | \widehat{W}_\varepsilon | x_0 \rangle$$

$$= \lim_{N_T \to \infty} C^{N_T} \int dx_0 ... dx_{N_T-1} \exp\left(-\varepsilon \sum_{j=0}^{N_T-1} \left(\frac{m}{2} \frac{(x_j - x_{j+1})^2}{\varepsilon^2} + U(x_j)\right)\right).$$

In this derivation we have inserted unit operators in the form $\int dx_i |x_i\rangle\langle x_i|$ and then used the result (1.34) with the abbreviation $C = \sqrt{m/(2\pi\varepsilon)}$.

Our expression (1.36) contains the limit $N_T \to \infty$. Since we are aiming at a numerical evaluation of our path integral with a computer, the number of steps $N_T$ we can use has to be finite. Hence we define an approximation of the partition function (we periodically identify $x_{N_T} \equiv x_0$)

$$Z_T^\varepsilon = C^{N_T} \int dx_0 \, ... \, dx_{N_T-1} \exp\left(-\varepsilon \sum_{j=0}^{N_T-1} \left(\frac{m}{2}\left(\frac{x_{j+1} - x_j}{\varepsilon}\right)^2 + U(x_j)\right)\right), \tag{1.37}$$

where the stepsize $\varepsilon$ and the Euclidean time are related by $T = \varepsilon N_T$.

The sum in the exponent has an interesting interpretation. Obviously, for smooth paths

$$\frac{x_{j+1} - x_j}{\varepsilon} = \dot{x}(t) + \mathcal{O}(\varepsilon) \text{ with } t = j\,\varepsilon,$$

$$\varepsilon \sum_{j=0}^{N_T-1} ... = \int_0^T dt ... + \mathcal{O}(\varepsilon) \text{ with } T = N_T\,\varepsilon. \tag{1.38}$$

Thus we can identify

$$\varepsilon \sum_{j=0}^{N_T-1} \left(\frac{m}{2}\left(\frac{x_{j+1} - x_j}{\varepsilon}\right)^2 + U(x_j)\right) = \int_0^T dt \left(\frac{m}{2}\dot{x}(t)^2 + U(x(t))\right) + \mathcal{O}(\varepsilon). \tag{1.39}$$

The expression on the right-hand side is called the Euclidean action $S_E$. It is obtained from the usual action $S$ when switching from real-time $\tau$ to imaginary-time $t = i\tau$ and rotating the contour of integration:

$$S[x, \dot{x}] = \int_0^T d\tau \left(\frac{m}{2}\dot{x}(\tau)^2 - U(x(\tau))\right) \to i \int_0^T dt \left(\frac{m}{2}\dot{x}(t)^2 + U(x(t))\right) = iS_E[x, \dot{x}]. \tag{1.40}$$

Let us summarize: We have found that we can construct an approximation $Z_T^\varepsilon$ of the partition function in the form of (1.37). In this approximation we

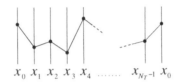

$x_0\ x_1\ x_2\ x_3\ x_4$ ........ $x_{N_T-1}\ x_0$

**Fig. 1.1.** A discretized path contributing in (1.37)

divide the Euclidean time interval $T$ into $N_T$ steps of size $\varepsilon$ (using periodic boundary conditions). At each step we insert a variable $x_j$ which we integrate from $-\infty$ to $+\infty$. The collection of values $x_j$ can be interpreted as a path (compare Fig. 1.1) and the integral is over all possible paths. The integrand is the exponential (1.37) of the Euclidean action for the discretized path.

In the next section we will develop a path integral representation for a system much closer to our target theory QCD. We will consider a scalar field theory. The technical steps of the construction – free case, infinitesimal timesteps, Trotter formula – will be the same. However, since we are dealing with a quantum field theory with many degrees of freedom, the calculation, in particular the corresponding notation, will be a little bit more involved.

## 1.3 The path integral for a scalar field theory

We now derive the key equation (1.2) for a scalar field theory. Although this is already a field theory, it has a simpler structure than QCD. This allows us to focus on the central points of the derivation of the path integral for a field theory without having to worry too much about technicalities. Although our presentation is self contained we recommend [8, 9] for an introductory reading about the canonical quantization of the scalar field, which is our starting point.

### 1.3.1 The Klein–Gordon field

Let us begin by recalling the classical action and equations of motion of the scalar field in Minkowski space. We consider a real scalar field $\Phi(t, \boldsymbol{x})$. The corresponding action $S$ is an integral over space-time

$$S = \int \mathrm{d}t\,\mathrm{d}^3x\; L\left(\Phi(t, \boldsymbol{x})\,,\, \partial_\mu \Phi(t, \boldsymbol{x})\right)\,, \tag{1.41}$$

with the Lagrangian density $L$ given by

$$\begin{aligned}
L(\Phi, \partial_\mu \Phi) &= \frac{1}{2}\,(\partial_\mu \Phi)(\partial^\mu \Phi) - \frac{m^2}{2}\,\Phi^2 - V(\Phi) \\
&= \frac{1}{2}\,\dot{\Phi}^2 - \frac{1}{2}\,(\nabla \Phi)^2 - \frac{m^2}{2}\,\Phi^2 - V(\Phi)\,. \tag{1.42}
\end{aligned}$$

This is a system of coupled oscillators. We stress again that this expression is in Minkowski space, i.e., in the second step we used the metric

$g_{\mu\nu} = \text{diag}(1, -1, -1, -1)$ to rewrite the kinetic term. In the action we also allow for a potential term $V(\Phi)$. A standard example is $V(\Phi(t, \boldsymbol{x})) = \lambda \Phi(t, \boldsymbol{x})^4$. Using the Euler–Lagrange equations

$$\partial_\mu \left( \frac{\partial L}{\partial(\partial_\mu \Phi)} \right) - \frac{\partial L}{\partial \Phi} = 0 \; , \tag{1.43}$$

we can derive the classical equation of motion

$$\partial_\mu(\partial^\mu \Phi) + m^2 \, \Phi + V'(\Phi) = 0 \; . \tag{1.44}$$

In the absence of a potential this is the Klein–Gordon equation.

As a next step we quantize the system using the canonical formalism. For this purpose we need the Hamiltonian which we obtain from the Lagrangian density. The canonical momentum $\Pi(t, \boldsymbol{x})$ is defined by

$$\Pi(t, \boldsymbol{x}) = \frac{\partial}{\partial \dot\Phi(t, \boldsymbol{x})} L\left(\Phi(t, \boldsymbol{x}) \, , \, \partial_\mu \Phi(t, \boldsymbol{x})\right) = \dot\Phi(t, \boldsymbol{x}) \; , \tag{1.45}$$

and the Hamiltonian function $H$ is obtained as the Legendre transform

$$H = \int d^3x \, \Pi(t, \boldsymbol{x}) \, \dot\Phi(t, \boldsymbol{x}) - \int d^3x \, L\left(\Phi(t, \boldsymbol{x}) \, , \, \partial_\mu \Phi(t, \boldsymbol{x})\right) \tag{1.46}$$

$$= \int d^3x \, \left( \frac{1}{2} \, \Pi(t, \boldsymbol{x})^2 + \frac{1}{2} \, (\nabla \Phi(t, \boldsymbol{x}))^2 + \frac{m^2}{2} \, \Phi(t, \boldsymbol{x})^2 + V\left(\Phi(t, \boldsymbol{x})\right) \right) \; .$$

The Hamiltonian function is the starting point for the quantization of the system in the continuum. Upon quantization, the Hamiltonian function $H$ and the fields $\Phi(t, \boldsymbol{x})$ and $\Pi(t, \boldsymbol{x})$ turn into Schrödinger operators $\widehat{H}, \widehat\Phi(\boldsymbol{x}), \widehat\Pi(\boldsymbol{x})$. The Hamiltonian operator reads (up to normal ordering)

$$\widehat{H} = \int d^3x \, \left( \frac{1}{2} \, \widehat\Pi(\boldsymbol{x})^2 + \frac{1}{2} \, \left(\nabla\widehat\Phi(\boldsymbol{x})\right)^2 + \frac{m^2}{2} \, \widehat\Phi(\boldsymbol{x})^2 + V\left(\widehat\Phi(\boldsymbol{x})\right) \right) \; . \tag{1.47}$$

All time arguments are gone, and the time evolution of operators is given by (1.23). The operators $\widehat\Phi$, $\widehat\Pi$ obey the canonical equal time commutation relations ($\hbar \equiv 1$)

$$\left[\widehat\Phi(\boldsymbol{x}), \widehat\Pi(\boldsymbol{y})\right] = i \, \delta(\boldsymbol{x} - \boldsymbol{y}) \; , \quad \left[\widehat\Phi(\boldsymbol{x}), \widehat\Phi(\boldsymbol{y})\right] = 0 \; , \quad \left[\widehat\Pi(\boldsymbol{x}), \widehat\Pi(\boldsymbol{y})\right] = 0 \; . \tag{1.48}$$

We now leave the canonical path and follow a different strategy by introducing the lattice as a cutoff before we quantize the system.

### 1.3.2 Lattice regularization of the Klein–Gordon Hamiltonian

In order to calculate something from a quantum field theory in a mathematically well-defined way, an ultraviolet regulator must be introduced. Such a

regulator is necessary to make the expressions finite. In continuum perturbation theory the regulator can, e.g., be introduced by using dimensional
regularization, Pauli–Villars regularization or a momentum cutoff. A different
approach is chosen in lattice field theory.

The central idea is to replace the continuous space by a 3D finite lattice
$\Lambda_3$:

$$x \quad \Rightarrow \quad a\,n\,, \quad n_i = 0, 1, \ldots, N-1 \quad \text{for} \quad i = 1, 2, 3\,. \tag{1.49}$$

The lattice constant $a$ has the physical dimension of length. The vector $n$ with
integer-valued components, $n_1, n_2, n_3$, labels the lattice sites. The operators
$\widehat{\Phi}(n)$ and $\widehat{\Pi}(n)$ now live only on the lattice sites and we stress this fact by
replacing the space argument $x$ by the label of the lattice site $n$. Since each
component $n_i$ runs from 0 to $N-1$, we have a total of $N^3$ lattice sites and
the lattice system thus has $2N^3$ degrees of freedom ($\widehat{\Phi}$ and $\widehat{\Pi}$ on each lattice
site). For any finite system one has to decide what to do at the boundary. We
choose periodic boundary conditions, i.e., we identify $n_j = N$ with $n_j = 0$.

For a lattice regularization of the Hamiltonian operator (1.47) we also
need to discretize the derivatives $\nabla\widehat{\Phi}(x)$. Using the Taylor series expansion of
$\widehat{\Phi}(x)$, it is easy to see that for small lattice constant $a$ the derivative may be
approximated as

$$\partial_j \widehat{\Phi}(x) = \frac{\widehat{\Phi}(n+\hat{j}) - \widehat{\Phi}(n-\hat{j})}{2a} + \mathcal{O}(a^2)\,, \tag{1.50}$$

where $\hat{j}$ denotes the unit vector in the $j$-direction. Using this definition for
the derivative, we obtain a lattice version of the Hamiltonian operator (1.47):

$$\widehat{H} = a^3 \sum_{n\in\Lambda_3} \left( \frac{1}{2}\widehat{\Pi}(n)^2 + \frac{1}{2}\sum_{j=1}^{3}\left(\frac{\widehat{\Phi}(n+\hat{j})-\widehat{\Phi}(n-\hat{j})}{2a}\right)^2 + \frac{m^2}{2}\widehat{\Phi}(n)^2 + V\left(\widehat{\Phi}(n)\right)\right). \tag{1.51}$$

The integral $\int d^3x$ has been replaced by the sum $a^3\sum_{n\in\Lambda_3}$. The Hamiltonian describes a system with a finite number of degrees of freedom $\widehat{\Phi}(n)$ and
their corresponding conjugate momenta $\widehat{\Pi}(n)$. The canonical quantization
condition is given by the commutators

$$\left[\widehat{\Phi}(n)\,,\,\widehat{\Pi}(m)\right] = \mathrm{i}\,a^{-3}\,\delta_{nm}\,, \tag{1.52}$$

$$\left[\widehat{\Phi}(n)\,,\,\widehat{\Phi}(m)\right] = 0\,, \quad \left[\widehat{\Pi}(n)\,,\,\widehat{\Pi}(m)\right] = 0\,. \tag{1.53}$$

When comparing the new quantization conditions (1.52) and (1.53) to their
continuum counterparts one sees that the Dirac-delta in (1.48) is replaced by
Kronecker deltas $\delta_{nm} = \delta_{n_1 m_1}\,\delta_{n_2 m_2}\,\delta_{n_3 m_3}$, as is usual for discrete degrees of
freedom. The factor $a^{-3}$ on the right-hand side of (1.52) is needed for getting

the dimensions right. When denoting the dimension of length by $\ell$ we find for the dimensions [...] of the quantities involved (using $\hbar \equiv 1$):

$$[S] = 1, \ [H] = \frac{1}{\ell}, \ [a] = \ell, \ [\widehat{\Phi}] = \frac{1}{\ell}, \ [\widehat{\Pi}] = \frac{1}{\ell^2}, \ [m] = \frac{1}{\ell} \ . \tag{1.54}$$

Thus the factor $a^{-3}$ in (1.52) is necessary in order to have objects of equal dimensions on both sides. This factor also matches the dimensions in the continuum formula (1.48) since the Dirac-delta there also has dimension $\ell^{-3}$.

The commutators (1.52) and (1.53) are obeyed when the momentum operators are represented as derivatives

$$\widehat{\Pi}(\boldsymbol{n}) = -\frac{i}{a^3} \frac{\partial}{\partial \Phi(\boldsymbol{n})} \ . \tag{1.55}$$

Using this representation we can write the Hamiltonian operator for the Klein–Gordon field on the lattice in its final form as

$$\widehat{H} = \widehat{H}_0 + \widehat{U} \ , \tag{1.56}$$

$$\widehat{H}_0 = a^3 \sum_{\boldsymbol{n} \in \Lambda_3} \frac{1}{2} \left( -\frac{i}{a^3} \frac{\partial}{\partial \Phi(\boldsymbol{n})} \right)^2 = -\frac{1}{2 a^3} \sum_{\boldsymbol{n} \in \Lambda_3} \frac{\partial^2}{\partial \Phi(\boldsymbol{n})^2} \ ,$$

$$\widehat{U} = a^3 \sum_{\boldsymbol{n} \in \Lambda_3} \left( \frac{1}{2} \sum_{j=1}^{3} \left( \frac{\widehat{\Phi}(\boldsymbol{n}+\hat{j}) - \widehat{\Phi}(\boldsymbol{n}-\hat{j})}{2a} \right)^2 + \frac{m^2}{2} \widehat{\Phi}(\boldsymbol{n})^2 + V\left( \widehat{\Phi}(\boldsymbol{n}) \right) \right) \ .$$

We have split the Hamilton operator into two parts: the free part $\widehat{H}_0$, which contains the derivatives, and the interaction part $\widehat{U}$. This splitting is useful for the calculation of the Euclidean correlators.

It is convenient to introduce a set of eigenstates of the field operators $\widehat{\Phi}(\boldsymbol{n})$, equivalent to the position states $|x\rangle$ of quantum mechanics. The condition (1.52) is a generalization of the condition $[\hat{x}, \hat{p}] = i\hbar$ in ordinary quantum mechanics. In a similar way the states $|\Phi\rangle$ are a generalization of the position states $|x\rangle$. The action of the operators $\widehat{\Phi}(\boldsymbol{n})$ on $|\Phi\rangle$ is defined as

$$\widehat{\Phi}(\boldsymbol{n}) |\Phi\rangle = \Phi(\boldsymbol{n}) |\Phi\rangle \ . \tag{1.57}$$

The application of the operator $\widehat{\Phi}(\boldsymbol{n})$ on an eigenstate $|\Phi\rangle$ thus gives as the eigenvalue the value $\Phi(\boldsymbol{n})$ of the field at the lattice point $\boldsymbol{n}$. Consequently, the state $|\Phi\rangle$ is labeled by the values of the field at all lattice points, i.e., by the set $\{\Phi(\boldsymbol{n}), \boldsymbol{n} \in \Lambda_3\}$. The states $|\Phi\rangle$ are orthogonal and complete:

$$\langle \Phi'|\Phi\rangle = \delta(\Phi' - \Phi) \equiv \prod_{\boldsymbol{n} \in \Lambda_3} \delta\left( \Phi'(\boldsymbol{n}) - \Phi(\boldsymbol{n}) \right) \ , \tag{1.58}$$

$$\mathbb{1} = \int_{-\infty}^{\infty} \mathcal{D}\Phi \, |\Phi\rangle\langle\Phi| \quad \text{with} \quad \mathcal{D}\Phi = \prod_{\boldsymbol{n} \in \Lambda_3} d\,\Phi(\boldsymbol{n}) \ . \tag{1.59}$$

### 1.3.3 The Euclidean time transporter for the free case

Having discussed the Hamiltonian of the scalar field and its lattice discretization, we can now return to our original problem, the calculation of the Euclidean correlators (1.12) and the partition function (1.13). In both these expressions we encounter Euclidean time transporters $\exp(-t\widehat{H})$. For the full Hamiltonian operator $\widehat{H}$ the calculation of matrix elements of the time transporter is not possible in closed form. However, for the free part $\widehat{H}_0$ this is an easy exercise which we perform using the eigenstates of $\widehat{H}_0$.

The free Hamiltonian operator $\widehat{H}_0$, as represented in (1.56), is a sum over second derivatives, i.e., a sum over squares of momentum operators. It is well-known from quantum mechanics that the eigenstates of a derivative operator are the momentum eigenstates $|\Pi\rangle$: plane waves.[2] Since $\widehat{H}_0$ is a sum over $N^3$ derivatives, the eigenstate is a product over $N^3$ plane waves, one for each field component $\Phi(n)$. Each plane wave is characterized by a wave number $\Pi(n)$. The whole eigenstate is then labeled by the set of all these wave numbers and we denote the eigenstate as $|\Pi\rangle$, which in field representation is given by

$$\langle\Phi|\Pi\rangle = \prod_{n\in\Lambda_3} \sqrt{\frac{a^3}{2\pi}} \, e^{i\,a^3\,\Pi(n)\,\Phi(n)} \,. \tag{1.60}$$

The eigenstates in (1.60) are normalized such that the unit operator reads

$$\mathbb{1} = \int_{-\infty}^{\infty} \mathcal{D}\Pi \, |\Pi\rangle\langle\Pi| \quad \text{with} \quad \mathcal{D}\Pi = \prod_{n\in\Lambda_3} \mathrm{d}\Pi(n) \,. \tag{1.61}$$

The action of the free Hamiltonian $\widehat{H}_0$ on $|\Pi\rangle$ gives rise to

$$\langle\Phi|\widehat{H}_0|\Pi\rangle = -\frac{1}{2\,a^3} \sum_{n\in\Lambda_3} \frac{\partial^2}{\partial\Phi(n)^2} \langle\Phi|\Pi\rangle = \frac{a^3}{2} \sum_{n\in\Lambda_3} \Pi(n)^2 \langle\Phi|\Pi\rangle \,. \tag{1.62}$$

Thus the eigenvalues of $\widehat{H}_0$ are given by $(a^3/2)\sum_n \Pi(n)^2$. It is important to realize that in both Eqs. (1.60) and (1.62) the quantities $\Phi(n)$ and $\Pi(n)$ are no longer operators, but real numbers. They are the values of the field and the wave number at the lattice site $n$.

With the help of the eigenstates (1.60) we can now compute matrix elements of the Euclidean time transporter of the free theory

$$\langle\Phi'|e^{-t\widehat{H}_0}|\Phi\rangle = \int \mathcal{D}\Pi \, \langle\Phi'|e^{-t\widehat{H}_0}|\Pi\rangle\langle\Pi|\Phi\rangle$$

$$= \int \mathcal{D}\Pi \, \langle\Phi'|\Pi\rangle\langle\Pi|\Phi\rangle \, e^{-t\,a^3/2\,\sum_n \Pi(n)^2} \,, \tag{1.63}$$

where in the first step we have inserted the unit operator written in the form (1.61) and in the second step the eigenvalue equation (1.62) was used. Inserting

---

[2] These are not plane waves in real space, but plane waves of the fields $\Phi(n)$.

(1.60) into the last expression, we have a product of Gaussian integrals that can be solved:

$$
\langle \Phi' | \mathrm{e}^{-t\widehat{H}_0} | \Phi \rangle = \prod_{n \in \Lambda_3} \frac{a^3}{2\pi} \int_{-\infty}^{\infty} \mathrm{d}\Pi(n)\, \mathrm{e}^{\mathrm{i}\, a^3\, \Pi(n)\left(\Phi'(n) - \Phi(n)\right)}\, \mathrm{e}^{-t\, a^3\, \Pi(n)^2/2} =
$$

$$
\prod_{n \in \Lambda_3} \sqrt{\frac{a^3}{2\pi t}}\, \mathrm{e}^{-a^3/(2t)\left(\Phi'(n) - \Phi(n)\right)^2} = \left(\frac{a^3}{2\pi t}\right)^{N^3/2} \mathrm{e}^{-a^3/(2t)\sum_n \left(\Phi'(n) - \Phi(n)\right)^2},
$$

$$(1.64)$$

where in the last step we used the Gaussian integral (1.31).

### 1.3.4 Treating the interaction term with the Trotter formula

When we switch from the free case to including the potential $\widehat{U}$ we again have to build up a finite step $t$ in Euclidean time from infinitesimal steps using the Trotter formula (1.35). Like for the quantum mechanical system of the last section we define

$$
\widehat{W}_\varepsilon = \mathrm{e}^{-\varepsilon \widehat{U}/2}\, \mathrm{e}^{-\varepsilon \widehat{H}_0}\, \mathrm{e}^{-\varepsilon \widehat{U}/2} . \tag{1.65}
$$

Using the Trotter formula (1.35) we find

$$
\langle \Phi' | \mathrm{e}^{-t\widehat{H}} | \Phi \rangle = \lim_{n_t \to \infty} \langle \Phi' | \widehat{W}_\varepsilon^{\,n_t} | \Phi \rangle = \tag{1.66}
$$

$$
\lim_{n_t \to \infty} \int \mathcal{D}\Phi_1 \mathcal{D}\Phi_2 \dots \mathcal{D}\Phi_{n_t-1}\, \langle \Phi' | \widehat{W}_\varepsilon | \Phi_{n_t-1} \rangle \langle \Phi_{n_t-1} | \widehat{W}_\varepsilon | \Phi_{n_t-2} \rangle \dots \langle \Phi_1 | \widehat{W}_\varepsilon | \Phi \rangle .
$$

In last step we have inserted the unit operator $n_t - 1$ times in the form (1.59).
    What remains to be done is the calculation of the matrix elements $\langle \Phi_{i+1} | \widehat{W}_\varepsilon | \Phi_i \rangle$ appearing in (1.66). We first note that, due to (1.56) and (1.57),

$$
\widehat{U} | \Phi \rangle = U[\Phi]\, | \Phi \rangle , \tag{1.67}
$$

with

$$
U[\Phi] = a^3 \sum_{n \in \Lambda_3} \left( \frac{1}{2} \sum_{j=1}^{3} \left( \frac{\Phi(n+\hat{j}) - \Phi(n-\hat{j})}{2a} \right)^2 + \frac{m^2}{2} \Phi(n)^2 + V(\Phi(n)) \right) . \tag{1.68}
$$

We stress that $U[\Phi]$ is a number, i.e., the classical interaction term evaluated for the classical fields $\Phi(n)$. Using (1.64) and (1.67) we obtain

$$
\langle \Phi_{i+1} | \widehat{W}_\varepsilon | \Phi_i \rangle = \langle \Phi_{i+1} | \mathrm{e}^{-\varepsilon \widehat{U}/2} \mathrm{e}^{-\varepsilon \widehat{H}_0} \mathrm{e}^{-\varepsilon \widehat{U}/2} | \Phi_i \rangle \tag{1.69}
$$

$$
= \mathrm{e}^{-\varepsilon\, U[\Phi_{i+1}]/2}\, \langle \Phi_{i+1} | \mathrm{e}^{-\varepsilon \widehat{H}_0} | \Phi_i \rangle\, \mathrm{e}^{-\varepsilon\, U[\Phi_i]/2}
$$

$$
= C^{N^3} \mathrm{e}^{-\varepsilon\, U[\Phi_i]/2\, -a^3/(2\varepsilon)\sum_n (\Phi(n)_i - \Phi(n)_{i+1})^2\, -\, \varepsilon\, U[\Phi_{i+1}]/2} .
$$

We have introduced the abbreviation $C$ for the factor $\sqrt{a^3/2\pi\varepsilon}$. Again we emphasize that the right-hand side of (1.69) is a real number. It is the matrix element of the Euclidean time transporter for infinitesimal time $\widehat{W}_\varepsilon$. These matrix elements may be combined in (1.66) to obtain the matrix element for a transport by a finite amount of Euclidean time.

### 1.3.5 Path integral representation for the partition function

Let us now use the two Eqs. (1.66) and (1.69) to compute the partition function $Z_T$ defined in (1.13). When the trace in this equation is evaluated using states $|\Phi_0\rangle$, we find

$$Z_T = \int \mathcal{D}\Phi_0 \, \langle \Phi_0 | e^{-T\widehat{H}} | \Phi_0 \rangle = \lim_{N_T \to \infty} \int \mathcal{D}\Phi_0 \, \langle \Phi_0 | \widehat{W}_\varepsilon^{N_T} | \Phi_0 \rangle \,, \qquad (1.70)$$

with $T = N_T \varepsilon$. We have already remarked that we are aiming at implementing the path integral in a numerical simulation and thus we cannot perform the limit $N_T \to \infty$, since a computer can only cope with a finite number of degrees of freedom. Therefore, we have to work with a finite resolution $\varepsilon$ and thus obtain only an approximation of the exact partition function. We denote this approximation by $Z_T^\varepsilon$. Inserting again $N_T - 1$ sets of complete states and using the expression (1.69) for the matrix elements $\langle \Phi_{i+1} | \widehat{W}_\varepsilon | \Phi_i \rangle$ we obtain

$$\begin{aligned} Z_T^\varepsilon &= \int \mathcal{D}\Phi_0 \, \langle \Phi_0 | \widehat{W}_\varepsilon^{N_T} | \Phi_0 \rangle \\ &= \int \mathcal{D}\Phi_0 \ldots \mathcal{D}\Phi_{N_T-1} \, \langle \Phi_0 | \widehat{W}_\varepsilon | \Phi_{N_T-1} \rangle \langle \Phi_{N_T-1} | \widehat{W}_\varepsilon | \Phi_{N_T-2} \rangle \ldots \langle \Phi_1 | \widehat{W}_\varepsilon | \Phi_0 \rangle \\ &= C^{N^3 N_T} \int \mathcal{D}\Phi_0 \ldots \mathcal{D}\Phi_{N_T-1} \, e^{-S_E[\Phi]} \,, \end{aligned} \qquad (1.71)$$

with

$$S_E[\Phi] = \frac{1}{2} \sum_{j=0}^{N_T-1} a^3 \sum_{\boldsymbol{n} \in \Lambda_3} \frac{1}{\varepsilon} \left( \Phi(\boldsymbol{n})_{j+1} - \Phi(\boldsymbol{n})_j \right)^2 + \varepsilon \sum_{j=0}^{N_T-1} U[\Phi_j] \,. \qquad (1.72)$$

We remark that in the first sum we have continued the index $j$ periodically by identifying $j = N_T$ with $j = 0$.

The expression for $S_E$ in (1.72) can be simplified by introducing a more compact notation $\Phi(\boldsymbol{n}, n_4) \equiv \Phi(\boldsymbol{n})_{n_4}$. This corresponds to introducing a 4D lattice $\Lambda$ defined as

$$\Lambda = \{ (\boldsymbol{n}, n_4) \mid \boldsymbol{n} \in \Lambda_3, \, n_4 = 0, 1, \ldots, N_T - 1 \} \,, \qquad (1.73)$$

where we have also relabeled $j$ to $n_4$. Note that the lattice $\Lambda$ is periodic in all four directions since it inherits the spatial periodicity of the spatial components of $\Lambda_3$. With this notation $S_E$ reads

$$S_E[\Phi] = \varepsilon a^3 \sum_{(\boldsymbol{n},n_4)\in\Lambda} \left( \frac{1}{2} \left( \frac{\Phi(\boldsymbol{n},n_4+1) - \Phi(\boldsymbol{n},n_4)}{\varepsilon} \right)^2 + \right. \tag{1.74}$$

$$\left. \frac{1}{2} \sum_{j=1}^{3} \left( \frac{\Phi(\boldsymbol{n}+\hat{j},n_4) - \Phi(\boldsymbol{n}-\hat{j},n_4)}{2a} \right)^2 + \frac{m^2}{2} \Phi(\boldsymbol{n},n_4)^2 + V\left( \Phi(\boldsymbol{n},n_4) \right) \right) .$$

As in the quantum mechanical problem of the last section, the expression (1.74) for $S_E$ again is a discretization of the Euclidean action: The first term on the right-hand side can be interpreted as the square of a discretized derivative in 4-direction. The sum over $\Lambda$ together with the factor $\varepsilon a^3$ is a discretization of an integral over space and Euclidean time. With these two identifications the expression (1.74) is very similar to the action of the Klein–Gordon field given in (1.41) and (1.42). The only differences are an overall minus sign and a relative sign between the time derivative in (1.42) and the rest of the terms. If one rotates the time in (1.42) into the imaginary axis, i.e., one replaces $t$ by $i\tau$, then this relative minus sign is gone. Thus $S_E[\Phi]$ is the action of the system for Euclidean time – the Euclidean action.

Also the measure in (1.71) can be written in simpler form as

$$\mathcal{D}[\Phi] = \prod_{(\boldsymbol{n},n_4)\in\Lambda} d\,\Phi(\boldsymbol{n},n_4) , \tag{1.75}$$

and the expression for the partition function turns into

$$Z_T^\varepsilon = C^{N^3 N_T} \int \mathcal{D}[\Phi] \, e^{-S_E[\Phi]} . \tag{1.76}$$

### 1.3.6 Including operators in the path integral

We may evaluate the right-hand side of (1.2) analogously to the method used for the partition function $Z_T$. In particular, we again use the Trotter formula to write the numerator of the Euclidean correlator as

$$\text{tr}\left[ e^{-(T-t)\hat{H}} \hat{O}_2 \, e^{-t\hat{H}} \, \hat{O}_1 \right] = \lim_{N_T\to\infty} \text{tr}\left[ \widehat{W}_\varepsilon^{N_T-n_t} \hat{O}_2 \, \widehat{W}_\varepsilon^{n_t} \, \hat{O}_1 \right] , \tag{1.77}$$

where the number of steps $N_T$ and $n_t$ are related to $T$ and $t$ by $T = \varepsilon N_T$ and $t = \varepsilon n_t$. Similar to the approximation $Z_T^\varepsilon$ we also introduce a Euclidean correlator with a finite temporal resolution $\varepsilon$. It reads

$$\langle O_2(t)\, O_1(0)\rangle_T^\varepsilon = \frac{1}{Z_T^\varepsilon} \text{tr}\left[ \widehat{W}_\varepsilon^{N_T-n_t} \hat{O}_2 \, \widehat{W}_\varepsilon^{n_t} \, \hat{O}_1 \right] \tag{1.78}$$

$$= \frac{1}{Z_T^\varepsilon} \int \mathcal{D}\Phi_0 \ldots \mathcal{D}\Phi_{N_T-1}\mathcal{D}\tilde{\Phi}_0\mathcal{D}\tilde{\Phi}_{n_t} \langle\Phi_0|\widehat{W}_\varepsilon|\Phi_{N_T-1}\rangle\langle\Phi_{N_T-1}|\widehat{W}_\varepsilon|\Phi_{N_T-2}\rangle \ldots$$

$$\langle\Phi_{n_t+1}|\widehat{W}_\varepsilon|\tilde{\Phi}_{n_t}\rangle\langle\tilde{\Phi}_{n_t}|\hat{O}_2|\Phi_{n_t}\rangle\langle\Phi_{n_t}|\widehat{W}_\varepsilon|\Phi_{n_t-1}\rangle \ldots \langle\Phi_1|\widehat{W}_\varepsilon|\tilde{\Phi}_0\rangle\langle\tilde{\Phi}_0|\hat{O}_1|\Phi_0\rangle .$$

This structure differs from the equivalent expression (1.71) for the normalization constant $Z_T^\varepsilon$ only by the two insertions of the matrix elements

$$\langle \widetilde{\Phi}_0 | \widehat{O}_1 | \Phi_0 \rangle \quad \text{and} \quad \langle \widetilde{\Phi}_{n_t} | \widehat{O}_2 | \Phi_{n_t} \rangle \,. \tag{1.79}$$

The operators $\widehat{O}_1$ and $\widehat{O}_2$ are expressions built from field operators and their conjugate momenta. Two typical examples are

$$\widehat{O}_A = \widehat{\Phi}(n_0)^\dagger \,, \quad \widehat{O}_B = \sum_{n \in \Lambda_3} \widehat{\Phi}(n)\, e^{-i\, a\, n\, p} \,, \tag{1.80}$$

where the first operator $\widehat{O}_A$ creates from the vacuum a field quantum at some position $n_0$, and the operator $\widehat{O}_B$ annihilates a quantum projected to momentum $p$. All operators $\widehat{O}_1$ and $\widehat{O}_2$ that can appear in (1.78) can be written as expressions of field operators and conjugate momenta (1.55). Thus, we obtain for matrix elements (1.79) of combinations of field operators

$$\langle \widetilde{\Phi} | \widehat{O} | \Phi \rangle = O[\Phi]\, \delta\left( \widetilde{\Phi} - \Phi \right) \,, \tag{1.81}$$

where we have made use of (1.57) and (1.58). The object $O[\Phi]$ is no longer an operator but a functional of the classical field variables $\Phi$. A functional maps a field configuration $\Phi$, specified by the set of all field values $\Phi(n), n \in \Lambda_3$, into the complex numbers. For the two operators $\widehat{O}_A$ and $\widehat{O}_B$ in the example of (1.80) above, these functionals read ($*$ denotes complex conjugation)

$$O_A[\Phi] = \Phi(n_0)^* \,, \quad O_B[\Phi] = \sum_{n \in \Lambda_3} \Phi(n)\, e^{-i\, a\, n\, p} \,. \tag{1.82}$$

Inserting the expression (1.81) for the matrix elements of the operators into (1.78) and integrating out the fields $\widetilde{\Phi}_0$ and $\widetilde{\Phi}_{n_t}$ we find

$$\langle O_2(t)\, O_1(0) \rangle_T^\varepsilon = \frac{C^{N^3 N_T}}{Z_T^\varepsilon} \int \mathcal{D}[\Phi]\, e^{-S_E[\Phi]}\, O_2[\Phi(.\,, n_t)]\, O_1[\Phi(.\,, 0)] \,. \tag{1.83}$$

In the last step we have collected and combined the different terms, just as we did for the partition function. In particular, we obtain the same expression for the action $S_E$ as given in (1.74) and use again the measure $\mathcal{D}[\Phi]$ as defined in (1.75). The functionals $O_1[\Phi(.\,, 0)]$ and $O_2[\Phi(.\,, n_t)]$ are the lattice transcriptions of the original operators $\widehat{O}_1$ and $\widehat{O}_2$ acting in Hilbert space. The functional $O_1[\Phi(.\,, 0)]$ is evaluated for the fields $\Phi(.\,, 0)$ with time argument 0, while the functional $O_2[\Phi(.\,, n_t)]$ is evaluated for the fields $\Phi(.\,, n_t)$ with time argument $n_t$, connected to the Euclidean time $t$ via $t = \varepsilon n_t$. The spatial arguments of the fields, used for evaluating the functionals, were replaced by a dot, since the functionals map the whole field configuration, defined by the set of all field values at the given time, into the complex numbers. We stress that the functionals $O_1[\Phi(.\,, 0)]$ and $O_2[\Phi(.\,, n_t)]$ as well as the Boltzmann factor

$\exp(-S_E[\Phi])$ are numbers, not operators, and therefore their relative order is no longer crucial. Let us finally remark that the factor $C^{N^3 N_T}$ in (1.83) is the same factor that also shows up in the expression (1.76) for the partition function. The two factors in the denominator and the numerator cancel and we can omit them from now on.

## 1.4 Quantization with the path integral

In the last section we completed the derivation of the path integral representation of the Euclidean correlators $\langle O_2(t) \, O_1(0) \rangle_T^\varepsilon$. In this section we now discuss several aspects of the equations we have derived and present some general concepts and strategies of lattice field theory. In particular, we present the idea of directly using the path integral as a method for quantizing a system. For further reading on this subject we recommend [7, 10, 11].

### 1.4.1 Different discretizations of the Euclidean action

Let us begin our discussion of the path integral with a closer look at the expression (1.74) for the Euclidean action. When inspecting the derivative terms – the first two terms on the right-hand side of (1.74) – one notes a difference between the two discretizations of the derivatives. The discretization for the temporal derivative uses forward differences, whereas for the spatial derivatives central differences are used. While the discretization of the spatial derivative was chosen in (1.50), the temporal derivative is a result of the stepwise transport in Euclidean time with the Trotter formula. Thus it contains only nearest neighbors. Such a discretization has, however, larger discretization errors than using central differences. From the Taylor expansion $f(x \pm \varepsilon) = f(x) \pm \varepsilon f'(x) + \varepsilon^2 f''(x)/2 \pm \varepsilon^3 f'''(x)/6 + \dots$ one finds that for the forward differences

$$\frac{f(x+\varepsilon) - f(x)}{\varepsilon} = f'(x) + \mathcal{O}(\varepsilon) , \tag{1.84}$$

while for the discretization with central differences

$$\frac{f(x+\varepsilon) - f(x-\varepsilon)}{2\,\varepsilon} = f'(x) + \mathcal{O}(\varepsilon^2) . \tag{1.85}$$

Due to the smaller discretization errors it is more advantageous to use the central difference formula also for the temporal derivative.

A second change that is usually implemented is to use the same temporal and spatial lattice constant, and to set

$$\varepsilon = a . \tag{1.86}$$

Although $\varepsilon$ and $a$ have a different origin, both of them should be small: The spatial lattice constant determines down to which length scale we can resolve the system, while the temporal constant determines how well the Trotter formula approximates the Euclidean time transporter. When using the central differences for time and implementing (1.86), the Euclidean action reads

$$
S_E[\Phi] = a^4 \sum_{n \in \Lambda} \left( \frac{1}{2} \sum_{\mu=1}^{4} \left( \frac{\Phi(n+\hat{\mu}) - \Phi(n-\hat{\mu})}{2a} \right)^2 + \frac{m^2}{2} \Phi(n)^2 + V\left(\Phi(n)\right) \right) .
$$
(1.87)

It is obvious that in this form the action treats the spatial directions and Euclidean time equally. Up to the fact that for Euclidean space we have no relative sign between temporal and spatial components of the metric tensor, the equal footing of space and time in (1.87) is the same as in the action (1.41) and (1.42) with the Minkowski metric. The metric tensor which is used in (1.42) to transform between covariant and contravariant indices is replaced by a Kronecker delta.

We finally remark that in Minkowski space the components run from $\mu = 0$ to 3, where $\mu = 0$ refers to time. To stress that Euclidean metric is used the index $\mu$ in (1.87) runs from 1 to 4, and $\mu = 4$ refers to Euclidean time.

### 1.4.2 The path integral as a quantization prescription

The action with central differences for time is not derived from first principles in the way presented above. The changes we performed in the action signals that we now change the philosophy for the quantization of the system. Let us discuss this new quantization prescription.

In the derivation of the last section the quantization of the system entered by enforcing the canonical commutation rules (1.52) and (1.53) for the fields and their conjugate momenta. Subsequently, the expression

$$
\langle O_2(t) \, O_1(0) \rangle_T = \frac{1}{Z_T} \int \mathcal{D}[\Phi] \, e^{-S_E[\Phi]} \, O_2[\Phi(.\,, n_t)] \, O_1[\Phi(.\,, 0)] \, ,
$$
(1.88)

for the Euclidean correlators as path integrals was derived (from now on we drop the superscripts $\varepsilon$ that indicate the discretization of time). Also the partition function is given as a path integral

$$
Z_T = \int \mathcal{D}[\Phi] \, e^{-S_E[\Phi]} .
$$
(1.89)

The measure

$$
\mathcal{D}[\Phi] = \prod_{n \in \Lambda} d\,\Phi(n) ,
$$
(1.90)

is a product measure of integration measures for the classical field variables at all points $n$ of the 4D lattice $\Lambda$.

Our new quantization prescription, which we introduce now, is based directly on Eqs. (1.88), (1.89), and (1.90). It states that the quantization is no longer implemented by enforcing canonical commutation rules for the commutators, but instead by a path integral over classical field variables. The steps involved in the quantization of a system are

- **Step 1:** The continuous space-time is replaced by a 4D Euclidean lattice with lattice constant $a$. The degrees of freedom are classical field variables $\Phi$ living on the lattice.

- **Step 2:** The Euclidean action $S_E[\Phi]$ of the system is discretized on the lattice such that in the limit $a \to 0$ the Euclidean continuum action is obtained. The Boltzmann weight factor is $\exp(-S_E[\Phi])$.

- **Step 3:** The operators that appear in the Euclidean correlator one wants to study are translated to functionals by replacing the field operators with the classical lattice field variables.

- **Step 4:** Euclidean correlation functions are computed by evaluating these functionals on some lattice field configuration, weighting them with the Boltzmann factor and integrating over all possible field configurations.

In the next chapters we will implement this prescription for QCD. In this approach there are, however, several different possibilities from which one can choose. We have already seen that various discretizations of derivatives can be used which differ in their discretization errors. An important question is whether different discretizations of a system all lead to a theory where the corresponding Euclidean correlators $\langle O_2(t)\, O_1(0)\rangle_T$ can be interpreted according to our first key equation

$$\lim_{T\to\infty} \langle O_2(t)\, O_1(0)\rangle_T = \sum_n \langle 0|\widehat{O}_2|n\rangle\langle n|\widehat{O}_1|0\rangle\, \mathrm{e}^{-t\,E_n} ,\qquad (1.91)$$

such that energy levels and physical matrix elements can be extracted. In other words: Does a given lattice discretization of the Euclidean action used in a path integral give rise to a proper quantum mechanical Hilbert space? If yes, how can the $n$-point functions with Minkowski metric be reconstructed from the Euclidean correlators? A very general answer to these questions is given by the so-called Osterwalder–Schrader reconstruction (see, e.g., [12] for a textbook presentation). If the Euclidean correlators calculated from the path integral obey a certain set of axioms, then the Hilbert space for the Minkowski theory can be obtained in a constructive way. For the Wilson formulation of lattice QCD, which we will present in Chaps. 2 and 5, this construction was generalized to lattice QCD in [13] (see also [14]). For an alternative construction on the lattice, based on the transfer matrix, the interested reader is referred to [15]. For our presentation here it is sufficient to conclude that we have some freedom in the discretization of QCD, and when the axioms are obeyed the physical Hilbert space can be constructed.

Once physical observables are calculated on the lattice one is interested in their values in the limit $a \to 0$, the so-called *continuum limit*. This limit can be understood in several ways. We have already encountered the so-called *naive* or *classical continuum limit* when we required the lattice discretization of the action to approach its continuum counterpart for $a \to 0$. However, the naive continuum limit for the action is only used as a guiding principle in the construction of the lattice theory. The fully quantized theory requires the evaluation of the path integral, giving rise to results for observables which are involved functions of $a$. In actual numerical calculations the continuum limit is approached in a different way: The couplings of the theory are driven to their critical values, i.e., to the values where the system undergoes a phase transition. When doing so, physical scales (e.g., the proton size) become large in lattice units. In other words, the resolution of our lattice becomes finer and finer. Certainly these few remarks are too short to understand this procedure which we will discuss in great detail in Sect. 3.5. This *true* continuum limit is based on concepts from statistical mechanics and we conclude this chapter by discussing the relation of lattice field theory to statistical mechanics.

### 1.4.3 The relation to statistical mechanics

As announced, there is a structural equivalence between our field theory on the lattice and statistical mechanics. A prototype system of statistical mechanics is a spin system. The degrees of freedom are classical spin variables $s_n$ located on some lattice which is typically 3D. The energy of the system is a functional $H[s]$ of the spins. In the canonical ensemble, i.e., for the system in a heat bath with temperature $T$, the probability $P[s]$ of finding the system in a particular configuration $s$ is given by

$$P[s] = \frac{1}{Z}\, e^{-\beta\, H[s]}\,, \tag{1.92}$$

where $\beta$ is the inverse temperature $\beta = 1/k_B T$, with $k_B$ denoting the Boltzmann constant. The partition function $Z$ is given by

$$Z = \sum_{\{s\}} e^{-\beta\, H[s]}\,, \tag{1.93}$$

where the sum runs over all possible spin configurations $\{s\}$. The expectation value of some observable $O$ is given by

$$\langle O \rangle = \frac{1}{Z} \sum_{\{s\}} e^{-\beta\, H[s]}\, O[s]\,. \tag{1.94}$$

The similarity between (1.88) and (1.94) is obvious. The Boltzmann factor $\exp(-\beta\, H)$ is replaced by the weight factor $\exp(-S_E)$ and the summation over all spin configurations by the integral over configurations of the classical field $\Phi$. Thus, the name "partition function" for the normalization factor of the path integral now appears natural. The structural equivalence between lattice field theory and statistical mechanics is an important observation also from a technical point of view: It allows one to apply analytical and numerical methods, originally developed in statistical mechanics, to lattice field theory. Some of these techniques will be discussed in subsequent chapters.

# References

1. C. Cohen-Tannoudji, B. Diu, and F. Laloë: *Quantum Mechanics* (John Wiley and Sons, New York, London, Sidney, Toronto 1977)
2. J. J. Sakurai: *Modern Quantum Physics* (Addison-Wesley, Reading, Massachusetts 1994)
3. A. Messiah: *Quantum Mechanics*, Vols. 1 and 2 (Dover Publications, Mineola, New York 1958)
4. J. W. Negele and H. Orland: *Quantum Many Particle Systems* (Westview Press – Advanced Book Classics, Boulder 1998)
5. E. Prugovečki: *Quantum Mechanics in Hilbert Space*, 2nd ed. (Academic Press, New York, London, Toronto, Sydney, San Francisco 1981)
6. M. Reed and B. Simon: *Methods of Modern Mathematical Physics*, revised and enlarged ed., Vol. 1: Functional Analysis (Academic Press, Boston, San Diego, New York, London, Sydney, Tokyo, Toronto 1980)
7. G. Roepstorff: *Path Integral Approach to Quantum Physics* (Springer, Berlin, Heidelberg, New York 1996)
8. C. Itzykson and J.-B. Zuber: *Quantum Field Theory* (McGraw-Hill, New York 1985)
9. M. E. Peskin and D. V. Schroeder: *An Introduction to Quantum Field Theory* (Addison-Wesley, Reading, Massachusetts 1995)
10. R. P. Feynman and A. Hibbs: *Quantum Mechanics and Path Integrals* (McGraw-Hill, New York 1965)
11. R. J. Rivers: *Path Integral Methods in Quantum Field Theory, Cambridge Monographs on Mathematical Physics* (Cambridge University Press, Cambridge, New York 1990)
12. J. Glimm and A. Jaffe: *Quantum Physics. A Functional Integral Point of View*, 2nd ed. (Springer, New York 1987)
13. K. Osterwalder and E. Seiler: Ann. Phys. **110**, 440 (1978)
14. E. Seiler: *Gauge Theories as a Problem of Constructive Quantum Field Theory and Statistical Mechanics*, Lect. Notes Phys. **159**. Springer, Berlin, Heidelberg, New York (1982)
15. M. Lüscher: Commun. Math. Phys. **54**, 283 (1977)

# 2

# QCD on the lattice – a first look

In Chap. 1 we derived the lattice path integral for a scalar field theory using the canonical approach for the quantization of the system. Subsequently, we changed our point of view and adopted the lattice path integral itself as a method of quantization. The steps involved are a lattice discretization of the classical Euclidean action and the construction of the measure for the integration over "all configurations" of the classical fields. In this chapter we begin with the implementation of these two steps for quantum chromodynamics.

Quantum chromodynamics (QCD) is the theory of strongly interacting particles and fields, i.e., the theory of quarks and gluons. In the first section of this chapter we will review the QCD action functional and its symmetries. This serves as a preparation for the subsequent discretization of QCD on the lattice. The discretization proceeds in several steps. We begin with the naive discretization of the fermionic part of the QCD action followed by discussing the lattice action for the gluons. In the end we write down a complete expression for the QCD lattice path integral. Already at this point we stress, however, that the expression for the path integral obtained in this chapter is not the final word. The nature of the quark fields in our path integral will have to be changed in order to incorporate Fermi statistics. This step will be taken in Chap. 5 where we reinterpret the quark fields as anti-commuting numbers, so-called *Grassmann numbers*.

## 2.1 The QCD action in the continuum

We begin our construction of the QCD lattice path integral with a review of the Euclidean continuum action. After introducing the quark and gluon fields, we develop the QCD continuum action starting with the discussion of the fermionic part of the action and its invariance under gauge transformations. Subsequently, we construct an action for the gluon fields that are invariant under these transformations.

Gattringer, C., Lang, C.B.: *QCD on the Lattice – A First Look.* Lect. Notes Phys. **788**, 25–41 (2010)
DOI 10.1007/978-3-642-01850-3_2      © Springer-Verlag Berlin Heidelberg 2010

### 2.1.1 Quark and gluon fields

The quarks are massive fermions and as such are described by Dirac 4-spinors

$$\psi^{(f)}(x)_{\alpha \atop c} \;, \quad \overline{\psi}^{(f)}(x)_{\alpha \atop c} \;. \tag{2.1}$$

These quark fields carry several indices and arguments. The space–time position is denoted by $x$, the Dirac index by $\alpha = 1, 2, 3, 4$, and the color index by $c = 1, 2, 3$. In general we will use Greek letters for Dirac indices and letters $a, b, c, \ldots$ for color. Each field $\psi^{(f)}(x)$ thus has 12 independent components. In addition the quarks come in several flavors called up, down, strange, charm, bottom, and top, which we indicate by a flavor index $f = 1, 2, \ldots, 6$. In many calculations it is sufficient to include only the lightest two or three flavors of quarks. Thus, our flavor index will run from 1 to $N_f$, the number of flavors. We remark that often we omit the indices and use matrix/vector notation instead.

In the Minkowskian operator approach the fields $\psi$ and $\overline{\psi}$ are related by $\overline{\psi} = \psi^\dagger \gamma_0$, where $\gamma_0$ is the $\gamma$-matrix related to time. In the Euclidean path integral one uses independent integration variables $\psi$ and $\overline{\psi}$.

In addition to the quarks, QCD contains gauge fields describing the gluons,

$$A_\mu(x)_{cd} \;. \tag{2.2}$$

These fields also carry several indices. As for the quark fields we have a space–time argument denoted by $x$. In addition, the gauge fields constitute a vector field carrying a Lorentz index $\mu$ which labels the direction of the different components in space–time. Since we are interested in the Euclidean action, the Lorentz index $\mu$ is Euclidean, i.e., we do not distinguish between covariant and contravariant indices. There is no metric tensor involved and $\mu = 1, 2, 3, 4$ simply label the different components. Finally, the gluon field carries color indices $c, d = 1, 2, 3$. For given $x$ and $\mu$, the field $A_\mu(x)$ is a traceless, hermitian $3 \times 3$ matrix at each space–time point $x$. We will discuss the structure of these matrices and their physical motivation in more detail below.

It is convenient to split the QCD action into a fermionic part, which includes quark fields and an interaction term coupling them to the gluons, and a gluonic part, which describes propagation and interaction of only the gluons.

### 2.1.2 The fermionic part of the QCD action

The fermionic part $S_F[\psi, \overline{\psi}, A]$ of the QCD action is a bilinear functional in the fields $\psi$ and $\overline{\psi}$. It is given by

$$S_F[\psi, \overline{\psi}, A] = \sum_{f=1}^{N_f} \int d^4x \, \overline{\psi}^{(f)}(x) \left( \gamma_\mu \left( \partial_\mu + i A_\mu(x) \right) + m^{(f)} \right) \psi^{(f)}(x)$$

$$= \sum_{f=1}^{N_f} \int d^4x \, \overline{\psi}^{(f)}(x) \underset{c}{\alpha} \left( (\gamma_\mu)_{\alpha\beta} \left( \delta_{cd} \partial_\mu + i A_\mu(x)_{cd} \right) \right. \tag{2.3}$$

$$\left. + m^{(f)} \delta_{\alpha\beta} \delta_{cd} \right) \psi^{(f)}(x) \underset{d}{\beta} \, .$$

In the first line of this equation we have used matrix/vector notation for the color and Dirac indices, while in the second line we write all indices explicitly. Note that we use the Einstein summation convention.

Equation (2.3) makes it obvious that the action is a sum of the actions for the individual flavors $f = 1, 2, \ldots, N_f$. The quarks with different flavor all couple in exactly the same way to the gluon field $A_\mu$ and only differ in their masses $m^{(f)}$. Of course, different flavors also have different electric charge and thus couple differently to the electromagnetic field. However, here we only discuss the strong interaction.

The color indices $c$ and $d$ of the quark fields $\overline{\psi}$, $\psi$ are summed over with the corresponding indices of the gauge field and, in this way, couple the quarks to the gluons. The coupling of the gauge field is different for each component $\mu$ since each component is multiplied with a different matrix $\gamma_\mu$. The $\gamma$-matrices are $4 \times 4$ matrices in Dirac space, and in the QCD action they mix the different Dirac components of the quark fields. They are the Euclidean versions of the (Minkowski) $\gamma$-matrices familiar from the Dirac equation. The Euclidean $\gamma$- matrices $\gamma_\mu$, $\mu = 1, 2, 3, 4$, obey the Euclidean anti-commutation relations

$$\{\gamma_\mu, \gamma_\nu\} = 2 \, \delta_{\mu\nu} \, \mathbb{1} \, . \tag{2.4}$$

In Appendix A.2 we discuss how to construct the Euclidean $\gamma$-matrices from their Minkowski counterparts and give an explicit representation. The different partial derivatives in the kinetic term of (2.3) mix the Dirac components in the same way as the gauge fields, i.e., the $\partial_\mu$ are also contracted with the matrices $\gamma_\mu$. The kinetic term is, however, trivial in color space. The mass term, finally, is trivial in both color and Dirac space.

Having discussed our notation in detail, we still should verify that the action (2.3) indeed gives rise to the relativistic wave equation for fermions, the Dirac equation. For a single flavor, the contribution to the action is given by (we drop the flavor index for the subsequent discussion and use matrix/vector notation for the color and Dirac indices)

$$S_F[\psi, \overline{\psi}, A] = \int d^4x \, \overline{\psi}(x) \left( \gamma_\mu \left( \partial_\mu + i A_\mu(x) \right) + m \right) \psi(x) \, . \tag{2.5}$$

The simplest way of applying the Euler Lagrange equations (1.43) in (2.5) is to differentiate the integrand of (2.5) with respect to $\overline{\psi}(x)$. This gives rise to

$$(\gamma_\mu \left(\partial_\mu + \mathrm{i}\, A_\mu(x)\right) + m)\, \psi(x) = 0 \,, \tag{2.6}$$

which is indeed the (Euclidean) Dirac equation in an external field $A_\mu$. Thus, we have verified that the action (2.5) has the correct form.

### 2.1.3 Gauge invariance of the fermion action

So far we have only discussed the different building blocks of QCD and how they are assembled in the fermionic part of the QCD action. Let us now dive a little bit deeper into the underlying structures and symmetries.

Up to the additional color structure, the action (2.5) is exactly the action of electrodynamics – when using matrix/vector notation this difference is not even explicit. As a matter of fact, the QCD action is obtained by generalizing the gauge invariance of electrodynamics.

In electrodynamics the action is invariant under multiplication of the fermion fields with an arbitrary phase at each space–time point $x$, combined with a transformation of the gauge field. In QCD we require invariance under local rotations among the color indices of the quarks. At each space-time point $x$ we choose an independent complex $3 \times 3$ matrix $\Omega(x)$. The matrices are required to be unitary, $\Omega(x)^\dagger = \Omega(x)^{-1}$, and to have $\det[\Omega(x)] = 1$. Such matrices are the defining representation of the *special unitary group*, denoted by SU(3) for the case of $3 \times 3$ matrices. It is easy to see that this set is closed under matrix multiplication. Furthermore the unit matrix is contained in this set and for every element there exists an inverse (the hermitian conjugate matrix). Thus the set of SU(3) matrices forms a group. We collect the basic equations showing these statements in Appendix A.1. Note, however, that the group operation – the matrix multiplication – is not commutative. Groups with a non-commutative group operation are called *non-abelian groups*. The idea of using non-abelian groups for a gauge theory was pursued by Yang and Mills [1], and such theories are often referred to as *Yang–Mills theories*.

Returning to our discussion of the QCD gauge invariance, we require that the action is invariant under the transformation

$$\psi(x) \to \psi'(x) = \Omega(x)\psi(x) \,, \ \overline{\psi}(x) \to \overline{\psi}'(x) = \overline{\psi}(x)\Omega(x)^\dagger \tag{2.7}$$

for the fermion fields and a yet unspecified transformation $A_\mu(x) \to A'_\mu(x)$ for the gauge fields. Invariance of the action means that we require

$$S_F[\psi', \overline{\psi}', A'] = S_F[\psi, \overline{\psi}, A] \,. \tag{2.8}$$

With (2.5) and (2.7), this gives

$$S_F[\psi', \overline{\psi}', A'] = \int \mathrm{d}^4 x\, \overline{\psi}(x)\Omega(x)^\dagger \left(\gamma_\mu \left(\partial_\mu + \mathrm{i}\, A'_\mu(x)\right) + m\right) \Omega(x)\psi(x) \,. \tag{2.9}$$

Using $\Omega(x)^\dagger = \Omega(x)^{-1}$, we see immediately that for the mass term the gauge transformation matrices cancel. For the other terms the situation is a bit more involved. From comparing (2.5) with (2.9) we obtain the condition

$$
\begin{aligned}
\partial_\mu + \mathrm{i}\, A_\mu(x) &= \Omega(x)^\dagger \left(\partial_\mu + \mathrm{i}\, A'_\mu(x)\right) \Omega(x) \\
&= \partial_\mu + \Omega(x)^\dagger \left(\partial_\mu \Omega(x)\right) + \mathrm{i}\Omega(x)^\dagger A'_\mu(x)\Omega(x) \, .
\end{aligned}
\tag{2.10}
$$

This is an equation for an operator acting on a function of $x$. Thus, due to the product rule, we find two terms with derivatives. We now can solve for $A'_\mu(x)$ (again we use $\Omega(x)^\dagger = \Omega(x)^{-1}$) and we arrive at the transformation property for the gauge field

$$
A_\mu(x) \to A'_\mu(x) = \Omega(x)A_\mu(x)\Omega(x)^\dagger + \mathrm{i}\left(\partial_\mu \Omega(x)\right)\Omega(x)^\dagger \, .
\tag{2.11}
$$

Note that also $A'_\mu(x)$ is hermitian and traceless as required for the gauge fields. For the first term on the right-hand side of (2.11), this follows from the fact that $A_\mu(x)$ is traceless and $\Omega(x)^\dagger = \Omega(x)^{-1}$. For the second term this is shown in Appendix A.1 (see (A.11) and (A.15)).

The requirement that the fermion action (2.5) remains invariant under the gauge transformation (2.7) of the fermions necessarily implies the presence of gauge fields $A_\mu(x)$ with the transformation properties given by (2.11).

In the next section, when we discretize QCD on the lattice, we again require the invariance of the lattice action under the local gauge transformations (2.7) for the quark fields. Gauge fields have to be introduced to achieve gauge invariance of the action.

### 2.1.4 The gluon action

Let us now discuss the action for the gluon fields $A_\mu(x)$. The gluon action $S_G[A_\mu]$ is a functional of only the gauge fields and is required to be invariant under the transformation (2.11):

$$
S_G[A'] = S_G[A] \, .
\tag{2.12}
$$

To construct an action with this property we define the *covariant derivative*

$$
D_\mu(x) = \partial_\mu + \mathrm{i}\, A_\mu(x) \, .
\tag{2.13}
$$

From our intermediate result in the first line of (2.10) we read off the transformation property for the covariant derivative as

$$
D_\mu(x) \to D'_\mu(x) = \partial_\mu + \mathrm{i}\, A'_\mu(x) = \Omega(x)D_\mu(x)\Omega(x)^\dagger \, .
\tag{2.14}
$$

These transformation properties ensure that $D_\mu(x)\psi(x)$ and $\psi(x)$ transform in exactly the same way – thus, the name "covariant derivative."

The covariant derivatives are now used to construct an action functional which is a generalization of the expression known from electrodynamics. We define the field strength tensor $F_{\mu\nu}(x)$ as the commutator

$$F_{\mu\nu}(x) = -\mathrm{i}[D_\mu(x), D_\nu(x)] = \partial_\mu A_\nu(x) - \partial_\nu A_\mu(x) + \mathrm{i}[A_\mu(x), A_\nu(x)] \,, \quad (2.15)$$

where the last term on the right-hand side does not vanish since $A_\mu(x)$ and $A_\nu(x)$ are matrices, i.e., objects where multiplication is a non-commutative operation. Up to this commutator, the field strength tensor $F_{\mu\nu}(x)$ has the same form as the field strength in electrodynamics.

The fact that the field strength tensor is the commutator of two covariant derivatives implies that it inherits the transformation properties (2.14), i.e., it transforms as

$$F_{\mu\nu}(x) \rightarrow F'_{\mu\nu}(x) = \Omega(x) F_{\mu\nu}(x) \Omega(x)^\dagger \,. \quad (2.16)$$

As a candidate for the gauge action we now consider the expression

$$S_G[A] = \frac{1}{2\,g^2} \int \mathrm{d}^4x \, \mathrm{tr}\,[F_{\mu\nu}(x) F_{\mu\nu}(x)] \,. \quad (2.17)$$

Taking the trace over the color indices ensures that (2.17) is invariant under gauge transformations. One can use (2.16), the invariance of the trace under cyclic permutations, and $\Omega(x)^\dagger = \Omega(x)^{-1}$ to verify this property. Furthermore, the summation over the Lorentz indices $\mu, \nu$ (summation convention) ensures that the action is a Lorentz scalar. We stress again that (2.17) is understood in Euclidean space.

From (2.15) and (2.17) it is obvious that our gauge action generalizes the action of electrodynamics. Up to the trace and the different overall factor, it exactly matches the action for the electromagnetic field. The trace is due to the fact that gluon fields are matrix valued. Below, we decompose the matrix-valued fields into components and in this way get rid of the trace, pushing the similarity to electrodynamics even further. The extra factor $1/g^2$ is just a convenient way to introduce the coupling. After rescaling the gauge fields

$$\frac{1}{g} A_\mu(x) \rightarrow A_\mu(x) \,, \quad (2.18)$$

the factor $1/g^2$ in (2.17) is gone and the gauge action assumes the more familiar form. Now, the gauge coupling shows up in the covariant derivative

$$D_\mu(x) \rightarrow \partial_\mu + \mathrm{i}\,g A_\mu(x) \,, \quad (2.19)$$

making obvious that $g$ is the coupling strength of the gauge fields to the quarks. On the lattice it is more convenient to have the gauge coupling as an overall factor of the gauge action, i.e., we work with the form (2.17).

### 2.1.5 Color components of the gauge field

We have introduced the gauge fields $A_\mu(x)$ as hermitian, traceless matrices and have shown that the gauge transformation (2.11) preserves these properties. Thus the $A_\mu(x)$ are in the Lie algebra su(3) and we can write

$$A_\mu(x) = \sum_{i=1}^{8} A_\mu^{(i)}(x)\, T_i \ . \tag{2.20}$$

The components $A_\mu^{(i)}(x), i = 1, 2, \ldots 8$, are real-valued fields, the so-called color components, and the $T_i$ are a basis for traceless hermitian $3 \times 3$ matrices (see Appendix A.1). We can use this representation (2.20) of the gauge field to write also the field strength tensor (2.15) in terms of its components. Inserting (2.20) into (2.15) we obtain

$$F_{\mu\nu}(x) = \sum_{i=1}^{8} \left(\partial_\mu A_\nu^{(i)}(x) - \partial_\nu A_\mu^{(i)}(x)\right) T_i \ + \ i \sum_{j,k=1}^{8} A_\mu^{(j)}(x) A_\nu^{(k)}(x)\, [T_j, T_k] \ . \tag{2.21}$$

The commutator on the right-hand side can be simplified further with the commutation relations (A.4) and one ends up with

$$F_{\mu\nu}(x) = \sum_{i=1}^{8} F_{\mu\nu}^{(i)}(x)\, T_i \ , \tag{2.22}$$

$$F_{\mu\nu}^{(i)}(x) = \partial_\mu A_\nu^{(i)}(x) - \partial_\nu A_\mu^{(i)}(x) - f_{ijk} A_\mu^{(j)}(x) A_\nu^{(k)}(x) \ . \tag{2.23}$$

This representation of the field strength can now be inserted in the expression (2.17) for the gauge action and, using (A.3) to evaluate the trace, we obtain

$$S_G[A] = \frac{1}{4\, g^2} \sum_{i=1}^{8} \int d^4x \, F_{\mu\nu}^{(i)}(x) F_{\mu\nu}^{(i)}(x) \ . \tag{2.24}$$

From this equation we see that the gauge action is a sum over color components and each term has the form of the action of electrodynamics. However, there appears a qualitatively new feature: From the right-hand side of (2.23) we see that the field strength color components are not linear in the gauge field $A_\mu^{(i)}(x)$ but have a quadratic piece which mixes the different color components of the gluon field. Thus, in the action (2.24) we not only encounter the term quadratic in the gauge fields which is familiar from electrodynamics, but also find cubic and quartic terms. These terms give rise to self-interactions of the gluons, making QCD a highly nontrivial theory. The self-interactions are responsible for confinement of color, the most prominent feature of QCD.

In Fig. 2.1 we show a schematic picture (a so-called tree-level Feynman diagram) illustrating the cubic and quartic interaction terms. The curly lines represent the gluons and the dots are the interaction vertices.

These remarks conclude our review of the continuum action and we have all the concepts and notations necessary to begin the lattice discretization of quantum chromodynamics. We stress, however, that besides gauge invariance, there are other important symmetries of the QCD action. We discuss these symmetries later as we need them.

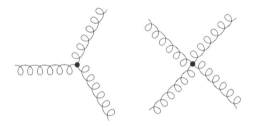

**Fig. 2.1.** Schematic picture of the cubic and quartic gluon self-interaction. The *wavy lines* represent the gluon propagators and the *dots* are the interaction vertices

## 2.2 Naive discretization of fermions

In this section we introduce the so-called *naive discretization* of the fermion action. Later, in Chap. 5, this discretization will be augmented with an additional term to remove lattice artifacts. Here, it serves to present the basic idea and, more importantly, to discuss the representation of the lattice gluon field which differs from the continuum form. We show that on the lattice the gluon fields must be introduced as elements of the gauge group and not as elements of the algebra, as is done in the continuum formulation.

### 2.2.1 Discretization of free fermions

As discussed in Sect. 1.4, the first step in the lattice formulation is the introduction of the 4D lattice $\Lambda$:

$$\Lambda = \{n = (n_1, n_2, n_3, n_4) \mid$$
$$n_1, n_2, n_3 = 0, 1, \ldots, N-1 ; \; n_4 = 0, 1, \ldots, N_T - 1\} . \quad (2.25)$$

The vectors $n \in \Lambda$ label points in space–time separated by a lattice constant $a$. In our lattice discretization of QCD we now place spinors at the lattice points only, i.e., our fermionic degrees of freedom are

$$\psi(n) , \; \overline{\psi}(n) , \;\; n \in \Lambda , \quad (2.26)$$

where the spinors carry the same color, Dirac, and flavor indices as in the continuum (we suppress them in our notation). Note that for notational convenience we only use the integer-valued 4-coordinate $n$ to label the lattice position of the quarks and not the actual physical space-time point $x = an$.

In the continuum the action $S_F^0$ for a free fermion is given by the expression (set $A_\mu = 0$ in (2.5))

$$S_F^0[\psi, \overline{\psi}] = \int d^4x \, \overline{\psi}(x) \left( \gamma_\mu \partial_\mu + m \right) \psi(x) . \quad (2.27)$$

When formulating this action on the lattice we have to discretize the integral over space–time as well as the partial derivative. The discretization is implemented as a sum over $\Lambda$, as we did for the scalar field theory in Chap. 1. The partial derivative is discretized with the symmetric expression

$$\partial_\mu \psi(x) \;\rightarrow\; \frac{1}{2a}\left(\psi(n+\hat{\mu}) - \psi(n-\hat{\mu})\right) . \tag{2.28}$$

Thus, our lattice version of the free fermion action reads

$$S_F^0[\psi,\overline{\psi}] = a^4 \sum_{n\in\Lambda} \overline{\psi}(n) \left(\sum_{\mu=1}^{4} \gamma_\mu \frac{\psi(n+\hat{\mu}) - \psi(n-\hat{\mu})}{2a} + m\,\psi(n)\right) . \tag{2.29}$$

This form is the starting point for the introduction of the gauge fields.

### 2.2.2 Introduction of the gauge fields as link variables

In the last section we showed that requiring the invariance of the action under the local rotation (2.7) of the color indices of the quark fields enforces the introduction of the gauge fields. On the lattice we implement the same transformation by choosing an element $\Omega(n)$ of SU(3) for each lattice site $n$ and transforming the fermion fields according to

$$\psi(n) \rightarrow \psi'(n) = \Omega(n)\,\psi(n) \; , \; \overline{\psi}(n) \rightarrow \overline{\psi}'(n) = \overline{\psi}(n)\,\Omega(n)^\dagger . \tag{2.30}$$

As for the continuum case, we find that on the lattice the mass term of (2.29) is invariant under the transformation (2.30). For the discretized derivative terms in (2.29), this is not the case. Consider, e.g., the term

$$\overline{\psi}(n)\psi(n+\hat{\mu}) \rightarrow \overline{\psi}'(n)\,\psi'(n+\hat{\mu}) = \overline{\psi}(n)\,\Omega(n)^\dagger\,\Omega(n+\hat{\mu})\,\psi(n+\hat{\mu}) . \tag{2.31}$$

This is not gauge-invariant. If, however, we introduce a field $U_\mu(n)$ with a directional index $\mu$, then

$$\overline{\psi}'(n)\,U'_\mu(n)\,\psi'(n+\hat{\mu}) \;=\; \overline{\psi}(n)\,\Omega(n)^\dagger\,U'_\mu(n)\,\Omega(n+\hat{\mu})\,\psi(n+\hat{\mu}) \tag{2.32}$$

is gauge-invariant if we define the gauge transformation of the new field by

$$U_\mu(n) \;\rightarrow\; U'_\mu(n) = \Omega(n)\,U_\mu(n)\,\Omega(n+\hat{\mu})^\dagger . \tag{2.33}$$

In order to make the fermionic action (2.29) gauge-invariant, we introduce the gauge fields $U_\mu(n)$ as elements of the gauge group SU(3) which transform as given in (2.33). These matrix-valued variables are oriented and are attached to the links of the lattice and thus are often referred to as *link variables*. $U_\mu(n)$ lives on the link which connects the sites $n$ and $n+\hat{\mu}$.

Since the link variables are oriented, we can also define link variables that point in negative $\mu$ direction. Note that these are not independent link variables but are introduced only for notational convenience. In particular, $U_{-\mu}(n)$

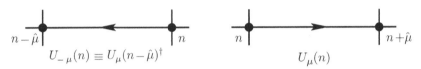

**Fig. 2.2.** The link variables $U_\mu(n)$ and $U_{-\mu}(n)$

points from $n$ to $n - \hat{\mu}$ and is related to the positively oriented link variable $U_\mu(n - \hat{\mu})$ via the definition

$$U_{-\mu}(n) \equiv U_\mu(n - \hat{\mu})^\dagger \ . \tag{2.34}$$

In Fig. 2.2 we illustrate the geometrical setting of the link variables on the lattice. From the definitions (2.34) and (2.33) we obtain the transformation properties of the link in negative direction

$$U_{-\mu}(n) \ \rightarrow \ U'_{-\mu}(n) = \Omega(n)\, U_{-\mu}(n)\, \Omega(n - \hat{\mu})^\dagger \ . \tag{2.35}$$

Note that we have introduced the gluon fields $U_\mu(n)$ as elements of the gauge group SU(3), not as elements of the Lie algebra which were used in the continuum. According to the gauge transformations (2.33) and (2.35) also the transformed link variables are elements of the group SU(3).

Having introduced the link variables and their properties under gauge transformations, we can now generalize the free fermion action (2.29) to the so-called *naive fermion action* for fermions in an external gauge field $U$:

$$S_F[\psi, \overline{\psi}, U] = a^4 \sum_{n \in \Lambda} \overline{\psi}(n) \left( \sum_{\mu=1}^{4} \gamma_\mu \frac{U_\mu(n)\psi(n+\hat{\mu}) - U_{-\mu}(n)\psi(n-\hat{\mu})}{2a} + m\,\psi(n) \right) \ . \tag{2.36}$$

Using (2.30), (2.33), and (2.35) for the gauge transformation properties of fermions and link variables, one readily sees the gauge invariance of the fermion action (2.36), $S_F[\psi, \overline{\psi}, U] = S_F[\psi', \overline{\psi}', U']$.

### 2.2.3 Relating the link variables to the continuum gauge fields

Let us now discuss the link variables in more detail and see how they can be related to the algebra-valued gauge fields of the continuum formulation. We have introduced $U_\mu(n)$ as the link variable connecting the points $n$ and $n + \hat{\mu}$. The gauge transformation properties (2.33) are consequently governed by the two transformation matrices $\Omega(n)$ and $\Omega(n + \hat{\mu})^\dagger$. Also in the continuum an object with such transformation properties is known: It is the path-ordered exponential integral of the gauge field $A_\mu$ along some curve $\mathcal{C}_{xy}$ connecting two points $x$ and $y$, the so-called *gauge transporter*:

$$G(x,y) = P \exp\left(i \int_{\mathcal{C}_{xy}} A \cdot ds\right) . \tag{2.37}$$

We may assume that a lattice is embedded in the continuum, where smooth gauge fields live. For a detailed discussion of the path-ordered exponential in the continuum see, e.g., [2]. We do not need the precise definition of the continuum gauge transporters and only use that under a gauge transformation (2.11) they transform as

$$G(x,y) \rightarrow \Omega(x) G(x,y) \Omega(y)^\dagger . \tag{2.38}$$

These transformation properties are the same as for our link variables $U_\mu(n)$ when $n$ and $n + \hat{\mu}$ are considered as end points of a path. Based on these transformation properties, we interpret the link variable $U_\mu(n)$ as a lattice version of the gauge transporter connecting the points $n$ and $n + \hat{\mu}$, i.e., we wish to establish $U_\mu(n) = G(n, n + \hat{\mu}) + \mathcal{O}(a)$. For that purpose we introduce algebra-valued lattice gauge fields $A_\mu(n)$ and write

$$U_\mu(n) = \exp\left(i\, a A_\mu(n)\right) . \tag{2.39}$$

When comparing (2.37) and (2.39) one sees that we have approximated the integral along the path from $n$ to $n + \hat{\mu}$ by $a A_\mu(n)$, i.e., by the length $a$ of the path times the value of the field $A_\mu(n)$ at the starting point.[1] This approximation is good to $\mathcal{O}(a)$ and no path ordering is necessary to that order. Since the link variables act as gauge transporters, we will often use this nomenclature instead of "link variable."

Based on the relation (2.39) we can now also connect the lattice fermion action (2.36) to its continuum counterpart (2.5). Since one of the guiding principles of our construction is the requirement that in the limit $a \rightarrow 0$ the lattice action approaches the continuum form, we expand (2.39) for small $a$,

$$U_\mu(n) = \mathbb{1} + i a A_\mu(n) + \mathcal{O}(a^2) , \quad U_{-\mu}(n) = \mathbb{1} - i a A_\mu(n - \hat{\mu}) + \mathcal{O}(a^2) , \tag{2.40}$$

where we use (2.34) and $A_\mu = A_\mu^\dagger$ for the expansion of $U_{-\mu}(n)$. Inserting these expanded link variables into expression (2.36) we find

$$S_F[\psi, \overline{\psi}, U] = S_F^0[\psi, \overline{\psi}] + S_F^I[\psi, \overline{\psi}, A] , \tag{2.41}$$

where $S_F^0$ denotes the free part of the action and the interaction part reads

$$S_F^I[\psi, \overline{\psi}, A] = i a^4 \sum_{n \in \Omega} \sum_{\mu=1}^{4} \overline{\psi}(n) \gamma_\mu \frac{1}{2} \left(A_\mu(n)\, \psi(n + \hat{\mu}) + A_\mu(n - \hat{\mu})\, \psi(n - \hat{\mu})\right)$$

$$= i a^4 \sum_{n \in \Omega} \sum_{\mu=1}^{4} \overline{\psi}(n) \gamma_\mu A_\mu(n) \psi(n) + \mathcal{O}(a) . \tag{2.42}$$

---

[1] We remind the reader that for notational convenience we denote the lattice points only by their integer-valued 4-coordinates $n$. For the current discussion it should, however, be kept in mind that the physical space–time coordinates are $a\,n$, i.e., neighboring lattice points are separated by the distance $a$.

In the second step we have used $\psi(n \pm \hat{\mu}) = \psi(n) + \mathcal{O}(a)$ and $A_\mu(n - \hat{\mu}) = A_\mu(n) + \mathcal{O}(a)$. The two Eqs. (2.41) and (2.42) establish that when expanding the lattice version (2.36) of the fermionic action for $a \to 0$, we indeed recover the continuum form (2.5).

Before we continue with discretizing the gauge part of the QCD action we stress an important conceptual point: The group-valued link variables $U_\mu(n)$ are not merely an auxiliary construction to sneak the Lie algebra-valued fields $A_\mu(x)$ of the continuum into the lattice formulation. Instead, the group elements $U_\mu(n)$ are considered as the fundamental variables which are integrated over in the path integral (see Chap. 3). This change from algebra-valued to (compact) group-valued fields has important consequences. In particular, the role of gauge fixing changes considerably. We will discuss these issues in detail once we have completed the construction of QCD on the lattice.

## 2.3 The Wilson gauge action

We have introduced the link variables as the basic quantities for putting the gluon field on the lattice. Now we construct a lattice gauge action in terms of the link variables and show that in the limit $a \to 0$ it approaches its continuum counterpart (assuming that the lattice gauge fields are embedded in a continuous background). This is the *naive continuum limit* in contradistinction to the continuum limit of the full, integrated quantum theory.

### 2.3.1 Gauge-invariant objects built with link variables

As a preparation for the construction of the gluon action let us first discuss the transformation properties of a string of link variables along a path of links. Let $\mathcal{P}$ be such a path of $k$ links on the lattice connecting the points $n_0$ and $n_1$. We define the ordered product

$$P[U] = U_{\mu_0}(n_0)U_{\mu_1}(n_0 + \hat{\mu}_0)\ldots U_{\mu_{k-1}}(n_1 - \hat{\mu}_{k-1}) \equiv \prod_{(n,\mu)\in\mathcal{P}} U_\mu(n) \ . \quad (2.43)$$

This object is the lattice version of the continuum gauge transporter (2.37). Note that the path $\mathcal{P}$ may contain link variables for both directions $\pm\mu$.

From the transformation properties for single link variables, (2.33) and (2.35), it follows that gauge rotations for all but the end points cancel: Consider the transformation of two subsequent link variables on the path, one ending at $n$ the other one starting from this point. The two transformation matrices $\Omega(n)^\dagger$ and $\Omega(n)$ cancel each other at $n$. Only the matrices at the two end points of the path, $n_0$ and $n_1$, remain. Thus the product $P[U]$ transforms according to

$$P[U] \ \to \ P[U'] = \Omega(n_0)\, P[U]\, \Omega(n_1)^\dagger \ . \quad (2.44)$$

Like for the single link term, from such a product of link variables $P[U]$ a gauge-invariant quantity can be constructed by attaching quark fields at the starting point and at the end point,

$$\overline{\psi}(n_0) \, P[U] \, \psi(n_1) \,. \tag{2.45}$$

An alternative way of constructing a gauge-invariant product of link variables is to choose for the path $\mathcal{P}$ and a closed loop $\mathcal{L}$ and to take the trace,

$$L[U] = \text{tr} \left[ \prod_{(n,\mu)\in\mathcal{L}} U_\mu(n) \right] \,. \tag{2.46}$$

According to (2.44), under a gauge transformation only the matrices $\Omega(n_0)$ and $\Omega(n_0)^\dagger$ at the point $n_0$ where the loop is rooted remain. These matrices then cancel when taking the trace. We find

$$L[U'] = \text{tr} \left[ \Omega(n_0) \prod_{(n,\mu)\in\mathcal{L}} U_\mu(n) \, \Omega(n_0)^\dagger \right] = \text{tr} \left[ \prod_{(n,\mu)\in\mathcal{L}} U_\mu(n) \right] = L[U] \,. \tag{2.47}$$

Thus, we have established that the trace over a closed loop of link variables is a gauge-invariant object. Such loops of link variables are used for the construction of the gluon action and later will also serve as physical observables.

### 2.3.2 The gauge action

For the gluon action it is sufficient to use the shortest, nontrivial closed loop on the lattice, the so-called *plaquette*. The plaquette variable $U_{\mu\nu}(n)$ is a product of only four link variables defined as

$$\begin{aligned} U_{\mu\nu}(n) &= U_\mu(n) \, U_\nu(n+\hat{\mu}) \, U_{-\mu}(n+\hat{\mu}+\hat{\nu}) \, U_{-\nu}(n+\hat{\nu}) \\ &= U_\mu(n) \, U_\nu(n+\hat{\mu}) \, U_\mu(n+\hat{\nu})^\dagger \, U_\nu(n)^\dagger \,. \end{aligned} \tag{2.48}$$

In the second formulation we have utilized the equivalence (2.34). We depict the plaquette in Fig. 2.3. As we have shown in the last paragraph, the trace of the plaquette variable is a gauge-invariant object.

We now present Wilson's form of the gauge action [3] – the first formulation of lattice gauge theory – and subsequently show that it indeed approaches the continuum form in the naive limit $a \to 0$. The *Wilson gauge action* is a sum over all plaquettes, with each plaquette counted with only one orientation. This sum can be realized by a sum over all lattice points $n$ where the plaquettes are located, combined with a sum over the Lorentz indices $1 \le \mu < \nu \le 4$,

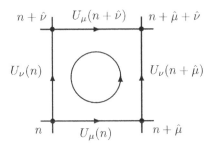

**Fig. 2.3.** The four link variables which build up the plaquette $U_{\mu\nu}(n)$. The *circle* indicates the order that the links are run through in the plaquette

$$S_G[U] = \frac{2}{g^2} \sum_{n\in\Lambda} \sum_{\mu<\nu} \mathrm{Re\ tr}\,[\mathbb{1} - U_{\mu\nu}(n)] \ . \tag{2.49}$$

The individual contributions are the real parts of traces over the unit matrix minus the plaquette variable. The factor $2/g^2$ is included to match the form of the continuum action (2.17) in the limit $a \to 0$. Let us now discuss this limit.

For establishing the correct limit we need to expand the link variables in the form (2.39) for small $a$. In order to handle the products of the four link variables in the plaquette in an organized way, it is useful to invoke the Baker–Campbell–Hausdorff formula for the product of exponentials of matrices:

$$\exp(A)\,\exp(B) = \exp\left(A + B + \frac{1}{2}[A, B] + \cdots\right), \tag{2.50}$$

where $A$ and $B$ are arbitrary matrices and the dots indicate powers of the matrices larger than 2 which are omitted. The formula (2.50) can be proven easily by expanding both sides in powers of $A$ and $B$. Inserting (2.39) into the definition (2.48) of the plaquette and applying (2.50) iteratively, we obtain

$$U_{\mu\nu}(n) = \ \exp\Big(\mathrm{i}\,aA_\mu(n) + \mathrm{i}\,aA_\nu(n+\hat\mu) - \frac{a^2}{2}[A_\mu(n), A_\nu(n+\hat\mu)] \tag{2.51}$$

$$-\mathrm{i}\,aA_\mu(n+\hat\nu) - \mathrm{i}\,aA_\nu(n) - \frac{a^2}{2}[A_\mu(n+\hat\nu), A_\nu(n)]$$

$$+\frac{a^2}{2}[A_\nu(n+\hat\mu), A_\mu(n+\hat\nu)] + \frac{a^2}{2}[A_\mu(n), A_\nu(n)]$$

$$+\frac{a^2}{2}[A_\mu(n), A_\mu(n+\hat\nu)] + \frac{a^2}{2}[A_\nu(n+\hat\mu), A_\nu(n)] \ + \ \mathcal{O}(a^3)\Big) \ .$$

In this expression we have gauge fields with shifted arguments such as $A_\nu(n+\hat\mu)$. We now perform a Taylor expansion for these fields, i.e., we set

$$A_\nu(n+\hat\mu) = A_\nu(n) + a\,\partial_\mu A_\nu(n) + \mathcal{O}(a^2)\,, \tag{2.52}$$

in all these terms and take into account contributions up to $\mathcal{O}(a^2)$. With this expansion several terms cancel and we obtain

$$U_{\mu\nu}(n) = \exp\left(i\,a^2\left(\partial_\mu A_\nu(n) - \partial_\nu A_\mu(n) + i[A_\mu(n), A_\nu(n)]\right) + \mathcal{O}(a^3)\right)$$
$$= \exp\left(i\,a^2 F_{\mu\nu}(n) + \mathcal{O}(a^3)\right) . \tag{2.53}$$

In the second step we use the continuum definition of the field strength given in (2.15). The form (2.53) can now be inserted in (2.49) for the Wilson gauge action. The exponential in (2.53) is expanded and we find

$$S_G[U] = \frac{2}{g^2} \sum_{n\in\Lambda}\sum_{\mu<\nu} \mathrm{Re}\,\mathrm{tr}\left[\mathbb{1} - U_{\mu\nu}(n)\right] = \frac{a^4}{2\,g^2} \sum_{n\in\Lambda}\sum_{\mu,\nu} \mathrm{tr}[F_{\mu\nu}(n)^2] + \mathcal{O}(a^2) . \tag{2.54}$$

The terms of $\mathcal{O}(a^2)$ that appear in the expansion of the exponential in (2.53) cancel when taking the real part of $\mathrm{tr}[\mathbb{1} - U_{\mu\nu}(n)]$ (one may use $\mathrm{tr}[U_{\mu\nu}(n)]^* = \mathrm{tr}[U_{\mu\nu}(n)^\dagger] = \mathrm{tr}[U_{\nu\mu}(n)]$ to see this). In a similar way also the $\mathcal{O}(a^3)$ terms in the expansion of (2.53) cancel, such that the Wilson action approximates the continuum form up to $\mathcal{O}(a^2)$, as stated in (2.54). Note that the factor $a^4$ together with the sum over $\Lambda$ is just the discretization of the space–time integral and thus $\lim_{a\to 0} S_G[U] = S_G[A]$. This completes our discussion of the naive continuum limit $a \to 0$ for the Wilson gauge action.

Concluding this section we remark that different lattice actions for the gauge fields have been proposed in order to reduce cutoff effects further. We return to this issue in Chap. 9.

## 2.4 Formal expression for the QCD lattice path integral

Having constructed the fermion and gauge field actions on the lattice we can now write down the complete expression for the lattice QCD path integral formula for Euclidean correlators. As already announced, the preliminary formulation presented here will be refined further in subsequent chapters. Nevertheless, we find such an intermediate summary helpful for understanding the line of the construction.

### 2.4.1 The QCD lattice path integral

Following the discussion of Sect. 1.4 we write Euclidean correlators as a lattice path integral in the form

$$\langle O_2(t)\,O_1(0)\rangle = \frac{1}{Z}\int \mathcal{D}\left[\psi,\overline\psi\right]\,\mathcal{D}[U]\,e^{-S_F[\psi,\overline\psi,U]-S_G[U]}\,O_2[\psi,\overline\psi,U]\,O_1[\psi,\overline\psi,U] , \tag{2.55}$$

where the partition function $Z$ is given by

$$Z = \int \mathcal{D}\left[\psi,\overline\psi\right]\,\mathcal{D}[U]\,e^{-S_F[\psi,\overline\psi,U]-S_G[U]} . \tag{2.56}$$

The quantization of the system in the path integral formalism is implemented as an integral over all field configurations. On the lattice the corresponding path integral measures are products of measures of all quark field components and products of measures for all link variables:

$$\mathcal{D}\left[\psi,\overline{\psi}\right] = \prod_{n\in\Lambda}\prod_{f,\alpha,c}d\psi^{(f)}(n)_\alpha^{\phantom{c}} d\overline{\psi}^{(f)}(n)_\alpha^{\phantom{c}} , \quad \mathcal{D}[U] = \prod_{n\in\Lambda}\prod_{\mu=1}^{4}dU_\mu(n) .$$

(2.57)

Both the fermion and gauge field measures shown here will be discussed in more detail in subsequent sections. For the fermion fields we have to include the Pauli's principle, turning the spinors $\psi$ and $\overline{\psi}$ into anticommuting variables. These so-called *Grassmann numbers* and the corresponding rules of integration will be discussed in Chap. 5. For the gauge fields we have denoted in (2.57) the measure for a single link variable as $dU_\mu(n)$, but not yet specified how we implement the integration over the group manifold of SU(3). This leads to the concept of *Haar measure* which we discuss in the next chapter. Despite these issues, the expressions in (2.57) already incorporate some of the essential features of lattice QCD, in particular the reduction of the original quantum fields to a countable number of classical variables.

As discussed in Sect. 1.4, in the path integral quantization (2.55) the configurations to be integrated over are weighted with the Boltzmann factor of the Euclidean action. The corresponding lattice versions of the fermion and gauge parts of the action have been derived as (compare (2.36) and (2.49))

$$S_F[\psi,\overline{\psi},U] = a^4\sum_{f=1}^{N_f}\sum_{n\in\Lambda}\left(\overline{\psi}^{(f)}(n)\sum_{\mu=1}^{4}\gamma_\mu\frac{U_\mu(n)\psi^{(f)}(n+\hat{\mu}) - U_{-\mu}(n)\psi^{(f)}(n-\hat{\mu})}{2a}\right.$$
$$\left. + m^{(f)}\overline{\psi}^{(f)}(n)\psi^{(f)}(n)\right) + \text{terms discussed in Chap. 5,} \quad (2.58)$$

where we now also sum over $N_f$ flavors of quarks. We stress again that the fermion action has to be augmented with another term in order to remove lattice artifacts (see Chap. 5). The gauge action, however, is ready to go and is taken over unchanged from (2.49):

$$S_G[U] = \frac{2}{g^2}\sum_{n\in\Lambda}\sum_{\mu<\nu}\text{Re tr}\left[\mathbb{1} - U_{\mu\nu}(n)\right] .$$

(2.59)

Let us remark that the functionals $O_1[\psi,\overline{\psi},U]$ and $O_2[\psi,\overline{\psi},U]$ are translations of the operators $\widehat{O}_1$ and $\widehat{O}_2$ acting in Hilbert space. The translation proceeds as expressed in (1.81) by evaluating the operators between field eigenstates. We stress that the functional $O_2$ depends only on the fields with time argument $n_t$ related to Euclidean time $t$ on the right-hand side of (2.55) via $t = a\,n_t$. The fields in the functional $O_1$ depend only on the fields with $t = 0$.

Equations (2.55), (2.56), (2.57), (2.58), and (2.59) comprise our current status in the construction of lattice QCD. So far we discussed the fundamental fields describing quarks and gluons and put them onto the lattice – quarks on the sites and the gauge fields on the links of the lattice. In order to allow for color rotations of the fermion fields in the same way as in the continuum, the gauge fields were introduced as group-valued link variables. Subsequently, we showed that products of link variables along closed paths are gauge-invariant and we have used the plaquette to construct Wilson's gauge action. Finally, we put together these ingredients according to the path integral quantization recipe of Sect. 1.4. Operators are translated into functionals of classical fields. These functionals are weighted with the Boltzmann factor and this product is then integrated over all possible field configurations. The precise definition of this integration – Grassmann integration for the fermions, Haar measure for the link variables – will be discussed in subsequent chapters.

# References

1. C. N. Yang and R. Mills: Phys. Rev. **96**, 191 (1954)
2. M. E. Peskin and D. V. Schroeder: *An Introduction to Quantum Field Theory* (Addison-Wesley, Reading, Massachusetts 1995)
3. K. G. Wilson: Phys. Rev. D **10**, 2445 (1974)

# 3

# Pure gauge theory on the lattice

In the last chapter we started the construction of QCD on the lattice. We discussed a particularly important feature of QCD, namely the self-interaction of gluons. This self-interaction makes pure gluodynamics, i.e., QCD without quarks, an interesting, highly nontrivial theory. Pure gluodynamics shows color confinement, an important property of full QCD. Since gluodynamics is much easier to handle than QCD with quarks, it is an important subject for studies of confinement and its underlying mechanisms.

In this chapter we complete the construction of gluodynamics on the lattice by discussing the integration measure for the link variables. After a section about gauge invariance and gauge fixing we discuss the potential between two static quark sources. The static quark potential can then be used to set the scale, i.e., to determine the value of the lattice spacing $a$ as a function of the gauge coupling $g$.

Before we come to these topics, let us briefly collect the formulas defining pure gauge theory on the lattice. Expectation values of observables $O$ (where $O$ may also be the product of several terms) are given by

$$\langle O \rangle = \frac{1}{Z} \int \mathcal{D}[U]\, e^{-S_G[U]}\, O[U] \,, \tag{3.1}$$

where the partition function $Z$ is defined as

$$Z = \int \mathcal{D}[U]\, e^{-S_G[U]} \,. \tag{3.2}$$

The integration measure for the link variables is the product measure

$$\int \mathcal{D}[U] = \prod_{n \in \Lambda} \prod_{\mu=1}^{4} \int dU_\mu(n) \,. \tag{3.3}$$

For the gauge field action we introduced the Wilson action

Gattringer, C., Lang, C.B.: *Pure Gauge Theory on the Lattice*. Lect. Notes Phys. **788**, 43–71 (2010)
DOI 10.1007/978-3-642-01850-3_3      © Springer-Verlag Berlin Heidelberg 2010

$$S_G[U] = \frac{\beta}{3} \sum_{n \in \Lambda} \sum_{\mu < \nu} \mathrm{Re}\,\mathrm{tr}\,[\mathbb{1} - U_{\mu\nu}(n)] \;, \tag{3.4}$$

but we again remark that other formulations are being used as well. In (3.4) we use a widely used abbreviation

$$\beta = \frac{6}{g^2} \;, \tag{3.5}$$

the so-called *inverse coupling*. Note that this is the definition of $\beta$ for the case of SU(3). The expression for the general case of SU($N$) is given in (3.93).

## 3.1 Haar measure

In the construction of the QCD lattice action, the link variables $U_\mu(n)$ are introduced as elements of SU(3). We now address the definition of the individual measures $dU_\mu(n)$ in the product measure of (3.3) which integrates the link variables $U_\mu(n)$ over the whole group manifold of SU(3). We consider the problem in a more general way and discuss the measure $dU$ with $U$ being an element of a Lie group, i.e., a group where the elements depend continuously on some parameters.

### 3.1.1 Gauge field measure and gauge invariance

An important restriction for the gauge field measure comes from the gauge invariance of our theory. In the last chapter we introduced the gauge transformation of the link variables as

$$U_\mu(n) \;\to\; U'_\mu(n) = \Omega(n)\,U_\mu(n)\,\Omega(n + \hat\mu)^\dagger \;. \tag{3.6}$$

The group-valued matrices $\Omega(n)$ can be chosen independently at each lattice site $n$. The action $S_G[U]$ for the gauge field is invariant under these transformations and we have

$$S_G[U'] = S_G[U] \;. \tag{3.7}$$

As for any integral, the result of a path integral should be invariant under a change of variables, in particular under the gauge transformation (3.6). For the partition function this requirement reads

$$Z = \int \mathcal{D}[U]\,e^{-S_G[U]} = \int \mathcal{D}[U']\,e^{-S_G[U']} = \int \mathcal{D}[U']\,e^{-S_G[U]} \;, \tag{3.8}$$

where in the last step we have used the gauge invariance (3.7) of the gauge action. Comparing the first and the final expression in (3.8) we obtain

$$\mathcal{D}[U] = \mathcal{D}[U'] \;. \tag{3.9}$$

Using the fact that $\mathcal{D}[U]$ is a product measure (compare (3.3)), we derive the condition

$$\mathrm{d}U_\mu(n) = \mathrm{d}U_\mu(n)' = \mathrm{d}\left(\Omega(n)U_\mu(n)\Omega(n+\hat{\mu})^\dagger\right) , \qquad (3.10)$$

for the integration over the individual link variables. This property is one of the defining properties of the *Haar measure*.

### 3.1.2 Group integration measure

The requirement of gauge invariance leads us to a structure which is well-known in the mathematical literature, the so-called *Haar measure*. This is a measure for integration over a continuous compact group $G$. We have derived the condition (3.10) for the integration over the single group variable $U_\mu(n)$. Since $\Omega(n)$ and $\Omega(n+\hat{\mu})$ can be chosen independently, the measure $\mathrm{d}U$ for a group element must be invariant under left- and right multiplication with another group element $V \in G$, i.e.,

$$\mathrm{d}U = \mathrm{d}(UV) = \mathrm{d}(VU) . \qquad (3.11)$$

The last equation, together with the normalization (integration over all of $G$)

$$\int \mathrm{d}U\, 1 = 1 , \qquad (3.12)$$

are the defining properties of the Haar measure for the integration over compact Lie groups. As announced, we discuss the measure not only for SU(3), but for more general compact Lie groups $G$, such as SU($N$) or U(1). For these cases we now give an explicit construction of the Haar measure. Let $U = U(\omega)$ be an element of the compact Lie group $G$ in the exponential representation (A.2), parameterized by the set of real numbers $\omega^{(k)}$. We have already discussed (see also Appendix A.1) that $\left(\partial/\partial\omega^{(k)}U(\omega)\right)U(\omega)^{-1}$ is in the Lie algebra of the group. Based on these Lie algebra-valued objects we can define a metric $\mathrm{d}s^2$ on the group as (we use summation convention for $n, m$)

$$\mathrm{d}s^2 \equiv \mathrm{tr}\left[\frac{\partial U(\omega)}{\partial\omega^{(n)}}U(\omega)^{-1}\left(\frac{\partial U(\omega)}{\partial\omega^{(m)}}U(\omega)^{-1}\right)^\dagger\right]\mathrm{d}\omega^{(n)}\mathrm{d}\omega^{(m)} = g(\omega)_{nm}\mathrm{d}\omega^{(n)}\mathrm{d}\omega^{(m)}.$$
$$(3.13)$$

From this equation we read off a metric tensor $g(\omega)$,

$$g(\omega)_{nm} = \mathrm{tr}\left[\frac{\partial U(\omega)}{\partial\omega^{(n)}}U(\omega)^{-1}\left(\frac{\partial U(\omega)}{\partial\omega^{(m)}}U(\omega)^{-1}\right)^\dagger\right] = \mathrm{tr}\left[\frac{\partial U(\omega)}{\partial\omega^{(n)}}\frac{\partial U(\omega)^\dagger}{\partial\omega^{(m)}}\right] .$$
$$(3.14)$$

In terms of the metric $g(\omega)$ we can define the measure $\mathrm{d}U$ as

$$\mathrm{d}U = c\sqrt{\det[g(\omega)]}\prod_k \mathrm{d}\omega^{(k)} . \qquad (3.15)$$

Because all groups we consider are compact groups, the parameters $\omega^{(k)}$ need only be integrated over finite intervals for $U(\omega)$ to cover all group elements. The constant $c$ in (3.15) can be used to fulfil the normalization condition (3.12). What is left to show is that the measure defined by (3.14) and (3.15) is indeed invariant under transformations within the group, i.e., that it obeys (3.11). Let $U(\widetilde{\omega})$ be the group element that is obtained from $U(\omega)$ by left- or right multiplication with some other group element $V$. The parameters $\widetilde{\omega}^{(k)}$ are related to the original parameters $\omega^{(k)}$ by some functional relation, i.e., we can write $\omega^{(k)} = \omega^{(k)}(\widetilde{\omega})$. The product measure for the $\omega^{(k)}$ transforms as

$$\prod_k d\omega^{(k)} = \det[J] \prod_k d\widetilde{\omega}^{(k)} \,, \tag{3.16}$$

where the Jacobi determinant is given by

$$J_{kn} = \frac{\partial \omega^{(k)}}{\partial \widetilde{\omega}^{(n)}} \,. \tag{3.17}$$

The metric tensor $g(\widetilde{\omega})$ for $U(\widetilde{\omega})$ is given by

$$
\begin{aligned}
g(\widetilde{\omega})_{nm} &= \operatorname{tr}\left[ \frac{\partial U(\widetilde{\omega})}{\partial \widetilde{\omega}^{(n)}} \frac{\partial U(\widetilde{\omega})^\dagger}{\partial \widetilde{\omega}^{(m)}} \right] \\
&= \operatorname{tr}\left[ \frac{\partial U(\omega)}{\partial \omega^{(k)}} \frac{\partial U(\omega)^\dagger}{\partial \omega^{(l)}} \right] \frac{\partial \omega^{(k)}}{\partial \widetilde{\omega}^{(n)}} \frac{\partial \omega^{(l)}}{\partial \widetilde{\omega}^{(m)}} = g(\omega)_{kl}\, J_{kn}\, J_{lm} \,.
\end{aligned}
\tag{3.18}
$$

In matrix notation the last equation reads ($T$ denotes transposition)

$$g(\widetilde{\omega}) = J^T\, g(\omega)\, J \,. \tag{3.19}$$

Using the factorization property of determinants we obtain from this equation

$$\det[g(\widetilde{\omega})] = \det[g(\omega)]\, \det[J]^2 \,. \tag{3.20}$$

Putting things together, we find that our group measure is indeed invariant,

$$
\begin{aligned}
d\widetilde{U} &= c\,\sqrt{\det[g(\widetilde{\omega})]}\, \prod_k d\widetilde{\omega}^{(k)} = c\,\sqrt{\det[g(\omega)]}\, \det[J] \prod_k d\widetilde{\omega}^{(k)} \\
&= c\,\sqrt{\det[g(\omega)]}\, \prod_k d\omega^{(k)} = dU \,.
\end{aligned}
\tag{3.21}
$$

In the first step we used (3.20) and in the second step (3.16). This completes the proof of invariance of the Haar measure.

### 3.1.3 A few integrals for SU(3)

In order to become familiar with the properties of the Haar measure, we now discuss a few SU(3) integrals which we will need subsequently. Our discussion is based on the invariance of the measure $dU$. The integrals we study are integrals over products of entries $U_{ab}$ of group elements $U$ in the fundamental representation. In particular, we consider the integrals

$$\int_{SU(3)} dU \, U_{ab} = 0 \, , \tag{3.22}$$

$$\int_{SU(3)} dU \, U_{ab}U_{cd} = 0 \, , \tag{3.23}$$

$$\int_{SU(3)} dU \, U_{ab}(U^\dagger)_{cd} = \frac{1}{3} \delta_{ad}\delta_{bc} \, , \tag{3.24}$$

$$\int_{SU(3)} dU \, U_{ab}U_{cd}U_{ef} = \frac{1}{6} \epsilon_{ace}\epsilon_{bdf} \, . \tag{3.25}$$

The basic tool for analyzing these integrals is the following equation for integrals over functions $f(U)$:

$$\int_{SU(3)} dU \, f(U) = \int_{SU(3)} dU \, f(V \, U) = \int_{SU(3)} dU \, f(U \, W) \, . \tag{3.26}$$

The elements $V$ and $W$ are arbitrary SU(3) matrices. This relation follows directly from the invariance (3.11) of the measure. Applying (3.26) to the left-hand side of the first integral (3.22) we obtain

$$\int_{SU(3)} dU \, U_{ab} = \int_{SU(3)} dU \, (V \, U)_{ab} = V_{ac} \int_{SU(3)} dU \, U_{cb} \, , \tag{3.27}$$

where the index $c$ is summed over. In order to fulfil this equation in a non-trivial way, we must have $V_{ac} = \delta_{ac}$. However, the last equation must hold for arbitrary group elements $V$, such that the integral itself must vanish, thus establishing (3.22). In an equivalent way one shows (3.23).

The Eq. (3.24) is somewhat more interesting. Let us for the moment set $b = c$ in the integrand and sum over $b$. We obtain

$$\int_{SU(3)} dU \, \left(U_{a1}(U^\dagger)_{1d} + U_{a2}(U^\dagger)_{2d} + U_{a3}(U^\dagger)_{3d}\right)$$

$$= \int_{SU(3)} dU \, (UU^\dagger)_{ad} = \int_{SU(3)} dU \, \delta_{ad} = \delta_{ad} \, . \tag{3.28}$$

In the second step we have used $U^\dagger = U^{-1}$ and in the last step the normalization of the Haar measure (3.12). For the left-hand side of (3.28) it is important to realize that the three contributions are entirely equivalent. For example the first term can be transformed into the second term by exchanging rows 1 and 2 as well as columns 1 and 2. This operation transforms an element of SU(3) into another group element, and both of these elements are summed over in the group integral. This equivalence implies that each of the three terms on the left-hand side contributes one-third to the result on the right-hand side of (3.28). Thus we have established formula (3.24) for the case $b = c$ (not summed). For $b \neq c$ we can again apply (3.26) to show that for this case the integral has to vanish. The essence of (3.26) is that the integral only gives a nonvanishing result if the integrand contributes to 1, i.e., the Haar measure projects out the contribution of the integrand to the so-called *trivial* or *singlet* representation of the group.

From (3.24) immediately follows a relation that will be useful later:

$$\int dU \ \text{tr}[V\,U] \ \text{tr}[U^\dagger\,W] = \frac{1}{3} \text{tr}[V\,W] \ . \tag{3.29}$$

This equation allows one to integrate the common link variable that occurs in a product of two plaquettes, resulting in a trace of link variables around the two plaquettes. This result is depicted in Fig. 3.1.

**Fig. 3.1.** Integrating out the common link of a product of two plaquettes

Essentially the same mechanism as in (3.24) is also at work in the integral (3.25): Again, different possible combinations of the integrand can be summed to produce a singlet. In particular, if one sets the second indices $b$, $d$, $f$ of the integrand to the values $b = 1$, $d = 2$, $f = 3$, and sums the first indices $a, c, e$ with the completely anti-symmetric $\epsilon$-tensor, one finds

$$\epsilon_{ace}\,U_{a1}\,U_{c2}\,U_{e3} = \det[U] = 1 \ , \tag{3.30}$$

where we have used the definition of the determinant and the fact that elements of SU(3) have $\det[U] = 1$. If two of the second indices in the integrand of (3.30) are equal, say $b = d = 1$, then summing with the $\epsilon$-tensor corresponds to a determinant with two equal columns, and therefore this sum vanishes. If two of the second indices are interchanged with respect to (3.30), e.g., $b = 2, d = 1$ then this corresponds to the interchange of two rows and the determinant acquires an extra minus sign. Thus, we obtain the following formula for summing the integrand of (3.30) with the $\epsilon$-tensor

$$\epsilon_{ace}\,U_{ab}\,U_{cd}\,U_{ef} = \epsilon_{bdf} \ . \tag{3.31}$$

Thus, whenever the set of the first coefficients $a, c, e$ and the set $b, d, f$ both give rise to nonvanishing $\epsilon$-tensor contributions ($\epsilon_{ace} \neq 0, \epsilon_{bdf} \neq 0$), the integrand gives a contribution to the singlet. The sign of this contribution is given by the combination of $\epsilon$-tensors on the right-hand side of (3.25). If this product of $\epsilon$-tensors vanishes for the given combination $a, b, c, d, e, f$, the integrand does not contribute to the singlet and the integral vanishes. Using the same argument as before, based on the simultaneous interchange of rows and columns, one finds that all the nonvanishing contributions have the same weight. The fact that there are six terms on the left-hand side of (3.31) then determines the factor $1/6$ on the right-hand side of (3.25).

This completes our excursion into group integrals. The integrals (3.22), (3.23), (3.24), (3.25), and (3.29) can easily be generalized to the case of $SU(N)$. An introductory account of more general $SU(N)$ group integrals can be found in the book by Creutz [1]. We finally remark that for a treatment of the strong coupling expansion in more detail than in our Sect. 3.4, the concept of character expansion is a powerful tool. However, character expansion, as well as the general representation theory of $SU(N)$, is beyond the scope of this book and we refer the reader to [2–5].

## 3.2 Gauge invariance and gauge fixing

We have seen that gauge invariance is one of the central principles in the construction of QCD. In this section we discuss the freedom of choosing a gauge in more detail and address the different roles of gauge fixing in the continuum and on the lattice. We argue that physical observables have to be gauge-invariant. We stress that all the gauge properties discussed here for pure gauge theory can be taken over to the full theory with fermions.

### 3.2.1 Maximal trees

Let us begin our discussion of gauge fixing with a kind of gauge that is particular for the lattice. We show that in a certain set of link variables, a so-called *maximal tree*, the link variables can be set to the identity element $\mathbb{1}$ and left out in the integration of the gauge field.

We start with an arbitrary configuration of link variables $U_\mu(n)$. In the beginning of our construction all gauge transformation matrices $\Omega(n)$ are set to the identity $\mathbb{1}$. We pick a single-link variable $U_{\mu_0}(n_0)$ and set the transformation matrix $\Omega(n_0 + \hat{\mu}_0)$ at the endpoint of the link to the value $U_{\mu_0}(n_0)$, keeping all other $\Omega(n)$ at $\mathbb{1}$. Our link $U_{\mu_0}(n_0)$ thus is transformed to

$$U_{\mu_0}(n_0)' = \Omega(n_0) U_\mu(n_0) \Omega(n_0 + \hat{\mu}_0)^\dagger = \mathbb{1} U_\mu(n_0) U_\mu(n_0)^\dagger = \mathbb{1} . \quad (3.32)$$

The transformation with the nontrivial $\Omega(n_0 + \hat{\mu}_0)$ will also affect all link variables starting at the site $n_0 + \hat{\mu}_0$, in particular

$$U_{\mu_1}(n_0 + \hat{\mu}_0)' = \Omega(n_0 + \hat{\mu}_0) U_{\mu_1}(n_0 + \hat{\mu}_0) . \quad (3.33)$$

For these transformed links we can repeat the step of (3.32) and choose the matrices $\Omega$ at their endpoints such that also these links are transformed to the

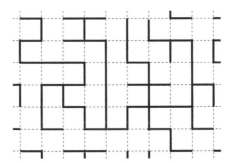

**Fig. 3.2.** Sketch of a maximal tree on a 2D sublattice. The *fat lines* represent link variables that are set to $\mathbb{1}$ and omitted in the integration

identity $\mathbb{1}$. The whole procedure can be repeated until we hit a link $U_{\mu^\star}(n^\star)$ which connects to another link that already has been transformed to $\mathbb{1}$ before. If we wanted to transform also this particular link $U_{\mu^\star}(n^\star)$, according to (3.33) we would transform the other link, which is already at $\mathbb{1}$, away from the identity. This restricts the set of links that can be transformed to $\mathbb{1}$ to a cluster of links which does not contain closed loops. This requirement defines a maximal tree as a maximal collection of link variables that can be transformed to $\mathbb{1}$ (note that there are many different maximal trees). In Fig. 3.2 we show an example of a maximal tree on a 2D sublattice. Note, however, that also smaller subsets than maximal trees can be gauged to $\mathbb{1}$.

Let us now discuss the consequences of the freedom to fix the gauge on a maximal tree or a subset of a maximal tree. We consider the vacuum expectation value of some gauge-invariant observable $O$ as defined in (3.1), (3.2), and (3.3). Gauge invariance of $O$ implies

$$O[U] = O[U'] \,, \tag{3.34}$$

where the link variables in the configurations $U$ and $U'$ are related by a gauge transformation (3.6). We will discuss the need for gauge invariance of our observables in more detail below. Also the action and the measure are gauge-invariant, i.e.,

$$S_G[U] = S_G[U'] \,, \quad \int \mathcal{D}[U] = \int \mathcal{D}[U'] \,. \tag{3.35}$$

Equations (3.34) and (3.35) imply that the whole integrand and the measure in (3.1) are unchanged when setting the links in a maximal tree to $\mathbb{1}$. Since the construction of a maximal tree works for any particular choice of link variables, we can keep the links in the maximal tree at $\mathbb{1}$ throughout the whole integration $\int \mathcal{D}[U]$. Since the Haar measure is normalized to 1 (see (3.12)), the links in the maximal tree can be omitted in the integration altogether. Thus, we can summarize the procedure of fixing the gauge to a maximal tree as follows: Select a maximal tree, or a smaller set of links without closed loops, and set the links in this set to $\mathbb{1}$. Subsequently, integrate over all link variables that are not contained in the chosen set. The expectation value of a gauge-invariant observable is unaffected, whether we fix the gauge or not.

The role of gauge fixing on the lattice is very different from the role it plays in the continuum. Fixing the gauge on the lattice is a step which we can implement to simplify some calculations or to make the interpretation of observables more transparent (see the discussion of the Wilson loop below). However, fixing the gauge is not a necessary step to make vacuum expectation values of operators well defined and computable. In the continuum the situation is different: There the kernel of the quadratic part of the gluon action (the inverse gluon propagator) has zero modes caused by pure gauges, i.e., gauge transformations of the trivial gauge field configuration. These modes give rise to singularities in the gluon propagator that have to be removed by choosing a particular gauge. Only after that step can the continuum theory be well defined in perturbation theory.

We have already mentioned that the procedure of fixing the gauge via a maximal tree is special to the lattice formulation. There is, however, a certain gauge, the so-called *temporal gauge*, which has a corresponding counterpart in the continuum. On a lattice with infinite temporal extent we can set all time-like links to $\mathbb{1}$, i.e., we can set

$$U_4(n) = \mathbb{1} \quad \forall\, n \,. \tag{3.36}$$

We remark that this is not a maximal tree but a smaller set. In the (Euclidean) continuum theory the temporal gauge is defined by $A_4(x) = 0$, which matches the lattice definition due to the relation $U_4 = \exp(\mathrm{i}\,a\,A_4)$. The temporal gauge is important for the Hamiltonian formulation of continuum gauge theories (see any textbook on quantum field theory, such as [6–8]). We will make use of the temporal gauge when we introduce and discuss the Wilson loop.

### 3.2.2 Other gauges

Other gauges that are used in the continuum also play a role on the lattice, in particular when one matches lattice results to continuum calculations done in a particular gauge (e.g., Landau gauge, Coulomb gauge). Also in lattice studies of the confinement mechanism special gauges, such as the maximally abelian gauge, are used. These gauges are not implemented via maximal trees but typically through extremizing some functional. As an example, which illustrates how such gauges are constructed, we briefly discuss the Landau gauge, which we will need in Chap. 11 when we discuss renormalization.

In the continuum the Landau gauge is defined by the condition

$$\partial_\mu A_\mu(x) = 0 \,. \tag{3.37}$$

The important step is to translate this prescription, which in the continuum is formulated in terms of the algebra-valued gauge field $A_\mu(x)$, into a prescription that can be used on the lattice, where we work with the group-valued link variables $U_\mu(n)$. In the first step we show that requiring the Landau gauge condition is equivalent to finding an extremal value of the functional

$$W[A] = \sum_{\mu=1}^{4} \int \mathrm{d}^4x \, \mathrm{tr}\left[A_\mu(x)^2\right] , \qquad (3.38)$$

with regard to gauge transformations. If $W[A]$ is at an extreme value, it has to be invariant when we apply an infinitesimal gauge transformation defined by $\Omega(x) = \exp(i\varepsilon H(x))$. Note that $H(x)$ is a traceless hermitian matrix, such that $\Omega(x)^\dagger = \exp(-i\varepsilon H(x))$. A line of algebra shows that, for our infinitesimal transformation, the gauge transformation (2.11) reduces to

$$A_\mu(x) \;\rightarrow\; A_\mu(x) + \varepsilon \, (i\,[H(x), A_\mu(x)] - \partial_\mu H(x)) + \mathcal{O}(\varepsilon^2) . \qquad (3.39)$$

Inserting the transformed gauge field into the functional $W[A]$, one finds that it changes according to

$$W[A] \rightarrow W[A] - 2\varepsilon \sum_{\mu=1}^{4} \int \mathrm{d}^4x \; \mathrm{tr}\left[A_\mu(x)\,\partial_\mu H(x)\right] + \mathcal{O}(\varepsilon^2)$$

$$= W[A] + 2\varepsilon \sum_{\mu=1}^{4} \int \mathrm{d}^4x \; \mathrm{tr}\left[(\partial_\mu A_\mu(x))\, H(x)\right] + \mathcal{O}(\varepsilon^2) , \quad (3.40)$$

where the periodicity of the trace makes the contribution of the commutator in (3.39) vanish. In the second step we performed an integration by parts (assuming that $H(x)$ vanishes for large $x$). If $W[A]$ is at an extremum, then the $\mathcal{O}(\varepsilon)$ term in (3.40) has to vanish. Since $H(x)$ can be chosen as an arbitrary traceless hermitian matrix, this statement is equivalent to (3.37).

Transforming the $W[A]$ into a corresponding lattice prescription is simple. We can consider the functional

$$W_{\mathrm{lat}}[U] = -a^2 \sum_{n} \sum_{\mu=1}^{4} \mathrm{tr}\left[U_\mu(n) + U_\mu(n)^\dagger\right] , \qquad (3.41)$$

which upon expansion of $U_\mu(n) = \exp\left(i\,a\,A_\mu(n)\right)$ in small $a$ turns into $W[A]$ (the additional constant is irrelevant).

The goal is to find a gauge transformation $\Omega$ which shifts $W_{\mathrm{lat}}[U]$ to a minimal value for a given gauge configuration. More explicitly we have to find the minimum of

$$F[\Omega] = -a^2 \sum_{n} \sum_{\mu=1}^{4} \mathrm{tr}\left[\Omega(n)U_\mu(n)\,\Omega(n+\hat\mu)^\dagger + \Omega(n+\hat\mu)\,U_\mu(n)^\dagger\,\Omega(n)^\dagger\right] ,$$

$$(3.42)$$

as a function of the matrices $\Omega(n)$ for fixed $U_\mu(n)$. This is a large-scale optimization problem which can be attacked with methods such as overrelaxation or simulated annealing (see, e.g., [9]). A particular problem is the constraint that the matrices $\Omega(n)$ have to be in SU(3). We discuss how such matrices are generated numerically in Chap. 4. We remark that a straightforward modification (sum $\mu$ from 1 to 3 only) of the method presented here for Landau gauge allows one to implement the Coulomb gauge condition $\sum_{j=1}^{3} \partial_j A_j = 0$ on the lattice. For a review of gauge-fixing procedures on the lattice see [10].

### 3.2.3 Gauge invariance of observables

In our discussion of maximal trees we already assumed that physical observables $O$ are gauge-invariant, i.e., that they obey (3.34). We now discuss this requirement in more detail.

By inspecting the integral (3.22) one sees that the attempt to use an individual link variable as an observable fails because its expectation value vanishes. Such an individual link variable is an example of a non-gauge-invariant functional $O[U]$, i.e., it obeys $O[U] \neq O[U']$ for configurations $U$ and $U'$ related by a gauge transformation. The expectation value of any non-gauge-invariant functional is equal to the expectation of its average over the gauge group. To see this we perform a gauge transformation $U_\mu(n) \rightarrow \Omega(n) U_\mu(n) \Omega(n + \hat{\mu})^\dagger$ and obtain an expectation value of an arbitrary observable

$$\langle O \rangle = \frac{1}{Z} \int \mathcal{D}[U] \, e^{-S_G[U]} O[U] = \frac{1}{Z} \int \mathcal{D}[\Omega U \Omega^\dagger] \, e^{-S_G[\Omega U \Omega^\dagger]} O[\Omega U \Omega^\dagger]$$
$$= \frac{1}{Z} \int \mathcal{D}[U] \, e^{-S_G[U]} O[\Omega U \Omega^\dagger] , \tag{3.43}$$

where in the last step we used both the gauge invariance of the action and the invariance of the Haar measure. Since the values $\Omega(n)$ can be chosen arbitrarily, we may even average over the gauge group:

$$\langle O \rangle = \frac{1}{Z} \int \mathcal{D}[U] \, e^{-S_G[U]} \left( \int \mathcal{D}[\Omega] \, O[\Omega U \Omega^\dagger] \right) = \int \mathcal{D}[\Omega] \, \langle O[\Omega U \Omega^\dagger] \rangle .$$
$$\tag{3.44}$$

Following these lines of argument, Elitzur [11] proved that local gauge symmetries cannot be broken spontaneously.

According to (3.44) we find for our example of a single-link variable

$$\langle U_\mu(n) \rangle = \int d\Omega(n) \, \langle \Omega(n) \, U_\mu(n) \rangle = \int d\Omega(n) \, \Omega(n) \, \langle U_\mu(n) \rangle = 0 . \tag{3.45}$$

We remark that for some calculations it is indeed advantageous to work with observables that are not gauge-invariant and to evaluate them after the gauge has been fixed. As mentioned, this procedure is useful when lattice results are compared to a continuum calculation done in a particular gauge.

## 3.3 Wilson and Polyakov loops

In this section we present observables which allow one to determine the potential between two static color sources. These observables are the so-called *Wilson* and *Polyakov loops* which we first introduce and only later give their interpretation.

### 3.3.1 Definition of the Wilson loop

In the last section we discussed that physical observables have to be gauge-invariant. A prototype of a gauge-invariant object, made from only the gauge fields, is the trace of a product of link variables along a closed loop which we introduced in (2.46),

$$L[U] = \text{tr} \left[ \prod_{(n,\mu)\in\mathcal{L}} U_\mu(n) \right] . \tag{3.46}$$

Here $\mathcal{L}$ is a closed loop of links on the lattice and the product in (3.46) runs over all these links. The Wilson loop which we introduce now is of that type.

A Wilson loop $W_{\mathcal{L}}$ is made from four pieces, two so-called *Wilson lines* $S(\boldsymbol{m},\boldsymbol{n},n_t)$, $S(\boldsymbol{m},\boldsymbol{n},0)$, and two temporal transporters $T(\boldsymbol{n},n_t), T(\boldsymbol{m},n_t)$. The Wilson line $S(\boldsymbol{m},\boldsymbol{n},n_t)$ connects the two spatial points $\boldsymbol{m}$ and $\boldsymbol{n}$ along some path $\mathcal{C}_{\boldsymbol{m},\boldsymbol{n}}$ with all link variables restricted to time argument $n_t$,

$$S(\boldsymbol{m},\boldsymbol{n},n_t) = \prod_{(\boldsymbol{k},j)\in\mathcal{C}_{\boldsymbol{m},\boldsymbol{n}}} U_j(\boldsymbol{k},n_t) . \tag{3.47}$$

The temporal transporter $T(\boldsymbol{n},n_t)$ is a straight line of $n_t$ link variables in time direction, all situated at spatial position $\boldsymbol{n}$,

$$T(\boldsymbol{n},n_t) = \prod_{j=0}^{n_t-1} U_4(\boldsymbol{n},j) . \tag{3.48}$$

Attaching the four pieces to each other gives a closed loop $\mathcal{L}$,

$$\mathcal{L}: \quad (\boldsymbol{m},n_t) \xrightarrow{S} (\boldsymbol{n},n_t) \xrightarrow{T^\dagger} (\boldsymbol{n},0) \xrightarrow{S^\dagger} (\boldsymbol{m},0) \xrightarrow{T} (\boldsymbol{m},n_t) . \tag{3.49}$$

The Wilson loop $W_{\mathcal{L}}$ is obtained by taking the trace,

$$W_{\mathcal{L}}[U] = \text{tr}\left[ S(\boldsymbol{m},\boldsymbol{n},n_t)T(\boldsymbol{n},n_t)^\dagger S(\boldsymbol{m},\boldsymbol{n},0)^\dagger T(\boldsymbol{m},n_t) \right] = \text{tr}\left[ \prod_{(\boldsymbol{k},\mu)\in\mathcal{L}} U_\mu(\boldsymbol{k}) \right] . \tag{3.50}$$

If the piece of loop $\mathcal{C}_{\boldsymbol{m},\boldsymbol{n}}$ used in $S(\boldsymbol{m},\boldsymbol{n},n_t)$ is a straight line we speak of a *planar Wilson loop*. Note that this can be the case only if $\boldsymbol{m}$ and $\boldsymbol{n}$ fall on a common coordinate axis. Otherwise the Wilson loop is called *nonplanar*. Figure 3.3 shows an example of a planar and a nonplanar loop.

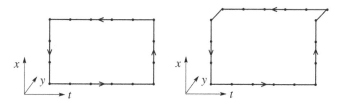

**Fig. 3.3.** Examples for a planar (*left-hand side plot*) and a nonplanar (*right-hand side*) Wilson loop. The horizontal direction is time

### 3.3.2 Temporal gauge

Before we can discuss the physical interpretation of the Wilson loop we must discuss a peculiarity of gauge theories. If one wants to evaluate the canonical momentum (1.45) for the gauge field action (2.17) one finds that for the temporal component $A_4$ the canonical momentum vanishes. The reason is that the field strength tensor $F_{\mu\nu}(x) = -i[D_\mu(x), D_\nu(x)]$ does not contain derivatives of $A_4$ with respect to time. A possible way out of this problem is to use a gauge where

$$A_4(x) = 0 , \tag{3.51}$$

i.e., the aforementioned temporal gauge. We remark that simply stating (3.51) does not by far do justice to the subtleties involved in the quantization of gauge theories, and we refer the reader to field theory books such as [6–8] on this issue. On the lattice, temporal gauge corresponds to the condition (3.36).

We stress that in the following we use the temporal gauge only to find the physical interpretation of the Wilson loop. For the actual computation of the expectation value we do not need to fix the gauge. The result for the expectation value of the Wilson loop is of course the same whether we fix the gauge or not.

### 3.3.3 Physical interpretation of the Wilson loop

In the temporal gauge (3.36), discussed in the last paragraph, the temporal transporters become trivial,

$$T(\boldsymbol{n}, n_t) = \prod_{j=0}^{n_t-1} U_4(\boldsymbol{n}, j) = \mathbb{1} , \tag{3.52}$$

and we obtain the following chain of identities

$$\langle W_{\mathcal{L}} \rangle = \langle W_{\mathcal{L}} \rangle_{\text{temp}} = \left\langle \text{tr} \left[ S(\boldsymbol{m}, \boldsymbol{n}, n_t)\, S(\boldsymbol{m}, \boldsymbol{n}, 0)^\dagger \right] \right\rangle_{\text{temp}} , \tag{3.53}$$

where in the first step we have used the fact that the expectation value of a gauge-invariant observable remains unchanged when fixing the gauge. In the second step we used $W_\mathcal{L}[U] = \text{tr}\left[S(\boldsymbol{m}, \boldsymbol{n}, n_t)\, S(\boldsymbol{m}, \boldsymbol{n}, 0)^\dagger\right]$, which follows from the definition of the Wilson loop in (3.50) and the triviality of the temporal transporter in temporal gauge (3.52).

The temporal gauge used in (3.53) makes explicit that the Wilson loop is the correlator of two Wilson lines $S(\boldsymbol{m}, \boldsymbol{n}, n_t)$ and $S(\boldsymbol{m}, \boldsymbol{n}, 0)$ situated at time slices $n_t$ and 0. Thus we can interpret this correlator using our first key equation in the form of (1.21). Accordingly, the correlator behaves for large total temporal extent $T$ of our Euclidean lattice as ($a, b$ are summed)

$$\left\langle \text{tr}\left[S(\boldsymbol{m}, \boldsymbol{n}, n_t) S(\boldsymbol{m}, \boldsymbol{n}, 0)^\dagger\right]\right\rangle_{\text{temp}} = \sum_k \langle 0|\widehat{S}(\boldsymbol{m}, \boldsymbol{n})_{ab}|k\rangle \langle k|\widehat{S}(\boldsymbol{m}, \boldsymbol{n})_{ba}^\dagger|0\rangle\, e^{-t E_k} ,$$

$$(3.54)$$

where the Euclidean time argument $t$ is related to $n_t$ via $t = a\, n_t$ with $a$ being the lattice spacing. The sum in (3.54) runs over all states $|k\rangle$ that have a nonvanishing overlap with $\widehat{S}(\boldsymbol{m}, \boldsymbol{n})^\dagger|0\rangle$.

In the next paragraph we will argue that the states $|k\rangle$ with nonvanishing overlap are states describing a static quark–antiquark pair located at spatial positions $\boldsymbol{m}$ and $\boldsymbol{n}$. Thus in (3.54) the term with the lowest energy $E_1$ is expected to be the state describing our static quark–antiquark pair. Higher states could be, e.g., this pair plus additional particle–antiparticle combinations with the quantum numbers of the vacuum. The energy $E_1$ is thus identified with the energy of the quark–antiquark pair, which is the static potential $V(r)$ at spatial quark separation $r$,

$$E_1 = V(r) \quad \text{with} \quad r = a\,|\boldsymbol{m} - \boldsymbol{n}| . \tag{3.55}$$

Combining (3.53), (3.54), and (3.55) we obtain

$$\langle W_\mathcal{L}\rangle \propto e^{-t V(r)}\left(1 + \mathcal{O}(e^{-t\,\Delta E})\right) = e^{-n_t\, a\, V(r)}\left(1 + \mathcal{O}(e^{-n_t\, a\,\Delta E})\right) . \tag{3.56}$$

Thus we find that we can calculate the static quark–antiquark potential from the large-$n_t$ behavior of the Wilson loop. The corrections in (3.56) are exponentially suppressed, where $\Delta E$ is the difference between $V(r)$ and the first excited energy level of the quark–antiquark pair.

We stress that Wilson loops are oriented. However, reversing the orientation, which on an algebraic level is complex conjugation (compare (2.34)), simply corresponds to the interchange of quark and antiquark. Thus both orientations serve equally well for a determination of the potential $V(r)$.

The Wilson loops we have introduced are not necessarily planar. In Fig. 3.3 we show two Wilson loops, a planar one (left-hand side plot) and a nonplanar loop (right-hand side). Both loops have $n_t = 5$ (the horizontal direction is time). The planar loop has $r = 3\, a$, the nonplanar loop has $r = \sqrt{3^2 + 1}\, a = \sqrt{10}\, a$. Thus with nonplanar Wilson loops we can calculate the potential $V(r)$ not only at distances $r$ that are integer multiples of $a$, but also at intermediate points. Nonplanar Wilson loops also allow one to study whether rotational invariance is eventually restored when approaching the continuum limit.

Wilson loops may also be used as operators for purely gluonic bound states, the so-called *glueballs*. The exponential decay of the corresponding Euclidean correlation functions allow to extract the mass spectrum of these states (cf. Chap. 6 and the discussion in [12]).

### 3.3.4 Wilson line and the quark–antiquark pair

To complete our physical interpretation of the Wilson loop we still need to show that the states $\widehat{S}(m, n)^\dagger_{ba}|0\rangle$ do indeed have overlap with a quark–antiquark pair. However, the true derivation that $S(m, n)$ is the correct expression to describe a quark–antiquark pair in the limit of large quark mass, will only be given in Chap. 5 after we have provided the final definition of the quark fields on the lattice. There we will show that the quark propagator reduces to $S(m, n)$ in the limit of infinitely heavy quarks. For now we are content with showing that $S(m, n)$ has the same transformation properties as a quark–antiquark pair under a gauge transformation.

According to our discussion in Sect. 2.1, a quark–antiquark pair at spatial positions $am$, $an$ is described by a product of fields

$$Q(m, n)_{ab} \equiv \psi(m)_{\substack{\alpha \\ a}} \, \overline{\psi}(n)_{\substack{\beta \\ b}} \, . \tag{3.57}$$

The quark fields carry spinor $(\alpha, \beta)$ and color $(a, b)$ indices. However, here we are not interested in the dependence of the potential on the spinor indices, and we ignore them in the definition of $Q(m, n)_{ab}$. $Q(m, n)_{ab}$ is not gauge-invariant. According to (2.7), it transforms under gauge transformations as

$$Q(m, n)_{ab} \to \Omega(m)_{aa'} \, Q(m, n)_{a'b'} \, \Omega(n)^\dagger_{b'b} \, . \tag{3.58}$$

From the discussion following the gauge transformation properties of link variables (2.33), we know that the products $P[U]$ of link variables, defined in (2.43), transform exactly as required (compare (2.44)),

$$S(m, n)_{ab} \to \Omega(m)_{aa'} \, S(m, n)_{a'b'} \, \Omega(n)^\dagger_{b'b} \, . \tag{3.59}$$

Thus we have at least verified that the Wilson line has the same transformation properties as the quark–antiquark pair.

### 3.3.5 Polyakov loop

Let us conclude this section with discussing a modification of the Wilson loop, the so-called *Polyakov loop* [13] (also called thermal Wilson line). Here we work with boundary conditions for the gauge fields that are periodic in the time direction. We make the temporal extent $n_t$ of the Wilson loop as large as possible on our lattice, i.e., we set $n_t = N_T$, where $N_T$ is the total number of lattice points in time direction. Then the spatial pieces of the

Wilson loop (the non-horizontal lines in Fig. 3.3) sit on top of each other but are oriented in opposite direction. Due to the periodic boundary conditions we cannot gauge-transform all temporal links to $\mathbb{1}$. We can, however, gauge the spatial pieces of our loop to $\mathbb{1}$. Then the Wilson loop reduces to the two disconnected paths (compare (3.48)) $T(\boldsymbol{m}, N_T), T(\boldsymbol{n}, N_T)^\dagger$ of temporal link variables, located in space at the two positions $\boldsymbol{m}$ and $\boldsymbol{n}$. Both these paths wind around the temporal direction of the lattice but have opposite orientations.

We can make this new observable gauge-invariant by taking the trace for each of the two loops individually. This is simply a rearrangement of the color indices and leaves the interpretation of the observable the same. In this way we introduce the so-called Polyakov loop

$$P(\boldsymbol{m}) = \operatorname{tr}\left[\prod_{j=0}^{N_T-1} U_4(\boldsymbol{m}, j)\right] , \qquad (3.60)$$

which as a trace over a closed loop is gauge-invariant. We now can abandon our special gauge and obtain ($r = a|\boldsymbol{m} - \boldsymbol{n}|$)

$$\langle P(\boldsymbol{m}) P(\boldsymbol{n})^\dagger \rangle \propto e^{-N_T a V(r)} \left(1 + \mathcal{O}(e^{-N_T a \Delta E})\right) . \qquad (3.61)$$

The numerical calculation of the static potential which we present in Chap. 4 is based on the Polyakov loop, i.e., on the two Eqs. (3.60) and (3.61). The identity of the potential as defined via Wilson or Polyakov loops, respectively, is not rigorously proven; there is a proof that the string tension derived from the Polyakov loop correlator is bounded from above by that of the Wilson loop [14].

An alternative way of motivating the introduction of the Polyakov loop is to couple a current $j_\mu$ to the field $A_\mu$ by defining an operator $O = \operatorname{tr}\left[P \exp(i \int d^4 z \, j_\mu(z) \, A_\mu(z))\right]$. Using the current corresponding to a static charge at position $\boldsymbol{x}$, given by $j_\mu(z) = (0, 0, 0, 1)\,\delta(\boldsymbol{z} - \boldsymbol{x})$ (Euclidean metric), on the lattice the operator $O$ translates to our Polyakov loop.

We finally remark that the vacuum expectation value $\langle P(\boldsymbol{n}) \rangle$ of a single Polyakov loop is also an important variable. As we will discuss in Chap. 12, it is an order parameter for the deconfinement transition in gluodynamics at finite temperature.

## 3.4 The static quark potential

Having introduced the Wilson loop as an observable for the static quark potential $V(r)$, we now discuss the general form of $V(r)$. As a first step we calculate the potential in the limit of strong coupling $g$ (corresponding to small $\beta$ – compare (3.5)) and find that this gives rise to a linearly rising term. Subsequently, we argue that, for small coupling $g$, we obtain the $1/r$ potential familiar from electrodynamics. Thus, we find that the static QCD potential can be parameterized by

$$V(r) = A + \frac{B}{r} + \sigma r \, . \tag{3.62}$$

Since the force between the quarks is the derivative of $V(r)$, the constant $A$ is only an irrelevant normalization of the energy. The second term in (3.62) is the Coulomb part of the potential with strength $B$. Finally, the third contribution is a linearly rising term and the real constant $\sigma$ is the so-called *string tension*. From QCD phenomenology one expects a value of $\sigma \approx 900$ MeV/fm. After providing the evidence for the individual terms in (3.62), we discuss the physical implications of the static QCD potential.

### 3.4.1 Strong coupling expansion of the Wilson loop

In order to demonstrate the presence of the linearly rising term in (3.62), we calculate the vacuum expectation value of the Wilson loop in the limit of strong coupling, i.e., large $g$ (small $\beta$). More explicitly we compute

$$\langle W_\mathcal{C} \rangle = \frac{1}{Z} \int \mathcal{D}[U] \exp\left( -\frac{\beta}{3} \sum_P \mathrm{Re}\,\mathrm{tr}[\mathbb{1} - U_P] \right) \mathrm{tr}\left[ \prod_{l \in \mathcal{C}} U_l \right] \, . \tag{3.63}$$

For this calculation we use a simplified notation: The sum runs over all plaquettes $P$, where each plaquette is counted with only one of the two possible orientations. The product over $l$ runs over all link variables contained in the contour $\mathcal{C}$ defining the Wilson loop. This expression can be rewritten as

$$\langle W_\mathcal{C} \rangle = \frac{1}{Z'} \int \mathcal{D}[U] \exp\left( \frac{\beta}{3} \sum_P \mathrm{Re}\,\mathrm{tr}[U_P] \right) \mathrm{tr}\left[ \prod_{l \in \mathcal{C}} U_l \right]$$

$$= \frac{1}{Z'} \int \mathcal{D}[U] \exp\left( \frac{\beta}{6} \sum_P \left( \mathrm{tr}[U_P] + \mathrm{tr}[U_P^\dagger] \right) \right) \mathrm{tr}\left[ \prod_{l \in \mathcal{C}} U_l \right] \, . \tag{3.64}$$

In the first step we separate the constant factor $\exp(-\beta/3 \sum_P \mathrm{Re}\,\mathrm{tr}[\mathbb{1}])$ from the Boltzmann factor $\exp(-S)$. Exactly the same constant factor appears in the partition function $Z$ and we cancel the two factors in the numerator and the denominator. The partition function without this factor is denoted by $Z'$. In the second step we use

$$\mathrm{Re}\,\mathrm{tr}[U_P] = \frac{1}{2} \left( \mathrm{tr}[U_P] + \mathrm{tr}[U_P^\dagger] \right) \, . \tag{3.65}$$

We stress that, according to (2.48), hermitian conjugation of the plaquette variable $U_P$ is equivalent to inverting the orientation of the plaquette. Thus in the second line of (3.64) we explicitly display both orientations of the plaquette variables $U_P$, which leads to an extra factor $1/2$.

In the form of (3.64) we can now discuss the expansion of the Wilson loop expectation value for strong coupling (small $\beta$). In particular we expand the Boltzmann factor of (3.64) in $\beta$ using the Taylor expansion for the exponential function,

$$\exp\left(\frac{\beta}{6}\sum_P\left(\text{tr}[U_P]+\text{tr}[U_P^\dagger]\right)\right) = \sum_{i,j=0}^\infty \frac{1}{i!j!}\left(\frac{\beta}{6}\right)^{i+j}\left(\sum_P\text{tr}[U_P]\right)^i\left(\sum_P\text{tr}[U_P^\dagger]\right)^j.$$

(3.66)

Note that in this expansion we have separated the contributions from clockwise oriented plaquettes $U_P^\dagger$ and counter-clockwise oriented plaquettes $U_P$. This is important since for the leading term in the expansion only those plaquettes oriented oppositely to the Wilson loop contribute.

For the normalization factor $Z'$ it is straightforward to determine the leading contribution in the small-$\beta$ expansion. Already the first term with $i = j = 0$ in (3.66) gives a nonvanishing contribution to the integral and we obtain (using the normalization of the Haar measure)

$$Z' = \int \mathcal{D}[U]\,\exp\left(\frac{\beta}{6}\sum_P\left(\text{tr}[U_P]+\text{tr}[U_P^\dagger]\right)\right) = \int \mathcal{D}[U]\,(1+\mathcal{O}(\beta)) = 1+\mathcal{O}(\beta^2).$$

(3.67)

The expansion of the numerator of (3.64) is less straightforward. If only the leading term in the expansion of the Boltzmann factor is kept, then the product of link variables building up the observable, $\prod_{l\in\mathcal{C}}U_l$, gives rise to integrals of the type (3.22) which all vanish. Thus, in order to find the leading nonvanishing term of the expansion we have to expand the Boltzmann factor in small $\beta$. This brings down additional link variables from the exponent and in this way we can saturate the integrals over the links to obtain nonvanishing contributions of the type (3.24). In sub-leading terms (3.25) also contributes.

If we consider the contour $\mathcal{C}$ of the Wilson loop to be a $n_r \times n_t$ rectangle of links, then the minimal area $\mathcal{A}_\mathcal{C}$ spanned by this contour contains $n_A = n_r\,n_t$ plaquettes (note that $n_r, n_t, n_A$ are positive integers). The physical area $\mathcal{A}_\mathcal{C}$ is related to the extension of the Wilson loop in physical units $a\,n_r, a\,n_t$ by $\mathcal{A}_\mathcal{C} = a^2\,n_A = a\,n_r\,a\,n_t$. Remembering (3.29) and Fig. 3.1 we find nonvanishing contributions only when each link variable $U_\mu(n)$ in the loop is paired with its conjugate partner $U_\mu(n)^\dagger$. Since we have plaquettes in our action, this must continue until we have filled the contour $\mathcal{C}$ with $n_A$ plaquettes obtained from the expansion of the Boltzmann factor. We depict this contribution in Fig. 3.4.

Note that the plaquettes used for filling the contour have to have the opposite orientation of the Wilson loop. Only in this way the contributions at each link have the form of the integrand in (3.24). Since we need at least $n_A = n_r\,n_t$ plaquettes from the exponent, the necessary term in the expansion (3.66) of the exponential is of order $n_A$. Explicitly this leading term reads (note that from the two orientations in (3.66) only the one opposite to the Wilson loop contributes)

$$\int \mathcal{D}[U] \frac{1}{n_A!} \left(\frac{\beta}{6}\right)^{n_A} \left(\sum_P \mathrm{tr}[U_P^\dagger]\right)^{n_A} \mathrm{tr}\left[\prod_{l \in \mathcal{C}} U_l\right]$$

$$= \left(\frac{\beta}{6}\right)^{n_A} \int \mathcal{D}[U] \prod_{P \in \mathcal{A}_\mathcal{C}} \mathrm{tr}[U_P^\dagger] \; \mathrm{tr}\left[\prod_{l \in \mathcal{C}} U_l\right]$$

$$= \mathrm{tr}[\mathbb{1}] \left(\frac{\beta}{6}\right)^{n_A} \left(\frac{1}{3}\right)^{n_A} = 3 \exp\left(n_A \ln\left(\frac{\beta}{18}\right)\right). \quad (3.68)$$

In the first step of (3.68) we expand the $n_A$-th power over the sum of plaquette variables with the correct orientation $(U_P^\dagger)$. We keep only the terms where each of the $n_A$ plaquettes $P$ inside the minimal area $\mathcal{A}_\mathcal{C}$ is occupied by the matching $U_P^\dagger$. There are exactly $n_A!$ such products and thus the factor $1/n_A!$ is canceled. All other terms in the expansion of the $n_A$-th power vanish since they give rise to integrals of the type (3.22). In the second step we evaluate the nonvanishing term using the integral (3.29) in the form depicted in Fig. 3.1. We glue together the plaquette variables inside the Wilson loop first into rows and then these rows into the full inner block of Fig. 3.4. Another gauge integral ties this block to the oppositely oriented outer contour of the Wilson loop. One finds that these steps give rise to the factor $(1/3)^{n_A}$ in (3.68). Combining (3.67) and (3.68) we find

$$\langle W_\mathcal{C} \rangle = 3 \exp\left(n_A \ln\left(\frac{\beta}{18}\right)\right) (1 + \mathcal{O}(\beta)) = 3 \exp\left(n_r\, n_t \ln\left(\frac{\beta}{18}\right)\right) (1 + \mathcal{O}(\beta)).$$
$$(3.69)$$

According to (3.56) this expression has to be compared to the asymptotic form, i.e., for large $t = a\, n_t$ we have

$$\langle W_\mathcal{C} \rangle \propto \exp\left(-a\, n_t\, V(r)\right). \quad (3.70)$$

Thus, we conclude that in the strong coupling limit (note that $r = a\, n_r$)

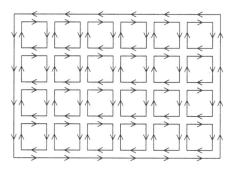

**Fig. 3.4.** Leading contribution in the strong coupling (small $\beta$) expansion of the Wilson loop. The outer, counter-clockwise-oriented *rectangle* is the Wilson loop, the smaller, clockwise-oriented *squares* are the single plaquette terms from the action

$$V(r) = \sigma\, r\,, \qquad (3.71)$$

where the string tension $\sigma$ is given by the leading order expression

$$\sigma = -\frac{1}{a^2}\, \ln\left(\frac{\beta}{18}\right)\, (1 + \mathcal{O}(\beta))\,. \qquad (3.72)$$

We remark that it is relatively easy to produce higher corrections to the string tension. However, since the strong coupling expansion does not play a central role in modern lattice gauge theory (we will soon see that we are actually interested in approaching weak coupling), we will not discuss the calculation of higher terms and the convergence properties of this series.[1] Instead, we stress once more that with a relatively simple expansion we have extracted from the lattice formulation a linearly rising potential. Below we discuss that such a term in the potential gives rise to the important feature of confinement. However, before we come to the discussion of the physical implications, let us first present the argument for the presence of the Coulomb-type term in the parameterization (3.62) of the static QCD potential.

### 3.4.2 The Coulomb part of the static quark potential

The presence of a Coulomb-type interaction can be readily seen from the behavior of the gluon action in the continuum for small coupling constant $g$. This argument is best presented in the normalization of the gauge field introduced in (2.18) and (2.19), where the gauge field $A_\mu$ was rescaled by a factor of $1/g$. In this form the Euclidean continuum gauge field action, written as a sum over the color components, reads (compare (2.23) and (2.24))

$$S_G[A] = \frac{1}{4}\sum_{i=1}^{8} \int \mathrm{d}^4x\, F_{\mu\nu}^{(i)}(x) F_{\mu\nu}^{(i)}(x)\,, \qquad (3.73)$$

where the field strength tensor for individual color components is given by

$$F_{\mu\nu}^{(i)}(x) = \partial_\mu A_\nu^{(i)}(x) - \partial_\nu A_\mu^{(i)}(x) - g\, f_{ijk} A_\mu^{(j)}(x) A_\nu^{(k)}(x)\,. \qquad (3.74)$$

From (3.74) it is obvious that the self-interaction terms that are specific for QCD are multiplied by the coupling constant $g$. Thus, when sending this coupling to zero, the field strength tensor (3.74) reduces to its abelian counterpart, i.e., it has the form known from QED. Thus, the action (3.73) turns into a sum over QED-type interactions for each color component. Since we know that in QED the static potential is of the Coulomb-type,[2] we conclude that the $1/r$ term should be present in the parameterization (3.62) of the static QCD potential.

---

[1]The original derivation by Wilson [15] has been rigorously proven in [16]. We refer the reader to Creutz' book [1] for an elementary presentation or to [17] for a more advanced calculation.

[2]See [18] for an elementary derivation of this fact in the path integral formalism.

### 3.4.3 Physical implications of the static QCD potential

Let us briefly sketch the physical implications of the static QCD potential, in particular the role of the linear term. The linearly rising term in the potential between a static quark–antiquark pair implies that the energy keeps rising linearly as one tries to pull the two constituents apart. Thus the quark and the antiquark are confined in a strongly bound meson state. Similarly, as we show later, also a combination of three quarks is bound, forming a baryon. The phenomenon that only color neutral combinations like hadrons are observable objects is one possible definition of confinement.

The physical mechanism which leads to the linearly rising term is the formation of a flux tube between the two sources. In QED, where we have no self-interaction of the gauge field, the field lines between a source and a sink spread out in space. In QCD the strong self-interaction of the gluons prevents this behavior, and the field is squeezed into a narrow tube or string producing the linear rise. Direct experimental evidence for the linearly rising potential is seen when the mass of hadrons is plotted as a function of their total spin and a linear behavior is found (see, e.g., the discussion in [19]). Since for a linearly rising potential the energy rises linearly with the angular momentum, this experimental finding confirms the linear term in (3.62).

So far our discussion is based only on the static potential obtained from pure gluodynamics. Certainly also the quarks, which we have not yet included in our discussion, play an important role. In the full theory with dynamical quarks, processes of particle–antiparticle creation and annihilation become important. In particular if the quark and antiquark are pulled sufficiently far apart, the energy becomes large enough to create a quark–antiquark pair, which may recombine with the two initial constituents to form two mesons. This phenomenon is called *string breaking* and can be studied on the lattice.

To summarize, we find that lattice QCD is very friendly to confinement. This property can easily be proven in the strong coupling limit, but as we shall see in Chap. 4, it is also relatively simple to extract the QCD potential in a numerical calculation at weaker coupling.

## 3.5 Setting the scale with the static potential

In the lattice formalism the action is given in units of $\hbar$ and all observables are dimensionless. Only by relating them to physical quantities we may introduce a scale parameter. An example of such a dimensionless quantity calculated on the lattice is the product $a\,M$ of the lattice spacing $a$ and some mass $M$, which may be determined from an exponential decay of a correlation function like in (1.21). Identifying $M$ with a physical mass allows us to determine the lattice constant $a$ in physical units.

In this section we discuss another method for setting the scale, which is based on the static quark potential $V(r)$. This allows to relate the lattice spacing $a$ to the inverse gauge coupling $\beta$. We define a certain physical distance $r_0$, the so-called Sommer parameter [20], which is a characteristic length scale tied to the static potential. In physical units this distance is $r_0 \simeq 0.5$ Fermi (1 Fermi = 1 fm = $10^{-15}$ m). This distance allows us to determine the lattice spacing $a$ from $V(r)$ simply by counting the number of lattice points between $r = 0$ and $r = r_0$. At the end of this section we will be able to discuss the true continuum limit of pure lattice gauge theory.

### 3.5.1 Discussion of numerical data for the static potential

Let us begin our presentation with a discussion of numerical data for the static potential. At the moment we assume that the data have already been calculated since we introduce the actual techniques for such a calculation only in Chap. 4. We assume that we have computed the expectation value $\langle W_{\mathcal{C}} \rangle$ for a planar Wilson loop of size $r \times t$, i.e., the contour $\mathcal{C}$ is an $r \times t$ rectangle. The spatial distance $r$ and the Euclidean time $t$ are related to integer numbers $n$ and $n_t$ through the lattice spacing $a$,

$$r = na , \quad t = n_t a . \tag{3.75}$$

According to (3.56) the vacuum expectation value of the Wilson loop is connected to the static potential via (we here omit the exponentially suppressed corrections shown explicitly in (3.56))

$$\langle W_{\mathcal{C}} \rangle = C \exp\left(-t\, V(r)\right) = C \exp\left(-n_t\, a V(na)\right) . \tag{3.76}$$

Since at this stage the lattice spacing $a$ is still unknown, all we can extract from our numerical data for $\langle W_{\mathcal{C}} \rangle$ are values for the product $aV(an)$ at different $n$. In practice this is done via a two-parameter fit of the data for different $n_t$ (but fixed $n$) according to (3.76). The fit parameters are $C$ and $aV(an)$. This procedure can be repeated for different values of $n$, and the result is a set of numerical data for $aV(an)$ as a function of $n$.

In Fig. 3.5 we show the results from a numerical calculation of $aV(an)$ as a function of $n$ for two different values of $\beta$. The actual numerical results are represented by the symbols which we connect with dotted lines to guide the eye. Since the data are from a numerical simulation, they come with (small) statistical errors represented by horizontal bars. Note that in the two plots different scales are used for the axes. The vertical scales are chosen such that the shape of the static potential is similar in the two plots.

The plots nicely show the linearly rising behavior at larger values of $n$ which we have already deduced from our strong coupling calculation. For short distances the data points also display a curvature which comes from the Coulomb part of the potential. This is more pronounced in the right-hand side plot which is for a larger value of $\beta$. In particular at small distances the lattice Coulomb potential deviates from the continuum form. The lattice Coulomb potential is related to the lattice-free boson propagator [23] and is often used to correct for this deviation [24–26].

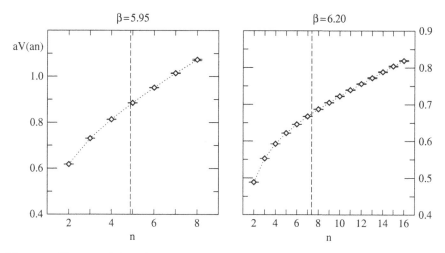

**Fig. 3.5.** Numerical data for the static quark potential computed with the Wilson gauge action at two different couplings $\beta$. The *symbols* are the numerical data which we connect to guide the eye. The *dashed vertical lines* are drawn at a distance that corresponds to the Sommer parameter. The data are taken from [21, 22]

### 3.5.2 The Sommer parameter and the lattice spacing

We have already announced that we introduce the Sommer parameter as a way to determine the lattice spacing $a$. Later in the book we discuss alternative methods to determine $a$, e.g., by using the masses of hadrons. Although the different methods provide values of $a$ that differ slightly at a given gauge coupling, in the continuum limit the final physical results should agree.

The Sommer parameter $r_0$ is a certain distance related to the shape of the static potential. Its physical value is $r_0 \simeq 0.5$ fm. Although we discuss the precise definition of the distance $r_0$ only below, we have already marked its position in the two plots of Fig. 3.5 using dashed vertical lines. Thus, the distance between $n = 0$ on the horizontal axis and the dashed line corresponds to 0.5 fm. Note that the variable $n$ on the horizontal axis is dimensionless – it simply is the number of lattice spacings in the spatial direction of the contour $\mathcal{C}$. Thus, what is marked by the vertical dashed lines is the Sommer scale in lattice units, i.e., $r_0/a$. We will discuss below how to extract the ratio $r_0/a$ from the numerical data for $aV(an)$.

Once the Sommer scale $r_0/a$ is computed, it is easy to determine the lattice spacing. When inspecting the two plots in Fig. 3.5, we find that the number of lattice points between the origin ($n = 0$) and the vertical dashed line at $r_0/a$ is different for the two values of $\beta$. In the left-hand side plot of Fig. 3.5, corresponding to $\beta = 5.95$, we find about $n = 4.9$ lattice points to the left of the dashed line. In the plot for $\beta = 6.20$ we find approximately $n = 7.4$ lattice points left of the vertical line. Since in both cases the distance between $n = 0$ and the vertical line corresponds to 0.5 fm, we find for the lattice spacing $a$:

$$a = 0.5\,\text{fm}\ /4.9\ \approx\ 0.102\,\text{fm}\quad \text{for}\ \beta = 5.95\ ,$$
$$a = 0.5\,\text{fm}\ /7.4\ \approx\ 0.067\,\text{fm}\quad \text{for}\ \beta = 6.20\ . \tag{3.77}$$

We finally need to discuss how the Sommer scale $r_0$ is defined [20]. The Sommer scale is not based directly on the potential $V(r)$, but instead on the force $F(r) = dV(r)/dr$ between the two static quarks.[3] For sufficiently heavy quarks, quark–antiquark bound states can be described by an effective nonrelativistic Schrödinger equation and the force $F(r)$ can be studied. From comparing with experimental data for the $\bar{b}b$ and $\bar{c}c$ spectra one finds that

$$F(r_0)\,r_0^2 = 1.65\quad \text{corresponds to}\quad r_0 \simeq 0.5\ \text{fm}\ . \tag{3.78}$$

Thus, we need to calculate from our numerical data for $V(r)$ the dimensionless product $F(r)\,r^2$ and determine the value $r = r_0$ where this product assumes the value 1.65. For an easy-to-digest presentation of this determination we discuss the Sommer parameter directly for the potential as parameterized in (3.62). We obtain for the force

$$F(r)\ =\ \frac{d}{dr}V(r)\ =\ \frac{d}{dr}\left(A + \frac{B}{r} + \sigma r\right)\ =\ -\frac{B}{r^2} + \sigma\ . \tag{3.79}$$

Thus, for the parameterized potential the condition (3.78) reads

$$F(r_0)\,r_0^2\ =\ -B + \sigma r_0^2\ =\ 1.65\ , \tag{3.80}$$

which implies $r_0 = \sqrt{(1.65 + B)/\sigma}$, or expressed in lattice units

$$\frac{r_0}{a}\ =\ \sqrt{\frac{1.65 + B}{\sigma\,a^2}}\ . \tag{3.81}$$

The ratio $r_0/a$ is exactly what we have used in Fig. 3.5 and the subsequent determination of $a$. The numbers $B$ and $\sigma a^2$ can be determined from the fit[4] of our data for $aV(an)$ to (this is equation (3.62) for $r = an$)

$$aV(an)\ =\ Aa + \frac{B}{n} + \sigma a^2\,n\ , \tag{3.82}$$

for different values of $n$. Thus, the determination of the lattice spacing can be summarized as follows: *Step 1:* Determine $B$ and $\sigma a^2$ from the numerical data for $aV(an)$. *Step 2:* Use (3.81) to calculate the dimensionless number $X = r_0/a$. *Step 3:* The lattice spacing is then given by $a = (0.5/X)$ fm.

---

[3]This definition of $F$ differs by a minus sign from the usual definition.

[4]We remark that such a fit already assumes that the potential has – at least locally – the form (3.62). An alternative approach is to directly determine the force $F$ by discretizing the derivative $d/dr$ on the lattice (see [20] for details). This approach has the further advantage that one does not need to fit $\sigma a^2$ and $B$, a fit which is dominated by large values of $n$ where the signal to noise ratio of the data for $aV(an)$ becomes poor.

In the discussion of Fig. 3.5 we have already addressed the fact that the lattice spacing decreases when increasing $\beta$ (see also (3.77)). In [22] the lattice spacing $a$ was determined for the Wilson action at several values of $\beta$ and the dependence of $a$ on $\beta$ was parametrized for $5.7 \leq \beta \leq 6.92$ as

$$a \; = \; r_0 \, \exp\left(-1.6804 - 1.7331(\beta - 6) + 0.7849(\beta - 6)^2 - 0.4428(\beta - 6)^3\right) . \tag{3.83}$$

The form of the parametrization in (3.83) is inspired by the renormalization group, an important concept which we briefly address now.

### 3.5.3 Renormalization group and the running coupling

Coupling constants like the gauge coupling $g$ or a quark mass $m$ as they enter the action functional are usually called *bare* parameters. These are not directly observable "physical" numbers. Only by computing observables such as hadron masses, the string tension, or the Sommer parameter, and by identifying those with experimental values one can find out the values of the bare parameters of the action in physical units.

Lattice actions may differ in various aspects. They may use different discretizations of derivatives or the lattice grid, which is usually taken to be hypercubic, may vary in its structure. However, when removing the lattice cutoff, i.e., in the limit $a \to 0$, physical observables should agree with the experimental value and become independent of $a$. In general this will imply that the bare parameters have a nontrivial dependence on the cutoff $a$, meaning that they are functions $g(a)$, $m(a)$, etc. As we send $a \to 0$ the values of the bare parameters will have to be changed in order to keep physics constant.

This *running* of the bare parameters is addressed by the so-called *renormalization group*. To simplify the discussion of this idea we consider pure gauge theory where we have only one bare coupling, the gauge coupling $g(a)$. Let $P\left(g(a), a\right)$ be a physical observable which in the limit $a \to 0$ obtains its physical value $P_0$,

$$\lim_{a \to 0} P\left(g(a), a\right) \; = \; P_0 . \tag{3.84}$$

Callan and Symanzik, following early suggestions for QED by Stückelberg, Peterman, Gell-Mann, and Low, formulated the requirement of constant physics in a differential equation:

$$\frac{dP(g, a)}{d \ln a} = 0 \quad \text{or, equivalently} \quad \left(\frac{\partial}{\partial \ln a} + \frac{\partial g}{\partial \ln a} \frac{\partial}{\partial g}\right) P(g, a) = 0 . \tag{3.85}$$

The equation relates to a semi-group of scale-changing transformations, hence the name renormalization group equation. (Actually and more precisely, the right-hand side of this equation is $\mathcal{O}\left((a/\xi)^2 \ln(a/\xi)\right)$ for a lattice system with correlation length $\xi$.) The coefficient function of the second term is called the $\beta$-function (not to be mixed up with our inverse gauge coupling),

$$\beta(g) \equiv -\frac{\partial g}{\partial \ln a} \, , \tag{3.86}$$

and determines, up to an integration constant, how the coupling $g$ depends on the cutoff $a$.

The $\beta$-function may be expanded in a power series around $g = 0$, with coefficients determined by perturbation theory. For $SU(N)$ and $n_f$ massless quarks the result reads

$$\beta(g) = -\beta_0 g^3 - \beta_1 g^5 + \mathcal{O}(g^7) \, ,$$
$$\beta_0 = \frac{1}{(4\pi)^2} \left( \frac{11}{3} N - \frac{2}{3} n_f \right) \, , \tag{3.87}$$
$$\beta_1 = \frac{1}{(4\pi)^4} \left( \frac{34}{3} N^2 - \frac{10}{3} N n_f - \frac{N^2 - 1}{N} n_f \right) \, .$$

For pure gauge $SU(3)$ we have to take $N = 3$ and $n_f = 0$. These first two coefficients of the expansion are universal, independent of the regularization scheme. In general, however, the $\beta$-function will depend on the details of the regularization.

The differential Eq. (3.86) with (3.87) can be solved, using separation of variables, and one obtains

$$a(g) = \frac{1}{\Lambda_{\mathrm{L}}} \left( \beta_0 g^2 \right)^{-\frac{\beta_1}{2\beta_0^2}} \exp \left( -\frac{1}{2\beta_0 g^2} \right) \left( 1 + \mathcal{O}(g^2) \right) \, . \tag{3.88}$$

The integration constant $\Lambda_{\mathrm{L}}$ is used to set the scale by fixing the value of $g$ at some $a$. Inverting the relation (3.88) one obtains the coupling $g$ as a function of the scale $a$, the so-called running coupling,

$$g(a)^{-2} = \beta_0 \ln \left( a^{-2} \Lambda_{\mathrm{L}}^{-2} \right) + \frac{\beta_1}{\beta_0} \ln \left( \ln \left( a^{-2} \Lambda_{\mathrm{L}}^{-2} \right) \right) + \mathcal{O} \left( 1/\ln \left( a^2 \Lambda_{\mathrm{L}}^2 \right) \right) \, . \tag{3.89}$$

Changing $a$ thus implies a corresponding change of $g$ such that physical observables remain independent of the scale-fixing procedure. The value of $\Lambda_{\mathrm{L}}$ depends on the regularization scheme. Different continuum or lattice actions have different values of $\Lambda_{\mathrm{L}}$; their ratios may be exactly related by a 1-loop perturbative calculation, however.

In fact, the validity of the perturbative expansion (3.87) has to be checked. We find that for shrinking lattice spacing the running coupling also decreases (for $n_f < 11 \, N/2$). Vanishing lattice spacing corresponds to vanishing coupling $g$ since $g = 0$ is indeed a zero of the $\beta$-function. This behavior is called *asymptotic freedom*. For honesty we have to point out that there are possible nonperturbative contributions of the type $\mathcal{O} \left( \exp(-1/g^2) \right)$ (nonvanishing for $g > 0$ but not contributing to the series expansion). It is generally expected that these are not relevant for sufficiently small $g$. Also, the order of performing the infinite volume limit and the continuum limit is an important issue, cf. [27] and references therein.

### 3.5.4 The true continuum limit

At the end of Chap. 1, we already briefly mentioned how the true continuum limit of a lattice field theory can be taken. However, at that point we did not have the prerequisites to address this limit in a proper way and promised a detailed discussion later. For the case of pure gauge theory, we can now provide this discussion: Since we have shown in (3.83) and (3.88), respectively, that the lattice spacing $a$ decreases with decreasing $g$ (increasing $\beta$), we conclude that we simply have to study the limit

$$\beta \;\rightarrow\; \infty \tag{3.90}$$

to obtain the true continuum limit $a \rightarrow 0$. There are, however, certain caveats to be considered in this procedure. If one performs the limit (3.90), then the physical volume of the box in which we study QCD is proportional to $a^4$ and thus shrinks to zero, unless we also increase the numbers of lattice points in the spatial ($N$ points) and temporal ($N_T$ points) directions of our lattice. In an ideal world one would first perform the so-called *thermodynamic limit*

$$N \;\rightarrow\; \infty\,, \quad N_T \;\rightarrow\; \infty\,, \tag{3.91}$$

and only after that step the continuum limit (3.90) would be taken. However, since in a numerical calculation this is not feasible, one is reduced to calculating the physical observables for a few values of $\beta$, giving rise to different values of $a$. The numbers of lattice points $N, N_T$ are always chosen such that the physical extension

$$L = a\,N\,, \quad T = a\,N_T\,, \tag{3.92}$$

of the box remains fixed for the different values of $a$. Studying the $a$-dependence of the results at fixed physical volume allows one to analyze the dependence on the scale $a$ and to extrapolate the results to $a \rightarrow 0$. The study of the $a$-dependence is often referred to as *scaling analysis*. The extrapolation to $a = 0$ can then be repeated for different physical sizes $L, T$ which in the end allows one to extrapolate the data to infinite physical volume.

## 3.6 Lattice gauge theory with other gauge groups

Although this book is dedicated to lattice QCD, where the gauge group is SU(3), we also briefly discuss lattice gauge theory with other gauge groups, in particular SU($N$) and U(1). SU(2) is of interest since it is simpler than SU(3) and is also a subgroup of SU(3), and it is therefore widely studied. Important ideas, such as topological charge and instantons, which are solutions of the classical equations of motion for the gauge field, can be formulated in SU(2). The groups SU($N$) with $N \geq 4$ are of interest since in the continuum the limit $N \rightarrow \infty$ is a limit appealing for analytic studies. Thus, on the lattice one can analyze the $N$-dependence of observables. For general $N$ the Wilson action reads

$$S_G[U] = \frac{\beta}{N} \sum_{n \in \Lambda} \sum_{\mu < \nu} \mathrm{Re\, tr}\, [\mathbb{1} - U_{\mu\nu}(n)] \ , \tag{3.93}$$

with $\beta = 2N/g^2$. The plaquette $U_{\mu\nu}(n)$ is the same ordered product (2.48) as for SU(3). All observables which we have discussed so far, in particular Wilson and Polyakov loops, can be taken over unchanged to general $N$.

The gauge group U(1) is interesting since it corresponds to QED. As an abelian group it is simpler to implement, and a numerical simulation of this case is a fairly easy exercise that can produce interesting results on a PC within a few hours. The main difference to the non-abelian groups is the fact that the elements of U(1) are not matrices but simply complex phases, i.e., we can write the link variables as $U_\mu(n) = \exp(\mathrm{i}\, A_\mu(n))$, where the $A_\mu(n)$ are real. This implies that the commutator of two link variables vanishes. Hence, the quadratic term in the field strength vanishes and one finds $F_{\mu\nu}(x) = \partial_\mu A_\nu(x) - \partial_\nu A_\mu(x)$. No self-interaction terms emerge. The Wilson action for U(1) is given by

$$S_G[U] = \beta \sum_{n \in \Lambda} \sum_{\mu < \nu} \mathrm{Re}\, (1 - U_{\mu\nu}(n)) \ . \tag{3.94}$$

No trace over color degrees of freedom is necessary and the plaquette is simply a product of complex numbers, $U_{\mu\nu}(n) = U_\mu(n)\, U_\nu(n+\hat{\mu})\, U_\mu(n+\hat{\nu})^*\, U_\nu(n)^*$ (the asterisk denotes complex conjugation). The coupling $\beta$ is related to the coupling $e$ of QED by $\beta = 1/e^2$. The Polyakov loop and Wilson loop observables can again be used, but no trace is needed.

An interesting observation can be made when applying the strong coupling (small-$\beta$) expansion to U(1) lattice gauge theory. One still finds a linearly rising term in the potential, indicating confinement. On the other hand this theory is expected to describe QED, which does not show confinement. The solution of this riddle is a phase transition in the U(1) lattice gauge theory. At a critical value of $\beta_{\mathrm{crit}} \approx 1.01$ the theory changes its behavior such that above $\beta_{\mathrm{crit}}$ the linearly rising term of the potential vanishes and only the Coulomb potential term remains. For SU($N$) lattice gauge theories it is generally believed that no such transition occurs and the theory remains confining for all $\beta$. However, for an alternative scenario see [28].

# References

1. M. Creutz: *Quarks, Gluons and Lattices* (Cambridge University Press, Cambridge, New York 1983)
2. H. Georgi: *Lie Algebras in Particle Physics* (Benjamin/Cummings, Reading, Massachusetts 1982)

3. H. F. Jones: *Groups, Representations and Physics* (Hilger, Bristol 1990)
4. M. Hamermesh: *Group Theory and Its Application to Physical Problems* (Addison-Wesley, Reading, Massachusetts 1964)
5. R. Gilmore: *Lie Groups, Lie Algebras, and Some of Their Applications*, Vol. 64 (John Wiley & Sons, New York 1974)
6. C. Itzykson and J.-B. Zuber: *Quantum Field Theory* (McGraw-Hill, New York 1985)
7. M. E. Peskin and D. V. Schroeder: *An Introduction to Quantum Field Theory* (Addison-Wesley, Reading, Massachusetts 1995)
8. S. Weinberg: *The Quantum Theory of Fields*, Vol. 1 and 2 (Cambridge University Press, Cambridge, New York 1996)
9. W. H. Press, S. A. Teukolsky, W. T. Vetterling, and B. P. Flannery: *Numerical Recipes in C*, 2nd ed. (Cambridge University Press, Cambridge, New York 1999)
10. L. Giusti et al.: Int. J. Mod. Phys. A **16**, 3487 (2001)
11. S. Elitzur: Phys. Rev. D **12**, 3978 (1975)
12. T. DeGrand and C. DeTar: *Lattice Methods for Quantum Chromodynamics* (World Scientific, Singapore 2006)
13. A. M. Polyakov: Phys. Lett. B **59**, 82 (1975)
14. C. Borgs and E. Seiler: Commun. Math. Phys. **91**, 329 (1983)
15. K. G. Wilson: Phys. Rev. D **10**, 2445 (1974)
16. K. Osterwalder and E. Seiler: Ann. Phys. **110**, 440 (1978)
17. J. M. Drouffe and J. B. Zuber: Phys. Rep. **102**, 1 (1983)
18. G. Roepstorff: *Path Integral Approach to Quantum Physics* (Springer, Berlin, Heidelberg, New York 1996)
19. D. H. Perkins: *Introduction to High Energy Physics*, 4th ed. (Cambridge University Press, Cambridge, New York 1999)
20. R. Sommer: Nucl. Phys. B **411**, 839 (1994)
21. M. Guagnelli, R. Sommer, and H. Wittig: Nucl. Phys. B **535**, 389 (1998)
22. S. Necco and R. Sommer: Nucl. Phys. B **622**, 328 (2002)
23. C. B. Lang and C. Rebbi: Phys. Lett. B **115**, 137 (1982)
24. C. Michael: Nucl. Phys. B **259**, 58 (1985)
25. R. G. Edwards, U. M. Heller, and T. R. Klassen: Nucl. Phys. B **517**, 377 (1998)
26. C. R. Allton et al.: Phys. Rev. D **65**, 054502 (2002)
27. E. Seiler: in *Applications of RG Methods in Mathematical Sciences* (RIMS, Kyoto University, Kyoto 2003)
28. A. Patrascioiu and E. Seiler: Phys. Rev. Lett. **74**, 1920 (1995)

# 4

---

# Numerical simulation of pure gauge theory

A basic simulation of pure SU(3) gauge theory is something that by now can be done already on a modern PC and certainly is a pedagogically valuable exercise. This section provides the techniques necessary for such a calculation. It also serves as an introduction to Monte Carlo simulations, although for the sake of having an easy-to-follow presentation, we concentrate on the simplest algorithms.

The vacuum expectation value of an observable in the quantized Euclidean gauge field theory on a lattice is formally given by the functional integral (cf. (3.1) and (3.2) in Chap. 3)

$$\langle O \rangle = \frac{1}{Z} \int \mathcal{D}[U]\, e^{-S_G[U]}\, O[U] \quad \text{with} \quad Z = \int \mathcal{D}[U]\, e^{-S_G[U]} . \tag{4.1}$$

However, this expression cannot be evaluated analytically, except for very small lattices. A Monte Carlo simulation approximates the integral by an average of the observable evaluated on $N$ sample gauge field configurations $U_n$, distributed with probability $\propto \exp(-S[U_n])$. The sum[1]

$$\langle O \rangle \approx \frac{1}{N} \sum_{\substack{U_n \text{ with} \\ \text{probability} \\ \propto\, e^{-S[U_n]}}} O[U_n] \tag{4.2}$$

is computed for sufficiently many configurations generated by Monte Carlo algorithms. In this chapter we discuss how such a sequence of configurations $U_n$ can be obtained as a so-called *Markov chain*. Usually, subsequently produced gauge configurations are not completely uncorrelated and we discuss methods how to deal with this problem and address the statistical analysis of the data. The observables we will consider prominently in this chapter are Wilson and Polyakov loops. At the end of this chapter the reader will be able to compute numerically the static potential with error bars.

---

[1]For the rest of this chapter we omit the subscript $G$ and denote the gauge action by $S[U]$.

Gattringer, C., Lang, C.B.: *Numerical Simulation of Pure Gauge Theory*. Lect. Notes Phys. **788**, 73–101 (2010)
DOI 10.1007/978-3-642-01850-3_4

## 4.1 The Monte Carlo method

### 4.1.1 Simple sampling and importance sampling

The Euclidean path integral for a finite lattice is a high-dimensional integral over the field variables: Scalar fields or fermions living on the sites and gauge fields attached to the links connecting neighboring sites.

Let us for a moment consider a similar but simpler situation. A 4D version of the Ising spin system of statistical physics has on its sites spin variables which may assume two values: $+1$ or $-1$ (compare (1.92), (1.93), and (1.94)). This system is a model of ferromagnetism where the variables represent microscopic magnets pointing up or down. One also may consider it as a simplification of a scalar quantum field theory where the integral over the continuous variable at each site is replaced by the sum over two values. A lattice with $N^4$ sites has $N^4$ such spin variables. Counting all possible combinations of values $\pm 1$ we have $2^{N^4}$ spin configurations. For a moderately large lattice we may have $N = 16$ and therefore $2^{65536} \approx 10^{19728}$ possible spin configurations. The exact evaluation of (1.94) corresponds to summing over all these configurations, clearly an impossible task. We have to find a way to provide an estimate for this sum.

Probability theory tells us that we may approximate an integral over a function by averaging the function values $f(x_n)$ at values $x_n$ randomly chosen according to the uniform distribution $\rho_u(x_n) = 1/(b-a)$:

$$\frac{1}{b-a} \int_a^b \mathrm{d}x \, f(x) = \langle f \rangle_{\rho_u} = \lim_{N \to \infty} \frac{1}{N} \sum_{n=1}^{N} f(x_n) . \tag{4.3}$$

The method replaces the exact mean by a sample mean, just like in opinion polls. Like in those polls, in actual calculations one is confined to a subsample and a finite number $N$. One can prove that the uncertainty in that estimate of the correct mean behaves like $\mathcal{O}\left(1/\sqrt{N}\right)$. Actually, even that is just a probabilistic statement, since this error has itself a statistical error and so on. However dubious this may seem at first sight, in actual calculations the method works amazingly well.

Monte Carlo sampling is easily applicable to higher dimensional integrals. All that changes is that now $x_n$ denotes a vector of random variables and one chooses random points in this multi-dimensional space. In the usual numerical quadrature methods the effort grows exponentially with the requested accuracy. For the Monte Carlo integration the estimated error is always $\propto 1/\sqrt{N}$: In order to improve the accuracy by a factor of 2 one has to take four times as many random points. Comparisons show that the Monte Carlo method becomes more efficient than quadrature for more than three dimensions.

However, in our path integral we have to take into account the *Boltzmann factor* $\exp(-S)$. Depending on the action $S$ it will give different importance to different field configurations. When summing over the configurations it is therefore more important to consider the configurations with larger weight than those with smaller weight. The central idea of the so-called *importance sampling* Monte Carlo method is to approximate the huge sum by a comparatively small subset of configurations, which are sampled according to the weight factor.

The expectation value of some function $f(x)$ with regard to a probability distribution with density $\rho(x)$ is given by

$$\langle f \rangle_\rho = \frac{\int_a^b \mathrm{d}x \, \rho(x) \, f(x)}{\int_a^b \mathrm{d}x \, \rho(x)} \, . \tag{4.4}$$

In the importance sampling Monte Carlo integration this expectation value is approximated by an average over $N$ values,

$$\langle f \rangle_\rho = \lim_{N \to \infty} \frac{1}{N} \sum_{n=1}^{N} f(x_n) \, , \tag{4.5}$$

for $x_n \in (a, b)$ each randomly sampled with the normalized probability density

$$\mathrm{d}P(x) = \frac{\rho(x)\mathrm{d}x}{\int_a^b \mathrm{d}x \, \rho(x)} \, . \tag{4.6}$$

Our path integral is of the form (4.4) and thus suitable for importance sampling. We may therefore write the expectation of an operator $O$ as

$$\langle O \rangle = \lim_{N \to \infty} \frac{1}{N} \sum_{n=1}^{N} O[U_n] \, , \tag{4.7}$$

with each of the $U_n$ sampled according to the probability distribution density

$$\mathrm{d}P(U) = \frac{\mathrm{e}^{-S[U]} \, \mathcal{D}[U]}{\int \mathcal{D}[U] \, \mathrm{e}^{-S[U]}} \, , \tag{4.8}$$

the so-called *Gibbs measure*. The gauge field configurations $U_n$ are our random variables. We approximate the integral using a sample of $N$ such configurations. In actual calculations this number may vary between a few hundreds up to several millions, depending on the available computer resources and the complexity of the problem. The statistical error of the result will be proportional to $1/\sqrt{N}$ and the exact value will be obtained for $N \to \infty$. The results are for finite size lattices, however. Control of the statistical error belongs to the central issues of Monte Carlo calculations.

## 4.1.2 Markov chains

How do we find field configurations $U_n$ following the probability distribution (4.8)? The idea is to start from some arbitrary configuration and then to

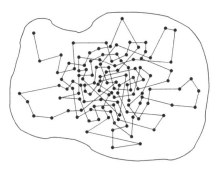

**Fig. 4.1.** Schematic sketch of a Markov chain in the space of all configurations

construct a stochastic sequence of configurations that eventually follows an equilibrium distribution $P(U)$. This is done with a so-called homogeneous *Markov chain* or *Markov process*

$$U_0 \longrightarrow U_1 \longrightarrow U_2 \longrightarrow \dots . \tag{4.9}$$

In this Markov chain configurations $U_n$ are generated subsequently. The index $n$ labels the configurations in the order they appear in the chain; it is often referred to as *computer time*, not to be mistaken with the Euclidean time of the 4D space–time. The change of a field configuration to a new one is called an *update* or a *Monte Carlo step*.

In Fig. 4.1 we show a schematic sketch of a Markov chain. The boundary delimits the space of all configurations. The dots represent configurations visited by our Markov chain and we connect them with straight lines to indicate that they are visited subsequently. The Markov chain in the figure starts in the upper left corner and then quickly moves toward the center of the blob, where we find a large density of dots. This corresponds to a region of configurations with large Boltzmann factor $\exp(-S)$ and thus with high probability. The Markov process is constructed such that it visits configurations with larger probabilities more often. A Markov process is characterized by a conditional transition probability (read: probability to get $U'$ if starting from $U$)

$$P(U_n = U' | U_{n-1} = U) = T(U'|U) . \tag{4.10}$$

This probability depends only on the configurations $U$ and $U'$ but not on the index $n$. The transition probabilities $T(U'|U)$ obey

$$0 \le T(U'|U) \le 1 , \quad \sum_{U'} T(U'|U) = 1 . \tag{4.11}$$

The inequality simply delimits the range of a probability. The sum states that the total probability to jump from some configuration $U$ to any target configuration $U'$ is equal to 1 (note that this includes also the case $U' = U$).

Let us now discuss an important restriction for $T(U'|U)$. Once it is in equilibrium, our Markov process cannot have sinks or sources of probability.

Thus the probability to hop into a configuration $U'$ at the step $U_{n-1} \to U_n$ has to be equal to the probability for hopping out of $U'$ at this step. The corresponding balance equation reads as

$$\sum_{U} T(U'|U)\, P(U) \stackrel{!}{=} \sum_{U} T(U|U')\, P(U') . \tag{4.12}$$

On the left-hand side we sum the transition probability $T(U'|U)$ leading into the final configuration $U'$ over all starting configurations $U$, weighted by the probability $P(U)$ that the system actually is in the configuration $U$. This expression gives the total probability to end up in $U'$ and has to equal the probability to hop out of $U'$, which we compute on the right-hand side. It is given by the probability $P(U')$ of finding the system in the configuration $U'$ times the sum of the transition probability $T(U|U')$ over all final configurations $U$ the system could jump into. Note that on both sides we also included the case where $U' = U$, i.e., the case without actual transition.

Before we discuss a solution of the balance equation (4.12), let us note an important property. The sum on the right-hand side can be calculated explicitly by using the normalization property (4.11). We find

$$\sum_{U} T(U'|U)\, P(U) = P(U') , \tag{4.13}$$

showing that the equilibrium distribution $P(U)$ is a fixed point of the Markov process. Once the equilibrium distribution is obtained, the system stays there upon applying $T$. Starting the process from an arbitrary start configuration $U_0$ with initial distribution $P^{(0)}(U) = \delta(U - U_0)$, one eventually obtains the equilibrium distribution $P(U)$ by applying the transition matrix iteratively:

$$P^{(0)} \stackrel{T}{\to} P^{(1)} \stackrel{T}{\to} P^{(2)} \stackrel{T}{\to} \ldots \stackrel{T}{\to} P \quad (= \text{equilibrium distribution}) . \tag{4.14}$$

For an elementary proof of this property see, e.g., [1].

Let us address an important point. For obtaining correct results, the Markov chain must be able to access all configurations. In other words, it must be possible to reach all points inside the blob of Fig. 4.1 in a finite number of steps. If the transition matrix $T(U'|U)$ is strictly positive for all pairs $U, U'$, then the process is aperiodic and every configuration can be eventually reached. This property is called *strong ergodicity*. In realistic simulations ergodicity and the related problem of relaxation are important questions. In particular, if there are topologically different sectors in configuration space, some Monte Carlo updating algorithms may have problems connecting them.

In an actual calculation one starts to calculate observables according to (4.7) only after a sufficient number of *equilibrating* Monte Carlo steps. The subtle question is when one can assume that the distribution of the considered configurations is already close enough to the equilibrium distribution. This decision is usually based on the measurement of certain observables and correlations. We will discuss this issue in more detail later in this chapter.

We now present a sufficient condition for a solution of the balance equation (4.12). On both sides we have sums over all configurations $U$, and these sums have to be equal. A solution can be obtained, by requiring that the equality holds term-wise,

$$T(U'|U)\, P(U) = T(U|U')\, P(U') \, . \tag{4.15}$$

This sufficient condition is known as the *detailed balance condition*. Although other solutions are known, most algorithms use the detailed balance condition. In the next section we will discuss the "mother of all Monte Carlo algorithms" based on (4.15), the *Metropolis algorithm*.

### 4.1.3 Metropolis algorithm – general idea

The Metropolis algorithm [2], which advances the Markov chain from a configuration $U_{n-1}$ to some new configuration $U_n$, consists of the following steps (we use $P(U) \propto \exp(-S[U])$):

**Step 1:** Choose some candidate configuration $U'$ according to some *a priori selection probability* $T_0(U'|U)$, where $U = U_{n-1}$.

**Step 2:** Accept the candidate configuration $U'$ as the new configuration $U_n$ with the acceptance probability

$$T_A(U'|U) = \min\left(1, \, \frac{T_0(U|U') \exp\left(-S[U']\right)}{T_0(U'|U) \exp\left(-S[U]\right)}\right) \, . \tag{4.16}$$

If a suggested change is not accepted, the unchanged configuration is considered again in the Markov chain and included in the measurements like the others.

**Step 3:** Repeat these steps from the beginning.

It is straightforward to see that the total transition probability $T = T_0 T_A$ fulfills the detailed balance condition

$$
\begin{aligned}
T(U'|U)&\exp\left(-S[U]\right) \\
&= T_0(U'|U) \min\left(1, \, \frac{T_0(U|U') \exp\left(-S[U']\right)}{T_0(U'|U) \exp\left(-S[U]\right)}\right) \exp\left(-S[U]\right) \\
&= \min\left(T_0(U'|U) \exp\left(-S[U]\right), \, T_0(U|U') \exp\left(-S[U']\right)\right) \\
&= T(U|U') \exp\left(-S[U']\right)
\end{aligned}
\tag{4.17}
$$

due to the positivity of all factors and the symmetry of the min operation. In many cases one uses a symmetric selection probability which obeys

$$T_0(U|U') = T_0(U'|U) \, . \tag{4.18}$$

In this case (4.16) simplifies to

$$T_A(U'|U) = \min\left(1, \, \exp\left(-\Delta S\right)\right) \quad \text{with} \quad \Delta S = S[U'] - S[U] \, . \tag{4.19}$$

In particular for symmetric $T_0$, the information necessary to decide on acceptance or rejection comes only from the change of the action $\Delta S$ with regard to the change of the configuration. If this change is local, e.g., just involves a single link variable $U_\mu(n)$, then $\Delta S$ may be determined from the field values in the local neighborhood.

### 4.1.4 Metropolis algorithm for Wilson's gauge action

Let us make the idea of the Metropolis algorithm more transparent by discussing its application in the $SU(N)$ Wilson gauge action (2.49) in four dimensions. Starting from some configuration $U$, our candidate configuration $U'$ for the Metropolis update differs from the configuration $U$ by the value of only a single link variable $U_\mu(n)'$. In four dimensions this link is shared by six plaquettes, and only these six plaquettes are affected when changing $U_\mu(n) \rightarrow U_\mu(n)'$. Their local contribution to the action is (compare (3.93))

$$S[U_\mu(n)']_{\text{loc}} = \frac{\beta}{N} \sum_{i=1}^{6} \text{Re tr} \left[ \mathbb{1} - U_\mu(n)' \, P_i \right] = \frac{\beta}{N} \, \text{Re tr} \left[ 6 \, \mathbb{1} - U_\mu(n)' \, A \right],$$

$$\text{with } A = \sum_{i=1}^{6} P_i = \sum_{\nu \neq \mu} \left( U_\nu(n{+}\hat{\mu}) \, U_{-\mu}(n{+}\hat{\mu}{+}\hat{\nu}) \, U_{-\nu}(n{+}\hat{\nu}) \right. \qquad (4.20)$$
$$\left. + \; U_{-\nu}(n{+}\hat{\mu}) \, U_{-\mu}(n{+}\hat{\mu}{-}\hat{\nu}) \, U_\nu(n{-}\hat{\nu}) \right).$$

Here the $P_i$ are products of the other three gauge link variables that build up the plaquettes together with $U_\mu(n)'$. These products are called *staples* and we have written explicitly the sum $A$ over all staples.

For the change of the action we obtain

$$\Delta S = S[U_\mu(n)']_{\text{loc}} - S[U_\mu(n)]_{\text{loc}} = -\frac{\beta}{N} \, \text{Re tr} \left[ (U_\mu(n)' - U_\mu(n)) \, A \right], \quad (4.21)$$

where $A$ is not affected by the change of $U_\mu(n)$.

An important part of the algorithm is the choice of the candidate link $U_\mu(n)'$. It should be an element of $SU(N)$ not too far away from the old link $U_\mu(n)$, such that the average acceptance probability (4.16) for the candidate does not become too small. A standard technique is to use

$$U_\mu(n)' = X \, U_\mu(n), \qquad (4.22)$$

where $X$ is a random element of the gauge group $SU(N)$ in the vicinity of $\mathbb{1}$. To achieve a symmetric selection probability $T_0$, $X$ and $X^{-1}$ have to be chosen with equal probability. How such matrices $X$ are constructed in practice will be discussed in Sect. 4.2.

Based on equations (4.20), (4.21), and (4.22), a realization of the Metropolis algorithm with single link variable updates and symmetric selection probability $T_0$ may be briefly summarized:

**Step 1:** Given some gauge field configuration, choose a site $n$ and direction $\mu$ and a candidate value $U_\mu(n)'$ according to some symmetric selection probability $T_0$, using, e.g., (4.22).

**Step 2:** Compute the sum over the staples and from this the change of the action $\Delta S$ according to (4.21). Compute a random number $r$ uniformly distributed in the interval $[0, 1)$. Accept the new variable $U_\mu(n)'$ if $r \leq \exp(-\Delta S)$ and reject it otherwise.

**Step 3:** Repeat these steps from the beginning.

We point out that the change in Step 2 is always accepted if the action decreases or remains invariant, i.e., $\exp(-\Delta S) \geq 1$. This alone would lead to a minimum of the action in configuration space, corresponding to a solution of the classical field equations. However, due to the random variable $r$ also configurations with increased action will be accepted every now and then. One could say that this feature reproduces the quantum fluctuations of the system.

The order in which one visits the links $(n, \mu)$ to update the corresponding link variables $U_\mu(n)$ can be chosen at one's discretion. However, some implementations prefer a certain order and visit all lattice points systematically, e.g., to utilize vector computation capabilities of the computer.

Furthermore, it is computationally economic to repeat the updating step a few times for the visited variable, since the computation of the sum of staples $A$ is costly. Thus, once $A$ is calculated, one offers the system candidate links $U_\mu(n)'$ repeatedly and accepts them according to Step 2. This modification is called *multi-hit Metropolis algorithm*. The number of candidate links can be tuned for efficiency. Theoretically, the infinite repetition leads to a method equivalent to the heat bath algorithm which we discuss in Sect. 4.3.

## 4.2 Implementation of Monte Carlo algorithms for SU(3)

In the last section we have presented the idea of the single link Metropolis update as a first approach to Monte Carlo algorithms. However, so far all the details of an actual implementation have been left out. These details, in particular the representation of the variables, boundary conditions, generation of a candidate link, and the generation of random numbers, will be provided in this section. Once we have presented these tools, we will be able to discuss two more Monte Carlo algorithms, namely the heat bath and overrelaxation methods.

Part of the material in this section is discussed for gauge groups U(1) and SU(2). The abelian group U(1) is a simple do-it-yourself example and the update of SU(2) subgroups is a building block for updating SU(3).

### 4.2.1 Representation of the link variables

We have already seen in the discussion of the Metropolis algorithm that matrix multiplication is the central operation of the update (compare (4.20), (4.21), and (4.22)). For this reason it is important to find a suitable and efficient representation for the link variables.

The defining representations of the gauge field variables are complex numbers for gauge group U(1), complex $2 \times 2$ matrices for SU(2), and complex $3 \times 3$ matrices for SU(3). However, due to the unitarity of the group elements there are further restrictions to be obeyed. The minimal number of parameters for the three groups is equal to the number of generators, 1, 3, and 8. Although one can think of representations of group elements that have just these minimal sets of parameters, in practical calculations it is often more convenient to use a redundant representation. This leads to faster evaluation of the multiplication of group elements. This operation is the most time-consuming part of the calculation, because it has to be done so frequently. For U(1) one uses a complex number $z$ with unit modulus $|z| = 1$. This implies that two real numbers instead of a single phase have to be stored.

For SU(2) it is convenient to store the first row $(a, b)$ of the matrix

$$U = \begin{pmatrix} a & b \\ -b^* & a^* \end{pmatrix} \quad \text{with} \quad |a|^2 + |b|^2 = 1 . \tag{4.23}$$

This corresponds to storing two complex numbers (four real numbers instead of the minimum of three). With $a = x_0 + i x_4$ and $b = x_3 + i x_2$ this is equivalent to using the vector $x = (x_0, \boldsymbol{x})$ of four real coefficients in the representation

$$U = x_0 \mathbb{1} + i \boldsymbol{x} \cdot \boldsymbol{\sigma} \quad \text{with} \quad \det[U] = |x|^2 = x_0^2 + |\boldsymbol{x}|^2 = \sum_{i=0}^{3} x_i^2 = 1 , \tag{4.24}$$

where $\boldsymbol{\sigma}$ denotes the vector of the three Pauli matrices (see (A.8)).

The group elements for SU(3) may be represented either by the complete complex matrix (9 complex numbers = 18 real numbers instead of the minimum of 8) or by the first two rows, corresponding to the complex 3 vectors $\boldsymbol{u}$ and $\boldsymbol{v}$ (6 complex numbers = 12 real numbers). One has to restrict these vectors to unit length, and they have to be orthogonal to each other:

$$|\boldsymbol{u}|^2 = \boldsymbol{u}^* \cdot \boldsymbol{u} = |u_1|^2 + |u_2|^2 + |u_3|^2 = 1 ,$$
$$|\boldsymbol{v}|^2 = \boldsymbol{v}^* \cdot \boldsymbol{v} = |v_1|^2 + |v_2|^2 + |v_3|^2 = 1 , \tag{4.25}$$
$$(\boldsymbol{u}, \boldsymbol{v}) = \boldsymbol{u}^* \cdot \boldsymbol{v} = u_1^* v_1 + u_2^* v_2 + u_3^* v_3 = 0 .$$

Due to the properties of SU(3) matrices, the third row of the matrix can be reconstructed from the first two rows:

$$U = \begin{pmatrix} \boldsymbol{u} \\ \boldsymbol{v} \\ \boldsymbol{u}^* \times \boldsymbol{v}^* \end{pmatrix} . \tag{4.26}$$

Thus in principle it is sufficient to store only two rows. However, due to the abundance of disc space and memory, nowadays mainly redundant representations of the group elements are used. Within the computer program the gauge field is then an array with indices for the space and time positions of the lattice sites and an index indicating the direction of the link variable. For non-abelian theories there are also color indices depending on the type of representation used (e.g., two if the matrix form is used).

Whatever representation one chooses, because of the accumulation of rounding errors in the multiplications of the group elements, the matrices have to be projected to unitarity regularly. The period depends on the number of digits chosen and has to be decided based on observation. Re-unitarization is done for $U(1)$ by dividing by the norm. For $SU(2)$ one normalizes the first row and then (if necessary in the chosen representation) reconstructs the second from the first. Equivalently, for $SU(3)$ one follows essentially the well-known Gram–Schmidt method for building orthonormal basis elements in vector spaces. The first row is normalized, the second constructed from the current values orthogonalized to the first row,

$$
\begin{aligned}
\boldsymbol{u}_{\text{new}} &= \boldsymbol{u}/|\boldsymbol{u}| \,, \\
\boldsymbol{v}_{\text{new}} &= \boldsymbol{v}'/|\boldsymbol{v}'| \quad \text{where} \quad \boldsymbol{v}' = \boldsymbol{v} - \boldsymbol{u}_{\text{new}} \left( \boldsymbol{v} \cdot \boldsymbol{u}_{\text{new}}^* \right) ,
\end{aligned}
\tag{4.27}
$$

and the third row is constructed from the first two rows as given in (4.26).

### 4.2.2 Boundary conditions

Since a numerical simulation works on a finite lattice, boundary conditions have to be implemented. For gauge fields one usually uses periodic boundary conditions (compare (A.27) for our definition of the lattice)

$$
\begin{aligned}
U_\mu(N, n_2, n_3, n_4) &= U_\mu(0, n_2, n_3, n_4) \,, \; U_\mu(n_1, N, n_3, n_4) = U_\mu(n_1, 0, n_3, n_4) \,, \\
U_\mu(n_1, n_2, N, n_4) &= U_\mu(n_1, n_2, 0, n_4) \,, \; U_\mu(n_1, n_2, n_3, N_T) = U_\mu(n_1, n_2, n_3, 0) \,.
\end{aligned}
\tag{4.28}
$$

For fermions one often chooses boundary conditions anti-periodic in one direction as will be discussed in Chap. 5. The boundary conditions define the topology of the underlying manifold. Periodic and anti-periodic boundary conditions correspond to a torus in four dimensions – each direction behaves like a circle. Choosing toroidal boundary conditions (as opposed to, e.g., Dirichlet boundary conditions) has the advantage of preserving the discrete translational symmetry of the lattice. Other boundary conditions and other lattice discretizations are possible and have been studied, but the majority of calculations are done with hypercubic lattices and (anti-) periodic boundary conditions.

When calculating the change of the action $\Delta S$ for a Monte Carlo step, one needs the addresses of nearest neighbors in the four space–time directions. This may seem trivial, but in actual calculations it takes some computer time. For this reason various acceleration techniques have been developed. Most of them use pre-calculated index arrays to implement neighbor calculations and boundary conditions at the same time.

### 4.2.3 Generating a candidate link for the Metropolis update

We still need to address how we generate the candidate link for the Metropolis algorithm presented in the last section.

In the Metropolis algorithm one has to suggest a candidate link variable $U_\mu(n)'$ in the vicinity of the old value $U_\mu(n)$. As already stated in (4.22) this can be done as

$$U_\mu(n)' = X U_\mu(n) , \qquad (4.29)$$

where $X$ is a randomly chosen element of the gauge group close to the unit element. The acceptance rate can be adjusted to reasonable values by tuning the spread of the matrices $X$ around unity. A high acceptance rate may seem desirable but usually means too small changes and slow motion in configuration space. Smaller acceptance is costly because many candidate configurations are generated but not accepted. As a rule of thumb an acceptance rate of 0.5 is reasonable, such that in average, one out of two suggested candidate configurations is accepted in the Monte Carlo step.

Choosing candidates for the Metropolis step is simple for $U(1)$ where one may choose $X = e^{i\varphi}$ with $\varphi \in (-\varepsilon, \varepsilon)$ randomly chosen with uniform distribution. Then $\varepsilon$ may be tuned for good acceptance. However, this requires computation of the sine function and one wants to avoid costly computations in the heart of the program. Thus it is more efficient to take $X = (1 + i\varphi)/\sqrt{1 + \varphi^2}$ instead, as long as $\varepsilon$ is not too large (less than 1). Further possibilities can be explored. One sees that this topic moves from science to art [3].

For the gauge groups SU(2) and SU(3) the determination of random group elements is more costly. One may use various methods to suggest elements around unity. For SU(2) one may choose four random numbers $r_i$ uniformly distributed in $(-1/2, 1/2)$. The SU(2) matrix $X$ then is constructed following (4.24) with $x_0$ and $\boldsymbol{x}$ given by

$$\boldsymbol{x} = \varepsilon\,\boldsymbol{r}/|\boldsymbol{r}| , \quad x_0 = \text{sign}(r_0)\,\sqrt{1 - \varepsilon^2} , \qquad (4.30)$$

where $\varepsilon$ is again the parameter that controls the spread of $X$ around $\mathbb{1}$.

For SU(3), updating matrices $X$ can be constructed from such SU(2) matrices embedded in $3 \times 3$ matrices according to

$$R = \begin{pmatrix} r_{11} & r_{12} & 0 \\ r_{21} & r_{22} & 0 \\ 0 & 0 & 1 \end{pmatrix} , \quad S = \begin{pmatrix} s_{11} & 0 & s_{12} \\ 0 & 1 & 0 \\ s_{21} & 0 & s_{22} \end{pmatrix} , \quad T = \begin{pmatrix} 1 & 0 & 0 \\ 0 & t_{11} & t_{12} \\ 0 & t_{21} & t_{22} \end{pmatrix} . \qquad (4.31)$$

A possible choice for $X$ is then given by the product

$$X = R\,S\,T . \qquad (4.32)$$

Since the matrices $R$, $S$, $T$ are close to $\mathbb{1}$, so is the product $X$.

In order to preserve the symmetry of the a priori selection probability $T_0(U'|U)$, $X$ and $X^{-1} = X^\dagger$ should be chosen with equal probability. A simple way of doing this is to construct a set of random SU(3) matrices close to unity (e.g., by using (4.32)) and to include in this set also the hermitian conjugate of each matrix. The matrices $X$ used to build the candidate link according to (4.29) are chosen randomly out of this set. The set is rebuilt with new group elements from time to time.

### 4.2.4 A few remarks on random numbers

The central step of the Monte Carlo procedure needs random numbers. In the computer programs these are so-called *pseudo random numbers*, generated reproducibly by algorithms. The statistical properties of the pseudo random numbers are very close to those of real random numbers. Typical Monte Carlo runs may need $\mathcal{O}(10^{12})$ random numbers. Therefore, utmost care has to be taken in selecting a proper generator. Standard implementations of random number generators are often not reliable enough and produce subtly correlated numbers with too small periods. High-quality generators use the so-called lagged Fibonacci method and there are generators with extremely long guaranteed periods $\mathcal{O}(10^{171})$ [4].

Pseudo random numbers are usually generated according to a uniform distribution in the interval $[0, 1)$. There exists a variety of algorithms to generate from these other distributions [3, 5]. This may be costly in terms of computer time, and finding efficient and "well-behaved" generators (in the statistical sense) is an important task of the Monte Carlo approach. Indeed, the whole gauge field configuration may be considered a vector of random numbers following a distribution: the Boltzmann distribution.

Any generator has to be initialized. In case one continues a long run one should also store the final state of the generator in order to be able to restart it at that position.

## 4.3 More Monte Carlo algorithms

Meanwhile there exists a collection of updating algorithms. If, such as in simple Metropolis, the changes of a Monte Carlo step affect only a few variables locally, one calls this a local algorithm. Unfortunately, local algorithms allow for only very small steps in the Markov chain and therefore are not very efficient. One has to perform many updating steps in order to obtain uncorrelated configurations. The situation is better for nonlocal algorithms, where large subsets of field variables are changed at once. Whereas for spin models and scalar field theories there are excellent nonlocal (e.g., cluster) algorithms, this is not the case for gauge fields, where, up to now, no efficient nonlocal updating method is known. The following algorithms have been applied to gauge fields:

**Heat bath**: Equivalent to an iterated Metropolis updating, optimizing the local acceptance rate [6].

**Overrelaxation**: A sometimes very efficient method to improve the step size in the Markov chain, exploiting symmetry properties of the action [7, 8]. To obtain ergodicity it has to be combined with other algorithms.

**Microcanonical**: Reformulates the updating in terms of a deterministic, discrete Hamiltonian evolution using a doubled number of variables: $(U, \Pi)$. A variant method introduces a microcanonical demon [9].

**Hybrid Monte Carlo**: Combines a number of microcanonical updates (a *trajectory*) with a final Monte Carlo acceptance step [10].

**Langevin**: Uses the stochastic differential equation for the construction of configurations [11–13]; equivalence to the microcanonical and the hybrid method may be demonstrated (in certain limits).

Different updating algorithms may have different performance in terms of the required computational effort and the resulting step size in the Markov chain.

Here we present in more detail the heat bath algorithm and overrelaxation. In Sect. 4.4 we will outline the steps necessary for a simulation of pure gauge theory based on these algorithms.

### 4.3.1 The heat bath algorithm

In the heat bath method one combines steps 1 and 2 of the single link Metropolis update into a single step and chooses the new value $U_\mu(n)'$ according to the local probability distribution defined by the surrounding staples,

$$dP(U) = dU \, \exp\left( \frac{\beta}{N} \, \mathrm{Re} \, \mathrm{tr} \, [U \, A] \right) . \tag{4.33}$$

The sum of staples $A$ is calculated according to (4.20) and all links, except for $U = U_\mu(n)'$, are held fixed and therefore $A$ is constant. Note that $dU$ denotes the Haar integration measure of the gauge group. This may be computationally quite demanding. It has the advantage, however, that the link variable always changes. The implementation depends on the details of the gauge group and of the action.

We first present in some detail the heat bath method for the gauge group SU(2) with Wilson action and generalize the method to SU(3) in the next paragraph. For the case of SU(2) there exists an efficient method to find a new link element. This group is special, since a sum of two SU(2) elements is proportional to another SU(2) matrix. We use this property and write the sum of staples $A$ from (4.21) in the form

$$A = a \, V \quad \text{with} \quad a = \sqrt{\det[A]} , \tag{4.34}$$

where it can be shown that $\det[A] \geq 0$. If $\det[A]$ vanishes one chooses a random $SU(2)$ matrix for $U$. Otherwise we find that $V = A/a$ is a properly normalized $SU(2)$ matrix. Inserting (4.34) in our probability distribution (4.33), we obtain (for $SU(2)$ we have set $N = 2$ in this equation)

$$dP(U) = dU \, \exp\left(\frac{1}{2} a\beta \, \mathrm{Re} \, \mathrm{tr} \, [U \, V]\right). \tag{4.35}$$

The Haar measure $dU$ is invariant under transformations of the origin in group space (cf. Chap. 3) and we may also write it as $d(U V)$. If we define a matrix $X$ by the product $X = UV$, the local probability distribution for $X$ is

$$dP(X) = dX \, \exp\left(\frac{1}{2} a\beta \, \mathrm{Re} \, \mathrm{tr} \, [X]\right). \tag{4.36}$$

If we generate a matrix $X$ distributed accordingly, the candidate link is obtained by

$$U_\mu(n)' = U = X \, V^\dagger = X \, A^\dagger \frac{1}{a}. \tag{4.37}$$

We therefore have reduced the problem to generating matrices $X$ distributed according to (4.36). The Haar measure in that equation may be written in terms of the real parameters used in the representation (4.24) of the group elements. For $X$ in representation (4.24) with $x \in \mathbb{R}^4, |x| = 1$, the Haar measure reads

$$dX = \frac{1}{\pi^2} \, d^4x \, \delta\left(x_0^2 + |\boldsymbol{x}|^2 - 1\right) \tag{4.38}$$

$$= \frac{1}{\pi^2} \, d^4x \, \frac{\theta(1 - x_0^2)}{2\sqrt{1 - x_0^2}} \left(\delta\left(|\boldsymbol{x}| - \sqrt{1 - x_0^2}\right) + \delta\left(|\boldsymbol{x}| + \sqrt{1 - x_0^2}\right)\right),$$

where in the second line we have used a well-known formula for the Dirac delta of functions. We rewrite the volume element as

$$d^4x = d|\boldsymbol{x}| \, |\boldsymbol{x}|^2 \, d^2\Omega \, dx_0, \tag{4.39}$$

where $d^2\Omega$ denotes the spherical angle element in the integration over the 3-vector $\boldsymbol{x}$. We can use the Dirac deltas to remove the $|\boldsymbol{x}|$ integration. Only the first Dirac delta in (4.38) contributes and from now on $|\boldsymbol{x}|$ is frozen to $\sqrt{1 - x_0^2}$. The Haar measure assumes the form

$$dX = \frac{1}{\pi^2} \, d^2\Omega \, dx_0 \, \frac{(1 - x_0^2) \, \theta(1 - x_0^2)}{2\sqrt{1 - x_0^2}} = \frac{1}{2\pi^2} \, d^2\Omega \, dx_0 \, \sqrt{1 - x_0^2} \, \theta(1 - x_0^2). \tag{4.40}$$

Note that in the matrix representation chosen for $X$ we have $|x_0| \leq 1$ and therefore we could have omitted the step function $\theta$. Due to $\mathrm{tr}[X] = 2x_0$, we end up with the distribution for $X$ in the form (using $d^2\Omega = d\cos\vartheta \, d\varphi$)

$$dP(X) = \frac{1}{2\pi^2} \, d\cos\vartheta \, d\varphi \, dx_0 \, \sqrt{1 - x_0^2} \, e^{a\beta x_0} , \tag{4.41}$$

with $x_0 \in [-1, 1]$, $\cos\vartheta \in [-1, 1]$, and $\varphi \in [0, 2\pi)$. In order to find a random matrix $X$ we have to determine random variables $x_0$, $\vartheta$, and $\varphi$ according to this distribution. Since the distribution for the three variables factorizes, we can generate them independently:

**Random variable $x_0$:** The task is to find values $x_0$ distributed according to $\sqrt{1 - x_0^2} \, e^{a\beta x_0}$. Following [14, 15] we introduce a variable $\lambda$

$$x_0 = 1 - 2\lambda^2 \quad \text{with} \quad x_0 \in [-1, 1] \quad \Rightarrow \tag{4.42}$$

$$dx_0 \, \sqrt{1 - x_0^2} \, e^{a\beta x_0} \propto d\lambda \, \lambda^2 \, \sqrt{1 - \lambda^2} \, e^{-2a\beta\lambda^2} \quad \text{with} \quad \lambda \in [0, 1] .$$

After this transformation we need to generate $\lambda$ with the polynomially modified Gaussian distribution density

$$p_1(\lambda) = \lambda^2 \, e^{-2a\beta\lambda^2} \tag{4.43}$$

and accept it with an accept/reject step using the square root function

$$p_2(\lambda) = \sqrt{1 - \lambda^2} . \tag{4.44}$$

Algorithms to compute random numbers with Gaussian distributions are well known [3, 5]. We proceed as follows [15]:

*Step 1:* One starts with a triplet of random numbers $r_i$, $i = 1, 2, 3$ uniformly distributed in $(0, 1]$ (the value 0 has to be avoided; since usual random number generators cover the interval $[0, 1)$ one just takes $(1 - r_i)$ instead). Then

$$\lambda^2 = -\frac{1}{2a\beta} \left( \ln(r_1) + \cos^2(2\pi r_2) \ln(r_3) \right) \tag{4.45}$$

follows the required distribution.

*Step 2:* We correct for the factor $p_2(\lambda)$ and thus accept only those values of $\lambda$ which obey

$$r \le \sqrt{1 - \lambda^2} \quad \text{or, better} \quad r^2 \le 1 - \lambda^2 , \tag{4.46}$$

where $r$ is a random variable uniformly distributed in $[0, 1)$. The accepted values give $x_0 = 1 - 2\lambda^2$ following the requested distribution.

**Random variable $|\boldsymbol{x}|$:** Actually this random variable was removed when we integrated it out using the Dirac delta of (4.38). However, in this step the length was frozen to $|\boldsymbol{x}| = \sqrt{1 - x_0^2}$ and we now can compute it from the $x_0$ determined in the last step.

**Random variables $\cos\vartheta$ and $\varphi$:** The angular variables correspond to the direction of $\boldsymbol{x}$ and are uniformly distributed. A possible method is to choose three random numbers $r_1$, $r_2$, and $r_3$ uniformly distributed in $[-1, 1)$ and to accept them when $r_1^2 + r_2^2 + r_3^2 \le 1$. This 3-vector is then normalized to length $|\boldsymbol{x}| = \sqrt{1 - x_0^2}$.

After these steps we end up with a vector $(x_0, \boldsymbol{x})$ and from that we can compute the matrix $X$ using representation (4.24). We summarize the steps for updating an SU(2) link variable with the heat bath algorithm:

1. Find the sum of staples $A$, compute $a = \sqrt{\det[A]}$, and set $V = A/a$.
2. Find a group element $X$ according to distribution (4.41) as discussed above.
3. The new link variable is $U = X V^\dagger$.

There is no heat bath algorithm which directly produces SU(3) link variables. However, one can apply a pseudo heat bath method by iterating the heat bath for the SU(2) subgroups of SU(3) [16]. Two such SU(2) subgroups would be sufficient to cover the whole group space. However, for symmetry reasons one chooses three such groups as given in (4.31). Each of the three matrices is determined with the heat bath for SU(2) as discussed.

For this aim let us consider again (4.33). Let us assume that we want to modify the link variable $U$ by left multiplication with, e.g., the first matrix $R$ of (4.31). The exponent in the local Boltzmann weight (4.33) then would read (now $N = 3$ since this is for SU(3))

$$\frac{\beta}{3} \operatorname{Re} \operatorname{tr}[R U A] . \tag{4.47}$$

Denoting $W = U A$ the trace contains the terms

$$\operatorname{tr}[R W] = r_{11} w_{11} + r_{12} w_{21} + r_{21} w_{12} + r_{22} w_{22} + \text{terms without } r_{ij} . \tag{4.48}$$

A heat bath algorithm for $R$ is therefore influenced only by those terms in $W$ that multiply the four nontrivial terms in $R$, the sub-block of $W$ corresponding to the relevant sub-matrix of $R$. We find that $W$ plays the role of $A$ in (4.33) in the discussion of the SU(2) heat bath algorithm. It is no group element of SU(2) but proportional to 1. Thus one can proceed like before finding the elements of $R$ in the "heat bath" of $W$.

We multiply the resulting $R$ with $W$ to obtain a new $W = R U A$, which now provides the SU(2) heat bath factors for $S$. The resulting $S$ again multiplies $W$ and leads to $T$. The three SU(2) heat bath factor matrices give the new value of the link variable:

$$U \quad \Rightarrow \quad U' = T S R U . \tag{4.49}$$

This pseudo heat bath update for the SU(2) subgroups may also be combined with overrelaxation steps as discussed in the next section.

### 4.3.2 Overrelaxation

This method tries to change the variables as much as possible in order to speed up the motion through configuration space. One utilizes the property that in the Metropolis algorithm new configurations are always accepted if they do not change the action. Like for the heat bath algorithm the starting point is the probability distribution (4.33) of a single link variable $U_\mu(n) = U$ in the background of its neighbors which we hold fixed with the sum of staples $A$ calculated according to (4.21). The idea of the overrelaxation method is to find a new value $U'$ which has the same probability weight as $U$ and thus is automatically accepted.

Let us first illustrate the idea for the gauge group U(1). In that case we can write $U = \exp(i\varphi)$ and for the sum of staples obtain $A = a \exp(i\alpha)$. The exponent for the local probability (4.33), the local action, can be written as (for U(1) the trace is gone and $N = 1$; compare (3.94))

$$\beta \operatorname{Re}(U A) = \beta a \operatorname{Re}\left(e^{i\varphi} e^{i\alpha}\right) = \beta a \cos(\varphi + \alpha). \tag{4.50}$$

Obviously the reflection of $(\varphi + \alpha) \to -(\varphi + \alpha)$ or, equivalently, the change $\varphi \to (2\pi - 2\alpha - \varphi)$ leaves the local action invariant and thus is always accepted.

For the non-abelian groups one suggests a change according to the ansatz

$$U \to U' = V^\dagger U^\dagger V^\dagger, \tag{4.51}$$

with a gauge group element $V$ chosen such that the action is invariant.

The choice of $V$ is nontrivial in the general case. However, as discussed in connection with (4.34), for the gauge group SU(2) the sum of staples $A$ is proportional to a group matrix and one constructs $V = A/a$ with the real number $a = \sqrt{\det[A]}$. The matrix $V$ is now unitary, i.e., $V^{-1} = V^\dagger$. We find

$$\operatorname{tr}[U' A] = \operatorname{tr}[V^\dagger U^\dagger V^\dagger A] = a \operatorname{tr}[V^\dagger U^\dagger] = \operatorname{tr}[A^\dagger U^\dagger] = \operatorname{tr}[U A]. \tag{4.52}$$

In the last step we have used the reality of the trace for SU(2) matrices. This choice for $U'$ indeed leaves the action invariant. Also the selection probability $T_0$ is symmetric, since $U' = V^\dagger U^\dagger V^\dagger$ implies $U = V^\dagger U'^\dagger V^\dagger$. In the rare case that $\det A$ vanishes, any random link variable is accepted.

Implementing overrelaxation for other gauge groups like SU(3) may be more involved and eventually not efficient. Usually only updating programs for the gauge groups U(1) and SU(2) make use of this method. For SU(3) overrelaxation has been studied in [17].

The overrelaxation algorithm alone is not ergodic. It samples the configuration space on the subspace of constant action. This is called the *microcanonical* ensemble. Since one wants to determine configurations according to the *canonical* ensemble, i.e., distributed according to the Boltzmann weight, one has to combine the overrelaxation steps with other updating algorithms, such as Metropolis or heat bath steps.

## 4.4 Running the simulation

After having introduced the techniques and algorithms for the update we can now discuss how to organize an actual simulation. A Monte Carlo simulation of a lattice gauge theory consists of several basic steps:

**Initialization:** The field configuration has to be initialized.

**Equilibration updates:** Sufficiently many updates have to be done, until the subsequently generated configurations represent the (approximate) equilibrium distribution.

**Evaluation of the observables and intermediate updates:** As soon as the system is in equilibrium, one can start to compute the observables on the Monte Carlo configurations. The configurations used for computing the observables should be separated by several intermediate updates.

These steps are discussed in some detail below. After the data are produced they have to be analyzed with regard to statistical properties (see Sect. 4.5). Only then one can study the physics content of the resulting data.

Table 4.1 shows a schematic listing demonstrating how the individual parts of the Monte Carlo program are combined in the computer code.

**Table 4.1.** Structure of a Monte Carlo program for pure gauge theory

---

**Start of program**

    **Declaration of parameters and variables:**
      *Initialize random number generator.*
      *Initialize tables (e.g., neighbor indices, factors, couplings).*
      *Initialize gauge field configuration.*

    **Loop over "equilibrating" iterations:**
      **for** $n = 1$ **until** $n_{\mathrm{equi}}$ :
        *Update the configuration.*
      **end for**

    **Loop over iterations with update and measurements:**
      **for** $n = 1$ **until** $n_{\mathrm{measure}}$ :
        **for** $i = 1$ **until** $n_{\mathrm{discarded}}$ :
          *Update the configuration.*
        **end for**
        *Measure the observables for the last of these configurations.*
        *Write the values to some data file. (Maybe also save the*
        *current configuration and the status of the random number*
        *generator to some file.)*
      **end for**

    *Write the final configuration and the status of the random*
    *number generator to some file for future restart.*
    *Write the final output to the data file.*
    *Close all open files.*

**End of program**

---

### 4.4.1 Initialization

Any field configuration can be chosen as initial configuration. After sufficiently many updates, eventually configurations distributed according to the equilibrium distribution will be produced due to the Markov chain property (4.14). Two typical start configurations are the so-called *cold start* and *hot start* configurations.

**Cold start:** All link variables are set to the unit element ($U_\mu(n) = \mathbb{1}$). This corresponds to trivial plaquette variables and minimal gauge action, a situation approximately expected for small gauge coupling $g$ (large $\beta$). In statistical models the weak coupling regime corresponds to low temperature, hence the name.

**Hot start:** The gauge field matrices are chosen randomly in group space. Usually one pays little care for a uniform distribution but just takes suitable normalized random vectors for that purpose. From these the matrix rows are built as discussed above in the unitarization procedure (4.26) and (4.27).

Some like it hot, some like it cold, some use mixed start configurations, like one half of the variables cold and the other half hot. Virtually any value of the total action can be mimicked this way, albeit without any internal structure. In any case one has to equilibrate the system before calculating observables.

Very long sequences of updates have to be split into several subsequent runs of the Monte Carlo program. In this case one starts with the last stored configuration and the corresponding state of the random number generator.

### 4.4.2 Equilibration updates

The central part of the program is the updating subroutine. Given a configuration one systematically updates all gauge links, suggesting new values and accepting them according to one of the methods discussed. Visiting all links once is often referred to as a *sweep* through the lattice. To optimize the performance, i.e., to have large steps in the Markov chain, often sweeps with different algorithms are combined. Such a combined sweep could consist of a heat bath sweep followed by two or three overrelaxation sweeps.

An important question is whether one has performed sufficiently many equilibration sweeps. Only after such an equilibration phase, the algorithm produces configurations with the correct distribution (compare (4.14)). A simple, first test for sufficient equilibration is given by comparing how a set of observables changes with the number of sweeps. In particular one can compare one such series of data obtained from a cold start to a series of data obtained from a hot start. As soon as the curves for the observables from the hot and the cold run approach each other, the system is nearing equilibrium.

In general the speed at which the system approaches the equilibrium will depend on the updating algorithm as well as on the gauge coupling $\beta$, the size of the lattice, and the type of action used. Large systems and large $\beta$

will require more equilibration steps than small lattices and small values of $\beta$. Also different observables will approach their equilibrium values at different rates.

In Sect. 4.5 we will put the determination of a sufficient number of equilibration steps on a sounder basis. There we will discuss the *autocorrelation time* which is a measure for how much two subsequent configurations in a Markov chain are correlated.

### 4.4.3 Evaluation of the observables

Once the system is in equilibrium, the configurations can be used for the evaluation of the observables. Depending on the updating algorithm, subsequently generated configurations will be more or less correlated. Thus several sweeps should be performed between evaluating the observables. We remark that often the evaluation of the observables is referred to as *measurement*.

A careful analysis of the statistical parameters, in particular the aforementioned autocorrelation time as discussed in Sect. 4.5, gives an indication of how many intermediate configurations should be discarded between two measurements. In practice this decision depends also on the cost for measuring the observables. If the measurement of observables is expensive (like it is for computing quark propagators on a gauge field background), one tries to have less correlation between subsequently measured configurations. In any case the final analysis should take into account that configurations may not be statistically independent.

Two key observables for simulations of pure gauge theory are Wilson and Polyakov loops (compare (3.50) and (3.60)). When performing the measurements one typically makes use of translation invariance (for periodic boundary conditions) and averages over all possible realizations of the observables in order to improve the statistics. This means that for the Wilson loop one averages over all positions where the loop can start from. In addition one can place the loop in different planes (corresponding to the symmetry operations of a hypercubic lattice) which gives for a quadratic loop an average over $6\,|\Lambda|$ terms (where $|\Lambda|$ denotes the number of sites) and an average over $12\,|\Lambda|$ contributions for rectangular loops. For the Polyakov loop one averages over all spatial positions $\boldsymbol{n}$ or, if products are considered, over all combinations $\boldsymbol{n}$, $\boldsymbol{m}$ for each distance $|\boldsymbol{n} - \boldsymbol{m}|$. This involves sums over very many terms and accumulation of rounding errors is then an issue, in particular if one has to subtract large contributions from each other. For that reason it is advisable to use higher precision accuracy for these observables.

Often the resulting mean values of observables may be computed "on the fly." However, it pays off to store as many of the intermediate results as possible in order to allow for a careful a posteriori analysis of the statistical properties.

## 4.5 Analyzing the data

The statistical analysis of the measured observables is the important final step of a Monte Carlo simulation. This analysis should also provide one with the information how many updating sweeps have to be discarded before configurations in equilibrium are produced and how many sweeps are necessary between two measurements. The final product of the statistical analysis is the average value which one quotes for an observable and an estimate for the corresponding statistical error.

### 4.5.1 Statistical analysis for uncorrelated data

We assume that we have computed the values $(x_1, x_2, \ldots, x_N)$ of some observable for a Markov sequence of Monte Carlo-generated configurations in equilibrium. Each of the values of the sample corresponds to a random variable $X_i$. All these variables have the same expectation value and variance:

$$\langle X_i \rangle = \langle X \rangle , \quad \sigma^2_{X_i} = \left\langle (X_i - \langle X_i \rangle)^2 \right\rangle = \sigma^2_X . \tag{4.53}$$

Candidates for unbiased estimators for these values are

$$\widehat{X} = \frac{1}{N} \sum_{i=1}^{N} X_i , \quad \widehat{\sigma}^2_X = \frac{1}{N-1} \sum_{i=1}^{N} \left( X_i - \widehat{X} \right)^2 . \tag{4.54}$$

If the $X_i$ are uncorrelated one finds for $i \neq j$,

$$\langle X_i X_j \rangle = \langle X_i \rangle \langle X_j \rangle = \langle X \rangle^2 \tag{4.55}$$

and the variance $\widehat{\sigma}^2_X$ allows one to determine the statistical error of $\widehat{X}$. To see this first note that the sample mean value $\widehat{X}$ is an estimator for the correct mean value: $\langle \widehat{X} \rangle = \langle X \rangle$. It is, however, itself a random variable, since its value may change from one set of $N$ configurations to another set. The variance of that estimator is

$$\sigma^2_{\widehat{X}} = \left\langle \left( \widehat{X} - \langle X \rangle \right)^2 \right\rangle = \left\langle \left( \frac{1}{N} \sum_{i=1}^{N} (X_i - \langle X \rangle) \right)^2 \right\rangle \tag{4.56}$$

$$= \frac{1}{N^2} \left\langle \sum_{i,j=1}^{N} (X_i - \langle X \rangle)(X_j - \langle X \rangle) \right\rangle = \frac{1}{N} \langle X^2 \rangle - \langle X \rangle^2 + \frac{1}{N^2} \sum_{i \neq j} \langle X_i X_j \rangle .$$

For uncorrelated $X_i$ the contributions from $i \neq j$ factorize due to (4.55) and

$$\sigma^2_{\widehat{X}} = \frac{1}{N} \sigma^2_X . \tag{4.57}$$

This is the well-known result for uncorrelated measurements. Thus, for the observable based on $N$ measurements, the statistical error, i.e., the *standard deviation* (s.d.), is $\sigma_{\widehat{X}}$. The value $\sigma_X$ on the right-hand side of (4.57) is approximated using $\widehat{\sigma}_X$ from (4.54). For the case of $N$ uncorrelated measurements one quotes the final result as

$$\widehat{X} \pm \sigma \quad \text{with} \quad \sigma = \frac{\widehat{\sigma}_X}{\sqrt{N}} \,. \tag{4.58}$$

The important message of this equation is that the statistical error decreases like $1/\sqrt{N}$ with the number $N$ of uncorrelated configurations.

### 4.5.2 Autocorrelation

Since in our case the data sample is the result of a (computer-)time series in our Monte Carlo simulation there is high chance that the observables are in fact correlated. This so-called *autocorrelation* leads to a nonvanishing *autocorrelation function*, which we define as

$$C_X(X_i, X_{i+t}) = \langle (X_i - \langle X_i \rangle)(X_{i+t} - \langle X_{i+t} \rangle) \rangle = \langle X_i X_{i+t} \rangle - \langle X_i \rangle \langle X_{i+t} \rangle \,. \tag{4.59}$$

For a Markov chain in equilibrium the autocorrelation function depends only on the (computer time) separation $t$ and we write

$$C_X(t) = C_X(X_i, X_{i+t}) \,. \tag{4.60}$$

Note that $C_X(0) = \sigma_X^2$. In a typical situation the normalized correlation function $\Gamma_X$ exhibits exponential behavior asymptotically for large $t$:

$$\Gamma_X(t) \equiv \frac{C_X(t)}{C_X(0)} \sim \exp\left(-\frac{t}{\tau_{X,\exp}}\right) \,, \tag{4.61}$$

and one calls $\tau_{X,\exp}$ the *exponential autocorrelation time* for $X$. The complete expression for $\Gamma_X(t)$ involves a sum over several such terms. In (4.61) we consider only the asymptotically leading term with the largest autocorrelation time. This number provides information on how strongly subsequent measurements are correlated. The exponential autocorrelation time $\tau_{\exp}$ is the supremum of the values $\tau_{X,\exp}$ for all possible observables $X$:

$$\tau_{\exp} = \sup_X \tau_{X,\exp} \,. \tag{4.62}$$

Autocorrelations lead to systematic errors which are $\mathcal{O}\left(\exp(-t/\tau_{\exp})\right)$ if the computer time between subsequent measurements is $t$.

For correlated random variables $X_i$ the terms with $i \neq j$ in the second line of (4.56) do not vanish and one can continue this equation to obtain for the correlated case

$$\sigma_{\hat{X}}^2 = \frac{1}{N^2} \sum_{i,j=1}^{N} C_X(|i-j|) = \frac{1}{N^2} \sum_{t=-(N-1)}^{N-1} \sum_{k=1}^{N-|t|} C_X(|t|)$$

$$= \sum_{t=-N}^{N} \frac{N-|t|}{N^2} C_X(|t|) = \frac{C_X(0)}{N} \sum_{t=-N}^{N} \Gamma_X(|t|) \left(1 - \frac{|t|}{N}\right) \quad (4.63)$$

$$\approx \frac{\sigma_X^2}{N} 2 \left(\frac{1}{2} + \sum_{t=1}^{N} \Gamma_X(|t|)\right) \equiv \frac{\sigma_X^2}{N} 2\tau_{X,\text{int}} ,$$

where we have introduced the *integrated autocorrelation time*

$$\tau_{X,\text{int}} = \frac{1}{2} + \sum_{t=1}^{N} \Gamma_X(t) . \quad (4.64)$$

This definition is motivated by the observation that for exponential behavior

$$\tau_{X,\text{int}} = \frac{1}{2} + \sum_{t=1}^{N} \Gamma_X(|t|) \approx \int_0^\infty dt \, e^{-t/\tau} = \tau \quad (\text{for large } \tau) . \quad (4.65)$$

In the last step of (4.63) we have neglected the factor $1 - |t|/N$ which is justified for large enough $N$ due to the exponential suppression of $\Gamma_X(|t|)$.

Computing $\tau_{X,\text{int}}$ in a realistic situation one has to cut off sum (4.64) at a value of $t$ where the values of $\Gamma(t)$ become unreliable. Usually one then assumes exponential behavior for the part not explicitly taken into account in the sum. Still, the determination of $\tau_{\text{exp}}$ or even $\tau_{\text{int}}$ is a delicate business. Usually one needs at least $1000\,\tau$ data values for estimates of $\tau$ itself. In order to judge whether the measured autocorrelation time is reliable, one therefore should start with small size lattices and high statistics and work oneself up to larger sizes, carefully checking the behavior and reliability of $C(t)$.

The variance $\sigma_{\hat{X}}^2$ computed in this way is larger than the variance computed from (4.57), which assumes an uncorrelated sample. The number of effectively independent data out of $N$ values is therefore

$$N_{\text{indep}} = \frac{N}{2\,\tau_{X,\text{int}}} \quad (4.66)$$

or

$$\sigma_{\hat{X},\text{corrected}}^2 = 2\,\tau_{X,\text{int}}\,\sigma_{\hat{X}}^2 . \quad (4.67)$$

For equilibration from some start configuration one should discard at least $20\,\tau$, for good statistical accuracy maybe $1000\,\tau$ configurations. When producing data with 1% errors one typically needs $> 10,000\,\tau$ values. For more detailed discussions, cf. [18–20].

Summing up our results we find that for the correlated case the result one quotes is given by

$$\hat{X} \pm \sigma \quad \text{with} \quad \sigma = \sqrt{\frac{1}{N}\,2\,\tau_{X,\text{int}}\,\hat{\sigma}_X^2} . \quad (4.68)$$

Finally let us briefly mention the issue of *critical slowing down*. The auto-correlation time depends on the updating algorithm but also on the parameters of the lattice system. For lattice field systems one expects that

$$\tau_{X,\text{int}} \sim (\xi_X)^z \,, \quad \tau_{\exp} \sim \xi^z \,, \tag{4.69}$$

where $\xi_X$ is the correlation length for the observable $X$ and $\xi$ the longest correlation length within the system. The correlation length is defined from the exponential decay of correlation functions between local observables measured at different points on the lattice, i.e., $\langle X(x)X(y) \rangle \sim e^{-|x-y|/\xi_X}$ for large $|x-y|$. The *dynamical critical exponent* $z \geq 0$ depends on the updating algorithm. At critical points $\xi$ approaches infinity, however, on finite lattices of linear size $L$ one has $\xi \leq L$. Thus, near a critical point, the computational effort grows like a power of the extension of the lattice:

$$\text{numerical cost} \propto L^z \,. \tag{4.70}$$

This behavior is called critical slowing down. For first-order phase transitions, where the system may tunnel between different phases, the autocorrelation time grows like $\exp(c\, L^{D-1})$ for a $D$-dimensional lattice.

We summarize: From the data one has to get an estimate of the autocorrelation time. This provides (a) information on the number of update sweeps to be discarded between measurements and (b) a correction factor to the statistical error derived naively as for statistically independent data.

### 4.5.3 Techniques for smaller data sets

If it is too expensive to compute the autocorrelation time – and unfortunately this is often the case in Monte Carlo calculations for quantum field theory problems – there are simpler statistical methods for obtaining at least some estimate for the correlation of the data.

**Data blocking methods:** One divides the data into sub-blocks of data of size $K$, computes the block mean values, and considers them as new variables $X_i$. The variance of these blocked $X_i$ then should decrease like $1/K$ if the original data were independent. One repeats this for a sequence of different values for $K$. As soon as the $1/K$ behavior is observed for large enough $K$ one may consider these block variables as statistically independent.

Once the data (or the block results) can be considered independent, one may determine the expectation values of the observables of interest and their errors. Often, however, the number of data is too small to get a reliable estimate of the variance of the computed expectation values. Another obstacle may be that error propagation is unreliable or impossible to determine. There are two efficient and easy-to-use methods dealing with both problems. Both assume that the data are not correlated.

**Statistical bootstrap:** Given a set of $N$ data, assume that we are interested in some observable $\theta$ which may be estimated from that set. This observable can also be, e.g., the result of a fit based on all $N$ raw data. Let us call the value of the observable obtained from the original data set $\widehat{\theta}$. One recreates from the sample repeatedly other samples by choosing randomly $N$ data out of the original set. This costs essentially nothing, since we just recycle the original data set for the building of new sets. Let us assume we have done this $K$ times and thus have $K$ sets of $N$ data values each. Of course some values will enter more than once in the new sets. For each of these sets one computes the observable $\theta$ resulting in values $\theta_k$ with $k = 1, \ldots, K$. Then one determines

$$\widetilde{\theta} \equiv \frac{1}{K} \sum_{k=1}^{K} \theta_k \, , \quad \sigma_{\theta}^2 \equiv \frac{1}{K} \sum_{k=1}^{K} \left( \theta_k - \widetilde{\theta} \right)^2 . \tag{4.71}$$

These are estimators for $\langle \theta \rangle$ and $\sigma_{\theta}^2$. They are not unbiased and therefore $\widetilde{\theta} \neq \widehat{\theta}$ for finite $K$. The difference is called bias and gives an idea on how far away the result may be from the true $\langle \theta \rangle$. As final result for the observable one quotes $\langle \theta \rangle = \widetilde{\theta} \pm \sigma_{\widetilde{\theta}}$.

**Jackknife:** We start with a data set of size $N$ and an observable $\theta$ like for the statistical bootstrap. The value of the observable computed for the original set is again called $\widehat{\theta}$. One now constructs $N$ subsets by removing the $n$th entry of the original set $(n = 1, \ldots, N)$ and determines the value $\theta_n$ for each set. Then

$$\sigma_{\theta}^2 \equiv \frac{N-1}{N} \sum_{n=1}^{N} \left( \theta_n - \widehat{\theta} \right)^2 . \tag{4.72}$$

The square root of the variance gives an estimate for the standard deviation of $\widehat{\theta}$. For the final result one quotes either $\langle \theta \rangle = \widehat{\theta} \pm \sigma_{\widehat{\theta}}$ or replaces $\widehat{\theta}$ by the unbiased estimator. The bias may be determined from

$$\widetilde{\theta} \equiv \frac{1}{N} \sum_{n=1}^{N} \theta_n \, , \tag{4.73}$$

leading to $\widehat{\theta} - (N-1) \left( \widetilde{\theta} - \widehat{\theta} \right)$ as the unbiased estimator for $\langle \theta \rangle$.

In a practical implementation both, statistical bootstrap and jackknife, may be combined with blocking by organizing the data in blocks and constructing subsets by removing blocks instead of only single values.

Another characteristic aspect of Monte Carlo simulations is the fact that often the observables, one is interested in, are not simple averages but quantities that are obtained from a fit. An example is the energy levels obtained from an exponential fit to a Euclidean correlator (see (1.1)). A single measurement of the correlator is fluctuating far too much for a reasonable fit. Thus one first has to average many measurements of the correlator before the fit can be performed. However, for a naive determination of the error, one would need many such sets of data for the correlator.

A powerful feature of the statistical bootstrap and the jackknife methods is the fact that they can be applied to the determination of the statistical error for fitted quantities. Also the randomly chosen samples of the bootstrap and the reduced samples of the jackknife method are large enough for performing a fit. Thus these two methods can be used for determining the errors for fitted quantities without the need for additional data or consideration of complicated error propagation.

### 4.5.4 Some numerical exercises

The data have been determined from the Monte Carlo-generated gauge field configurations. Now one can start to explore physics. In this last section we briefly discuss two simple observables – the average plaquette and the static potential as obtained from the correlator of Polyakov loops. These observables are simple, can be calculated easily, and are well suited for a first simulation of a lattice gauge theory in an introductory course. We stress that the data and methods presented here are not meant to be state of the art but illustrate what can be achieved in a basic simulation.

Repeated Monte Carlo runs for different gauge couplings and lattice sizes provide an idea about the functional behavior of observables. As our first example let us consider the sum of plaquette variables as it enters the Wilson gauge action (3.93) for the gauge group SU($N$),

$$S_P[U] = \frac{1}{N} \sum_P \mathrm{Re\ tr}\,[U_P] \,, \tag{4.74}$$

where the sum runs over all plaquettes, counting only one orientation. Its expectation value is

$$\langle S_P \rangle = \frac{\int \mathcal{D}[U]\ \exp\left(\beta\, S_P[U]\right) S_P[U]}{\int \mathcal{D}[U]\ \exp\left(\beta\, S_P[U]\right)} \,. \tag{4.75}$$

Here we have used that the action $S$ is related to $S_P$ by $S[U] = \beta\,(6|\Lambda| - S_P[U])$ (see (3.93)). This relation allows one to cancel the constant term of the Boltzmann factor in the numerator with the same factor appearing in the normalization $1/Z$. As a function of the inverse gauge coupling $\beta$ one knows $\langle S_P \rangle$ for small $\beta$ from the strong coupling expansion and for large $\beta$ from the weak coupling expansion. Usually one plots the normalized quantity $E_P \equiv \langle S_P \rangle/(6|\Lambda|)$ which assumes values between 0 and 1 as in Fig. 4.2. The derivative $dE_P/d\beta$ is related to the second and first moments of $S_P$, since from (4.75) one finds

$$\frac{d\langle S_P \rangle}{d\beta} = \langle S_P^2 \rangle - \langle S_P \rangle^2 \,. \tag{4.76}$$

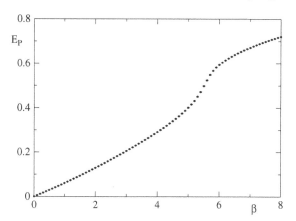

**Fig. 4.2.** The plaquette expectation value $E_P \equiv \langle S_P \rangle/(6|\Lambda|)$ as a function of $\beta$ for SU(3) gauge theory with Wilson gauge action on a $12^4$ hypercubic lattice. The data have been generated as discussed in the text, with increasing $\beta$ in small steps. For each step 50 equilibrating iterations have been done, followed by 200 groups of one discarded update and one update with measurement. Although there appears to be a maximum of the slope near $\beta \approx 5.7$, further analysis shows that this is not a signal of a phase transition. As far as numerics shows, the confinement phase for pure SU(3) lattice gauge theory case extends from $\beta = 0$ to the highest values studied and presumably up to infinity

This quantity is called the specific heat in analogy to statistical spin systems. From these we know that the specific heat grows with the system's correlation length. The quantity therefore is an indicator whether one approaches a critical point in the phase landscape. Studying (4.75) and (4.76) for U(1) gauge theory exhibits a (first-order) phase transition near $\beta \approx 1$ (for Wilson's action). For SU(2) and SU(3) no such signal has been seen although one does observe a maximum at intermediate coupling values. This indicates a nearby singularity in the multi-dimensional coupling space of actions.

The correlation function of Polyakov loops is related to the static quark potential (compare (3.61)),

$$\langle P(\boldsymbol{m})P(\boldsymbol{n})^\dagger \rangle \propto e^{-N_T a V(r)} \left( 1 + \mathcal{O}(e^{-N_T a \Delta E}) \right) , \tag{4.77}$$

where $r = |\boldsymbol{m} - \boldsymbol{n}|$. Up to an irrelevant overall constant the potential is

$$aV(r) = -\ln\left( \langle P(\boldsymbol{m})P(\boldsymbol{n})^\dagger \rangle \right)/N_T . \tag{4.78}$$

The result of a determination of $aV$, based on this relation, is shown in the left-hand side plot of Fig. 4.3. We present the result of a straightforward evaluation of the Polyakov loop correlation function with relatively low statistics, which can be reproduced within a few hours on a PC.

We stress that state-of-the-art calculations are based on expectation values of Wilson loops. Results with much higher statistics and more sophisticated

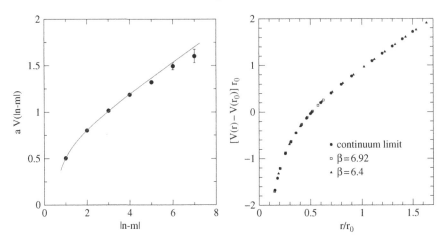

**Fig. 4.3.** The static potential $V$ for SU(3) gauge theory in a pure gauge field simulation with Wilson action. *Left-hand side*: values from the correlation of Polyakov loops as obtained on a $16^3 \times 6$ lattice at $\beta = 5.7$ from 50,000 subsequent iterations; this is shown just to demonstrate that main features can be seen already in such a very simple analysis. Much better and more sophisticated methods have been used. An example is shown in the *right-hand side* figure, where the values are derived from the expectation values of Wilson loops [21]. Both $r$ and $V$ were made dimensionless using the Sommer parameter $r_0$ (compare Chap. 3). (*Right-hand side* figure reprinted from [21] with kind permission from Silvia Necco)

methods of analysis (e.g., [22, 23]) exist [21, 24, 25]. As an example of such a calculation we show in the right-hand side plot the result from [21], where different lattice sizes and gauge couplings were used. The relation between the results for different gauge couplings is derived from the data and gives information on the scaling properties (see Sect. 3.5 and Chap. 9).

Furthermore, using nonplanar Wilson loops (compare Fig. 3.3) one can find out whether the lattice system, although on a hypercubic grid, does recover rotational invariance as it should when approaching the continuum limit (small lattice spacing, corresponding to large $\beta$). Figure 4.3 demonstrates that this is indeed the case: Also points from off-axis distance vectors agree with the overall shape. There one has chosen data for sufficiently large lattice sizes and values of $\beta$. Choosing smaller lattices and smaller $\beta$ produces noticeable deviations.

# References

1. H. J. Rothe: *Lattice Gauge Theories – An Introduction* (World Scientific, Singapore 1992)
2. N. Metropolis et al.: J. Chem. Phys. **21**, 1087 (1953)

3. D. E. Knuth: *The Art of Computer Programming*, 2nd ed., Vol. 2 (Addison-Wesley, Reading, Massachusetts 1981)
4. M. Lüscher: Comput. Phys. Commun. **79**, 100 (1994)
5. W. H. Press, S. A. Teukolsky, W. T. Vetterling, and B. P. Flannery: *Numerical Recipes in C*, 2nd ed. (Cambridge University Press, Cambridge, New York 1999)
6. M. Creutz: Phys. Rev. D **21**, 2308 (1980)
7. S. L. Adler: Phys. Rev. D **23**, 2901 (1981)
8. C. Whitmer: Phys. Rev. D **29**, 306 (1984)
9. M. Creutz: Phys. Rev. Lett. **50**, 1412 (1983)
10. D. J. E. Callaway and A. Rahman: Phys. Rev. D **28**, 1506 (1983)
11. G. Parisi and Y.-S. Wu: Sci. Sin. **24**, 483 (1981)
12. J. R. Klauder: in *Recent Developments in High-Energy Physics; Acta Phys. Austriaca Suppl. 25*, edited by H. Mitter and C. B. Lang, p. 251 (Springer, Wien, New York 1983)
13. P. H. Damgaard and H. Hüffel: Phys. Rep. **152**, 227 (1987)
14. K. Fabricius and O. Haan: Phys. Lett. B **143**, 459 (1984)
15. A. D. Kennedy and B. J. Pendleton: Phys. Lett. B **156**, 393 (1985)
16. N. Cabibbo and E. Marinari: Phys. Lett. B **119**, 387 (1982)
17. R. Petronzio and E. Vicari: Phys. Lett. B **248**, 159 (1990)
18. B. A. Berg: *Introduction to Markov Chain Monte Carlo Simulations and Their Statistical Analysis* (World Scientific, Singapore 2004)
19. A. D. Sokal: *Monte Carlo Methods in Statistical Physics: Foundations and New Algorithms, Lecture Notes Cours de Troisieme Cycle de la Physique en Suisse Romande, Lausanne, Switzerland* (unpublished, 1989)
20. U. Wolff: In: H. Gausterer, C. B. Lang (eds.) *Computational Methods in Field Theory*, Lect. Notes Phys. **409**, 127. Springer, Berlin, Heidelberg, New York (1992)
21. S. Necco: Ph.D. thesis, Humboldt Universität zu Berlin, Germany, arXiv:heplat/0306005 (2003)
22. M. Lüscher and P. Weisz: JHEP **0109**, 010 (2001)
23. M. Lüscher and P. Weisz: JHEP **0207**, 049 (2002)
24. G. Bali: Phys. Rep. **343**, 1 (2001)
25. P. Majumdar: Nucl. Phys. B **664**, 213 (2003)

# 5

# Fermions on the lattice

In our first look at QCD on the lattice in Chap. 2 we have already announced that our formulation of lattice fermions has not reached its final form. In particular we still need to take into account the correct statistics. In the first section of this chapter we will show that the required Fermi statistics can be implemented by using anti-commuting numbers for the quark fields, so-called *Grassmann numbers*. We will discuss the rules for calculating with Grassmann numbers, and we will derive a few key formulas for Gaussian integrals with Grassmann numbers.

The lattice formulation of the fermions as presented in Chap. 2 still suffers from certain lattice artifacts, so-called *doublers*. We will identify the doublers by analyzing the Fourier transform of the lattice Dirac operator. To remove the doublers we will add an extra term to the fermion action and in this way arrive at the *Wilson fermion action*.

In a short section we will present the so-called *hopping expansion* for fermionic observables. Although this expansion is useful only for very heavy quarks, it provides an interesting conceptual insight: We will be able to interpret the fermions as paths of link variables connecting points in space–time. Finally we will discuss the symmetries of the Wilson action which we will need later when we construct hadron interpolators in Chap. 6.

## 5.1 Fermi statistics and Grassmann numbers

### 5.1.1 Some new notation

Before we come to presenting the arguments for the need of Grassmann numbers, let us introduce some new notation. For the discussion in this chapter it is convenient to separate the fermionic part $\langle\ldots\rangle_F$ and the gauge field part $\langle\ldots\rangle_G$ of the path integral and to write (compare (2.55), (2.56), and (2.57))

$$\langle O \rangle = \langle\, \langle O \rangle_F \,\rangle_G \,. \tag{5.1}$$

Gattringer, C., Lang, C.B.: *Fermions on the Lattice*. Lect. Notes Phys. **788**, 103–122 (2010)
DOI 10.1007/978-3-642-01850-3_5          © Springer-Verlag Berlin Heidelberg 2010

The fermionic part $\langle \dots \rangle_F$ of the path integral is defined by

$$\langle A \rangle_F = \frac{1}{Z_F[U]} \int \mathcal{D}\left[\psi, \overline{\psi}\right] \, e^{-S_F[\psi, \overline{\psi}, U]} \, A[\psi, \overline{\psi}, U] \,. \tag{5.2}$$

In this integration $U$ is an external field. The fermionic partition function

$$Z_F[U] = \int \mathcal{D}\left[\psi, \overline{\psi}\right] \, e^{-S_F[\psi, \overline{\psi}, U]} \,, \tag{5.3}$$

depends on the gauge field through the gauge field dependence of the fermion action $S_F[\psi, \overline{\psi}, U]$. Later we will identify $Z_F[U]$ as the so-called *fermion determinant*, i.e., the determinant of the Dirac operator. The abbreviation $\langle \dots \rangle_G$ for the gauge field part of the path integral is defined as

$$\langle B \rangle_G = \frac{1}{Z} \int \mathcal{D}[U] \, e^{-S_G[U]} \, Z_F[U] \, B[U] \,. \tag{5.4}$$

Here $B[U]$ could be $\langle A \rangle_F$, but also other integrands, such as observables made from the gauge fields only, are possible. As we will discuss below, the nature of the fermionic part and the gauge field part of the path integral is quite different. Thus the separation of the two parts will turn out to be convenient.

### 5.1.2 Fermi statistics

Let us now discuss the fermionic vacuum expectation value of a product of fermion fields ($\alpha_i, \beta_i$ are Dirac indices, $a_i, b_i$ refer to color, $n_i, m_i$ are the space–time arguments and $f_i, g_i$ flavor labels)

$$\left\langle \psi^{(f_1)}(n_1)_{\substack{\alpha_1 \\ a_1}} \psi^{(f_2)}(n_2)_{\substack{\alpha_2 \\ a_2}} \dots \psi^{(f_k)}(n_k)_{\substack{\alpha_k \\ a_k}} \overline{\psi}^{(g_1)}(m_1)_{\substack{\beta_1 \\ b_1}} \dots \overline{\psi}^{(g_k)}(m_k)_{\substack{\beta_k \\ b_k}} \right\rangle_F \,. \tag{5.5}$$

As it stands, this product of fermion fields is not gauge-invariant in general. Factors built from products of link variables might be needed to make it gauge-invariant. However, here we are focusing on the fermionic part of the path integral and such gauge factors can be added later before the integration over the gauge field is performed.

One of the defining properties of fermions is the requirement that they obey Fermi statistics. For our vacuum expectation value (5.5) this implies antisymmetry under the interchange of quantum numbers. If we interchange the quantum numbers of any two of the fermions, say we interchange

$$f_1 \leftrightarrow f_2 \,, \quad n_1 \leftrightarrow n_2 \,, \quad \alpha_1 \leftrightarrow \alpha_2 \,, \quad a_1 \leftrightarrow a_2 \,, \tag{5.6}$$

the vacuum expectation value (5.5) has to acquire a minus sign. The interchange (5.6) is equivalent to commuting the first two fermion variables in (5.5), and the requirement of producing a minus sign can be accounted for by demanding an extra minus sign under this commutation. Thus we require the fermion fields to behave as anti-commuting numbers for any combination of the indices $f, f', n, n', \alpha, \alpha', a, a'$,

$$\psi^{(f)}(n)_{\substack{\alpha \\ a}} \, \psi^{(f')}(n')_{\substack{\alpha' \\ a'}} \; = \; - \psi^{(f')}(n')_{\substack{\alpha' \\ a'}} \, \psi^{(f)}(n)_{\substack{\alpha \\ a}} \; . \tag{5.7}$$

Also the $\overline{\psi}$ have to anti-commute among each other and furthermore the $\psi$ have to anti-commute with the $\overline{\psi}$. Hence in addition to (5.7) we demand

$$\overline{\psi}^{(f)}(n)_{\substack{\alpha \\ a}} \, \overline{\psi}^{(f')}(n')_{\substack{\alpha' \\ a'}} \; = \; - \overline{\psi}^{(f')}(n')_{\substack{\alpha' \\ a'}} \, \overline{\psi}^{(f)}(n)_{\substack{\alpha \\ a}} \; , \tag{5.8}$$

$$\psi^{(f)}(n)_{\substack{\alpha \\ a}} \, \overline{\psi}^{(f')}(n')_{\substack{\alpha' \\ a'}} \; = \; - \overline{\psi}^{(f')}(n')_{\substack{\alpha' \\ a'}} \, \psi^{(f)}(n)_{\substack{\alpha \\ a}} \; . \tag{5.9}$$

Thus all fermionic degrees in our lattice theory anti-commute with each other.

We remark that the fermionic path integral in terms of anti-commuting numbers may be derived from the canonical anti-commutation relations for fermions by introducing coherent states. For this approach we refer the reader to [1, 2] for an introductory presentation.

### 5.1.3 Grassmann numbers and derivatives

After having motivated the use of anti-commuting numbers, so-called *Grassmann numbers*, for the fermionic path integral, we need to learn how to calculate with Grassmann numbers.[1] We consider a set of Grassmann numbers $\eta_i, i = 1, \ldots, N$, obeying

$$\eta_i \, \eta_j \; = \; - \eta_j \, \eta_i \; , \tag{5.10}$$

for all $i, j$. This equation implies that the $\eta_i$ are nilpotent, i.e., they obey $\eta_i^2 = 0$. Hence the power series for a function of the $\eta_i$ terminates after a finite number of terms, and the only relevant class of functions are polynomials,

$$A = a + \sum_i a_i \eta_i + \sum_{i<j} a_{ij} \eta_i \eta_j + \sum_{i<j<k} a_{ijk} \eta_i \eta_j \eta_k + \ldots + a_{12\ldots N} \eta_1 \eta_2 \cdots \eta_N \; ,$$

$$\tag{5.11}$$

with complex coefficients $a, a_i, a_{ij} \ldots a_{12\ldots N}$. These polynomials can be added and multiplied and form a so-called *Grassmann algebra*. The Grassmann numbers $\eta_i$ are referred to as the *generators* of the Grassmann algebra.

One may differentiate elements of the Grassmann algebra with respect to the generators. In order to develop the rules for such Grassmann derivatives, we consider a simple example. For a Grassmann algebra with only $N = 2$ generators the most general function is

$$A = a + a_1 \, \eta_1 + a_2 \, \eta_2 + a_{12} \, \eta_1 \eta_2 \; . \tag{5.12}$$

---

[1]An introductory text on Grassmann numbers, going beyond our short presentation, can be found in [3].

We define the left derivative of $A$ with respect to $\eta_1$ by

$$\frac{\partial}{\partial \eta_1} A = a_1 + a_{12}\,\eta_2 \,. \tag{5.13}$$

However, using (5.10) we can interchange the order of the generators in the last term of (5.12) and write our polynomial $A$ also in the form

$$A = a + a_1\,\eta_1 + a_2\,\eta_2 - a_{12}\,\eta_2\eta_1 \,. \tag{5.14}$$

Thus, in order to get consistent results, we have to assign the Grassmann property also to the derivative operator:

$$\frac{\partial}{\partial \eta_1}\eta_2 = -\eta_2\frac{\partial}{\partial \eta_1} \,. \tag{5.15}$$

Furthermore, if we apply another derivative $\partial/\partial \eta_2$ to $\partial/\partial \eta_1 A$, we find that for consistency also the derivatives have to anti-commute with each other. Thus we define the following rules for our derivatives:

$$\frac{\partial}{\partial \eta_i}1 = 0 \,, \quad \frac{\partial}{\partial \eta_i}\eta_i = 1 \,, \quad \frac{\partial}{\partial \eta_i}\frac{\partial}{\partial \eta_j} = -\frac{\partial}{\partial \eta_j}\frac{\partial}{\partial \eta_i} \,, \quad \frac{\partial}{\partial \eta_i}\eta_j = -\eta_j\frac{\partial}{\partial \eta_i} \quad (\text{for } i \neq j) \,. \tag{5.16}$$

### 5.1.4 Integrals over Grassmann numbers

In addition to differentiation, we also want to find a consistent definition of integration over Grassmann numbers. The guiding principle of our construction is to implement the properties of the integration in $\mathbb{R}^N$. More specifically, we consider the integral over a domain $\Omega \subset \mathbb{R}^N$, with an integrand $f$ that vanishes at the boundary $\partial\Omega$,

$$\int_\Omega d^N x\, f(x) \;=\; \int_\Omega dx_1 \,\dots\, dx_N\, f(x_1, x_2, \dots, x_N) \,, \tag{5.17}$$

as a linear functional of the function $f$. We want to construct a linear functional $\int d^N\eta$ acting on Grassmann polynomials $A$ with equivalent properties. The requirement that the integral is a complex linear functional reads

$$\int d^N\eta\, A \in \mathbb{C} \,, \quad \int d^N\eta\,(\lambda_1 A_1 + \lambda_2 A_2) \;=\; \lambda_1 \int d^N\eta A_1 + \lambda_2 \int d^N\eta A_2 \,, \tag{5.18}$$

where $\lambda_1, \lambda_2$ are complex numbers. The second defining equation of the Grassmann integral is

$$\int d^N\eta\, \frac{\partial}{\partial \eta_i}A \;=\; 0 \,. \tag{5.19}$$

This requirement is equivalent to demanding for the integration over $\Omega \subset \mathbb{R}^N$ that the integrand $f(x_1, \dots, x_N)$ vanishes at the boundary $\partial\Omega$. The corresponding formula equivalent to (5.19) is

$$\int_\Omega d^N x \, \frac{\partial}{\partial x_i} f(x_1, \ldots, x_N) \; = \; 0 \,. \tag{5.20}$$

Formula (5.19) implies that whenever a polynomial $A$ can be written as a derivative of some other polynomial $A'$, the integral over $A$ vanishes. Thus we conclude that the integral over $A$ must be proportional to the coefficient $a_{12 \ldots N}$ for the highest power of the generators $\eta_i$. In order to complete the definition of our integral we demand the normalization

$$\int d^N \eta \, \eta_1 \eta_2 \, \cdots \, \eta_N \; = \; 1 \,, \quad \text{which implies} \quad \int d^N \eta \, A \; = \; a_{12 \ldots N} \,. \tag{5.21}$$

We can push the definition of the Grassmann integral even closer to integration in $\mathbb{R}^N$, by writing the measure $d^N \eta$ as a product

$$d^N \eta \; = \; d\eta_N \, d\eta_{N-1} \, \cdots \, d\eta_1 \,, \tag{5.22}$$

where the individual measures $d\eta_i$ obey

$$\int d\eta_i \, 1 = 0 \,, \quad \int d\eta_i \, \eta_i = 1 \,, \quad d\eta_i \, d\eta_j = - \, d\eta_j \, d\eta_i \,. \tag{5.23}$$

The last property is required by the freedom to interchange the $\eta_i$ in the integrand. Using (5.23) it is possible to define the integration over only a subset of the generators $\eta_i$, analogous to the case of $\mathbb{R}^N$. It is interesting to note that the measures $d\eta_i$ and the derivatives $\partial/\partial \eta_i$ obey the same algebraic properties (compare (5.23) and (5.16)).

Finally we need to discuss how the measure $d^N \eta$ transforms under a linear change of variables defined by

$$\eta_i' \; = \; \sum_{j=1}^{N} M_{ij} \eta_j \,, \tag{5.24}$$

where $M$ is a complex $N \times N$ matrix. Applying this change of variables to the normalization integral (5.21) we find

$$\int d^N \eta \, \eta_1 \ldots \eta_N = \int d^N \eta' \, \eta_1' \ldots \eta_N' = \int d^N \eta' \sum_{i_1, \ldots, i_N} M_{1 i_1} \ldots M_{N i_N} \, \eta_{i_1} \ldots \eta_{i_N}$$

$$= \int d^N \eta' \sum_{i_1, \ldots, i_N} M_{1 i_1} \ldots M_{N i_N} \, \epsilon_{i_1 i_2 \ldots i_N} \, \eta_1 \ldots \eta_N = \det[M] \int d^N \eta' \, \eta_1 \ldots \eta_N \,. \tag{5.25}$$

In the second line of this equation we have reordered the product of Grassmann numbers $\eta_{i_1} \ldots \eta_{i_N}$. This product vanishes if two of the indices $i_k$ are equal. The sign from the reordering is given by the completely anti-symmetric tensor $\epsilon_{i_1 i_2 \ldots i_N}$. The summation of the matrix elements $M_{ij}$ with this tensor gives the determinant. Comparing the first and the last expression in (5.25) one reads off the transformation properties of the measure in the Grassmann integration

$$d^N \eta = \det[M] \, d^N \eta' \,. \tag{5.26}$$

This transformation is "opposite" to the transformation of the integration measure for $\mathbb{R}^N$, where $\det[M]$ would appear on the left-hand side.

### 5.1.5 Gaussian integrals with Grassmann numbers

We finally derive formulas for Gaussian integrals with Grassmann numbers which we will need for our treatment of lattice fermions. In particular we consider a Grassmann algebra with $2N$ generators $\eta_i, \overline{\eta}_i, i = 1, 2, \ldots, N$. We stress that all these $2N$ generators anti-commute with each other, i.e., we have

$$\eta_i \eta_j = -\eta_j \eta_i \,, \quad \overline{\eta}_i \overline{\eta}_j = -\overline{\eta}_j \overline{\eta}_i \,, \quad \overline{\eta}_i \eta_j = -\eta_j \overline{\eta}_i \,. \tag{5.27}$$

The first integral we prove is the so-called *Matthews–Salam formula* [4, 5]

$$Z_F = \int d\eta_N d\overline{\eta}_N \, \ldots \, d\eta_1 d\overline{\eta}_1 \, \exp\left( \sum_{i,j=1}^N \overline{\eta}_i M_{ij} \eta_j \right) = \det[M] \,, \tag{5.28}$$

where $M$ is a complex $N \times N$ matrix. A possible minus sign in the exponent, as it appears in (5.2), can be absorbed in the definition of $M$. When setting $M = -D$, (5.28) establishes that the fermionic partition function (5.3) is indeed a determinant, called *fermion determinant*.

To show the result (5.28) we define transformed integration variables $\eta'_j$

$$\eta'_j = \sum_{k=1}^N M_{jk} \eta_k \,. \tag{5.29}$$

From (5.26) we find for the transformation of the measure

$$d\eta_N d\overline{\eta}_N \, \ldots \, d\eta_1 d\overline{\eta}_1 = \det[M] \, d\eta'_N d\overline{\eta}_N \, \ldots \, d\eta'_1 d\overline{\eta}_1 \,. \tag{5.30}$$

Using the transformation (5.29) and (5.30) we prove formula (5.28):

$$Z_F = \det[M] \int \prod_{i=1}^N d\eta'_i d\overline{\eta}_i \, \exp\left( \sum_{j=1}^N \overline{\eta}_j \eta'_j \right) = \det[M] \prod_{i=1}^N \int d\eta'_i d\overline{\eta}_i \, \exp\left( \overline{\eta}_i \eta'_i \right)$$

$$= \det[M] \prod_{i=1}^N \int d\eta'_i d\overline{\eta}_i \, (1 + \overline{\eta}_i \eta'_i) = \det[M] \,. \tag{5.31}$$

In (5.31) we have used the fact that pairs of Grassmann objects such as $\eta'_i \overline{\eta}_i$ and $d\eta'_j d\overline{\eta}_j$ commute with other pairs. In the second line we have expanded the individual exponential functions in power series. Due to the nilpotency of Grassmann numbers these series terminate after the second term.

Let us now generalize the integral (5.28) by embedding the $2N$-dimensional Grassmann algebra generated by $\eta_i, \overline{\eta}_i, i = 1, 2, \ldots, N$ into a $4N$-dimensional algebra generated by $\eta_i, \overline{\eta}_i, \theta_i, \overline{\theta}_i, i = 1, 2, \ldots, N$. All these $4N$ Grassmann numbers anti-commute with each other. However, we integrate only over the $\eta_i, \overline{\eta}_i$ and the other generators $\theta_i, \overline{\theta}_i$ serve as source terms. The integral we consider is the so-called *generating functional for fermions*. It is given by

$$W\left[\theta, \overline{\theta}\right] = \int \prod_{i=1}^{N} \mathrm{d}\eta_i \mathrm{d}\overline{\eta}_i \exp\left(\sum_{k,l=1}^{N} \overline{\eta}_k M_{kl} \eta_l + \sum_{k=1}^{N} \overline{\theta}_k \eta_k + \sum_{k=1}^{N} \overline{\eta}_k \theta_k\right)$$

$$= \det[M] \exp\left(-\sum_{n,m=1}^{N} \overline{\theta}_n \left(M^{-1}\right)_{nm} \theta_m\right). \tag{5.32}$$

For proving this result we first complete the square in the exponent and write the exponent as (we use summation convention for all indices):

$$\left(\overline{\eta}_i + \overline{\theta}_j \left(M^{-1}\right)_{ji}\right) M_{ik} \left(\eta_k + \left(M^{-1}\right)_{kl} \theta_l\right) - \overline{\theta}_n \left(M^{-1}\right)_{nm} \theta_m. \tag{5.33}$$

Then we perform a transformation of variables

$$\eta_k' = \eta_k + \left(M^{-1}\right)_{kl} \theta_l \ , \ \ \overline{\eta}_i' = \overline{\eta}_i + \overline{\theta}_j \left(M^{-1}\right)_{ji}. \tag{5.34}$$

From (5.23) it follows that the integration measure remains invariant under this transformation. Applying the transformation (5.33) to (5.32) we find

$$W\left[\theta, \overline{\theta}\right] = \exp\left(-\sum_{n,m=1}^{N} \overline{\theta}_n \left(M^{-1}\right)_{nm} \theta_m\right) \int \prod_{i=1}^{N} \mathrm{d}\eta_i' \mathrm{d}\overline{\eta}_i' \exp\left(\sum_{k,l=1}^{N} \overline{\eta}_k' M_{kl} \eta_l'\right)$$

$$= \det[M] \exp\left(-\sum_{n,m=1}^{N} \overline{\theta}_n \left(M^{-1}\right)_{nm} \theta_m\right), \tag{5.35}$$

where in the last step we have used (5.28), thus completing the proof.

### 5.1.6 Wick's theorem

Using the generating functional (5.32) we can now discuss a key formula for calculating the fermionic expectation value $\langle \ldots \rangle_F$ for products of Grassmann numbers. The formula is known as *Wick's theorem* and reads

$$\langle \eta_{i_1} \overline{\eta}_{j_1} \cdots \eta_{i_n} \overline{\eta}_{j_n} \rangle_F = \frac{1}{Z_F} \int \prod_{k=1}^{N} \mathrm{d}\eta_k \mathrm{d}\overline{\eta}_k \ \eta_{i_1} \overline{\eta}_{j_1} \cdots \eta_{i_n} \overline{\eta}_{j_n} \ \exp\left(\sum_{l,m=1}^{N} \overline{\eta}_l M_{lm} \eta_m\right)$$

$$= (-1)^n \sum_{P(1,2,\ldots,n)} \mathrm{sign}(P) \left(M^{-1}\right)_{i_1 j_{P_1}} \left(M^{-1}\right)_{i_2 j_{P_2}} \cdots \left(M^{-1}\right)_{i_n j_{P_n}}, \tag{5.36}$$

where the sum in the second line runs over all permutations $P(1, 2, \ldots, n)$ of the numbers $1, 2, \ldots, n$, and $\text{sign}(P)$ is the sign of the permutation $P$. The expectation values in Wick's theorem are often referred to as $n$-*point functions*.

The formula can be proven by noting that from the definition of $W\left[\theta, \overline{\theta}\right]$ in the first line of (5.32) follows

$$\langle \eta_{i_1} \overline{\eta}_{j_1} \cdots \eta_{i_n} \overline{\eta}_{j_n} \rangle_F = \frac{1}{Z_F} \frac{\partial}{\partial \theta_{j_1}} \frac{\partial}{\partial \overline{\theta}_{i_1}} \cdots \frac{\partial}{\partial \theta_{j_n}} \frac{\partial}{\partial \overline{\theta}_{i_n}} W\left[\theta, \overline{\theta}\right] \bigg|_{\theta, \overline{\theta} = 0} . \quad (5.37)$$

Using the explicit form of the generating functional $W\left[\theta, \overline{\theta}\right]$, given in the second line of (5.32), and performing the derivatives with respect to the sources, one arrives at the result (5.36). In a similar way it is easy to show that expectation values with different numbers of $\eta_i$ and $\overline{\eta}_j$ vanish.

This completes our discussion of computing with Grassmann numbers, and we now have the algebraic tools ready for working with fermions.


## 5.2 Fermion doubling and Wilson's fermion action

In the last section we have introduced Grassmann numbers in order to incorporate Fermi statistics in our formulation of QCD on the lattice. In this section we will first rewrite the naive fermion action presented in Chap. 2 as a quadratic form. This step makes explicit that Wick's theorem, shown in the last section, can be applied. Based on this form we will then consider the free case and analyze the Fourier transform of the lattice Dirac operator. We will identify the aforementioned doublers and add a term to the action in order to remove these unwanted degrees of freedom in the continuum limit.


### 5.2.1 The Dirac operator on the lattice

The naive fermion action presented in Sect. 2.2 reads (see (2.36)):

$$S_F[\psi, \overline{\psi}, U] = a^4 \sum_{n \in \Lambda} \overline{\psi}(n) \left( \sum_{\mu=1}^{4} \gamma_\mu \frac{U_\mu(n)\psi(n+\hat{\mu}) - U_{-\mu}(n)\psi(n-\hat{\mu})}{2a} + m\, \psi(n) \right) . \quad (5.38)$$

Here we discuss the case of only a single flavor for notational convenience and thus no sum over the flavor index occurs. The flavors differ only by the value of the mass parameter $m$ which is irrelevant for the discussion below.

Since the action is bilinear in $\overline{\psi}$ and $\psi$, we can write it in the form

$$S_F[\psi, \overline{\psi}, U] = a^4 \sum_{n, m \in \Lambda} \sum_{a, b, \alpha, \beta} \overline{\psi}(n)_{\alpha\atop a} D(n|m)_{\alpha\beta \atop a b} \psi(m)_{\beta \atop b} . \quad (5.39)$$

The naive Dirac operator on the lattice is then given by

$$D(n|m)_{\substack{\alpha\beta\\ab}} = \sum_{\mu=1}^{4}(\gamma_\mu)_{\alpha\beta}\frac{U_\mu(n)_{ab}\,\delta_{n+\hat\mu,m} - U_{-\mu}(n)_{ab}\,\delta_{n-\hat\mu,m}}{2a} + m\,\delta_{\alpha\beta}\,\delta_{ab}\,\delta_{n,m}\,.$$

$$(5.40)$$

In (5.39) we have rewritten the fermion action in exactly the form which is used in Wick's theorem (5.36) when setting $M = -a^4\,D$. The only notational difference is that in (5.39) we sum over several different indices, while in (5.36) we use only a single index to label the different Grassmann numbers. However, this does not change the algebraic content or the applicability of the formula.

### 5.2.2 The doubling problem

Let us now compute the Fourier transform of the lattice Dirac operator $D(n|m)$ for trivial gauge fields $U_\mu(n) = \mathbb{1}$, i.e., for the case of free lattice fermions. In Appendix A.3 we collect and discuss the formulas necessary for Fourier transformation on the lattice. Fourier transformation is applied independently to the two space–time arguments $n$ and $m$. In order to have a unitary similarity transformation the second index (here $m$) of a matrix is Fourier transformed using the complex conjugate phase $\exp(iq \cdot ma)$. The Fourier transform of the Dirac operator (5.40) for trivial gauge field reads (since $U_\mu(n) = \mathbb{1}$ we omit the color indices for notational convenience and use vector/matrix notation in Dirac space)

$$\widetilde{D}(p|q) = \frac{1}{|\Lambda|}\sum_{n,m\in\Lambda}\mathrm{e}^{-ip\cdot na}\,D(n|m)\,\mathrm{e}^{iq\cdot ma}$$

$$= \frac{1}{|\Lambda|}\sum_{n\in\Lambda}\mathrm{e}^{-i(p-q)\cdot na}\left(\sum_{\mu=1}^{4}\gamma_\mu\frac{\mathrm{e}^{+iq_\mu a} - \mathrm{e}^{-iq_\mu a}}{2a} + m\mathbb{1}\right)$$

$$= \delta(p-q)\,\widetilde{D}(p)\,,\qquad\qquad(5.41)$$

where $|\Lambda|$ is the total number of lattice points (see Appendix A.3) and the Fourier transform of the lattice Dirac operator is defined by

$$\widetilde{D}(p) = m\mathbb{1} + \frac{i}{a}\sum_{\mu=1}^{4}\gamma_\mu\sin(p_\mu a)\,.\qquad\qquad(5.42)$$

From the last line in (5.41) it is obvious that in the new basis, where the matrix elements are labeled by $p, q$, the Dirac operator is diagonal in the momenta. Thus, in order to calculate the inverse $D^{-1}(n|m)$ of the Dirac operator in real space, we simply need to compute the inverse of the $4 \times 4$ matrix $\widetilde{D}(p)$ and then invert the Fourier transformation. The matrix $\widetilde{D}(p)^{-1}$ can be calculated easily with the help of (A.26),

$$\widetilde{D}(p)^{-1} = \frac{m\mathbb{1} - ia^{-1}\sum_\mu\gamma_\mu\sin(p_\mu a)}{m^2 + a^{-2}\sum_\mu\sin(p_\mu a)^2}\,,\qquad\qquad(5.43)$$

and we obtain upon inverting the Fourier transformation

$$D^{-1}(n|m) = \frac{1}{|\Lambda|} \sum_{p \in \tilde{\Lambda}} \tilde{D}(p)^{-1} e^{ip \cdot (n-m)a} , \qquad (5.44)$$

where we have eliminated one of the momentum sums using $\delta(p-q)$ from (5.41). Thus for free fermions we have calculated the inverse $D^{-1}(n|m)$ of the lattice Dirac operator. This inverse is referred to as the *quark propagator.*

According to the Wick's theorem (5.36), the quark propagator governs the behavior of $n$-point functions and therefore it is important to analyze it. For free fermions this analysis is best done in momentum space, i.e., we study the momentum space propagator $\tilde{D}(p)^{-1}$. Of particular interest is the case of massless fermions and we set $m = 0$ in (5.43). First we remark that for fixed $p$ the propagator has the correct naive continuum limit,

$$\tilde{D}(p)^{-1}\Big|_{m=0} = \frac{-ia^{-1}\sum_{\mu}\gamma_{\mu}\sin(p_{\mu}a)}{a^{-2}\sum_{\mu}\sin(p_{\mu}a)^2} \xrightarrow{a \to 0} \frac{-i\sum_{\mu}\gamma_{\mu}p_{\mu}}{p^2} . \qquad (5.45)$$

In the continuum the momentum space propagator for massless fermions (right-hand side of (5.45)) has a pole at

$$p = (0, 0, 0, 0) . \qquad (5.46)$$

This pole corresponds to the single fermion which is described by the continuum Dirac operator. On the lattice the situation is different. There the propagator for free fermions (center term in (5.45)) has additional poles: Whenever all components are either $p_{\mu} = 0$ or $p_{\mu} = \pi/a$, we find a pole. The momentum space contains all momenta $p_{\mu} \in (-\pi/a, \pi/a]$, with $-\pi/a$ and $\pi/a$ identified (compare Appendix A.3), and we cannot simply exclude the value $p_{\mu} = \pi/a$. Thus, as it stands, our lattice Dirac operator has unphysical poles at

$$p = (\pi/a, 0, 0, 0) , (0, \pi/a, 0, 0) , \dots , (\pi/a, \pi/a, \pi/a, \pi/a) . \qquad (5.47)$$

It is easy to see that this gives rise to 15 unwanted poles, the so-called *doublers.* We now discuss a way to remove the doublers from our theory, at least in the continuum limit.

### 5.2.3 Wilson fermions

In order to remove the doublers we need to distinguish between the proper pole with all $p_{\mu} = 0$ and the doublers that contain momentum components $p_{\mu} = \pi/a$. A possible solution was suggested by Wilson. He proposed to add an extra term, such that the momentum space Dirac operator reads

$$\tilde{D}(p) = m\mathbb{1} + \frac{i}{a}\sum_{\mu=1}^{4}\gamma_{\mu}\sin(p_{\mu}a) + \mathbb{1}\frac{1}{a}\sum_{\mu=1}^{4}(1 - \cos(p_{\mu}a)) . \qquad (5.48)$$

The extra term, the so-called *Wilson term*, is exactly what we need. For components with $p_\mu = 0$ it simply vanishes. For each component with $p_\mu = \pi/a$ it provides an extra contribution $2/a$. This term acts like an additional mass term, and the total mass of the doublers is given by

$$m + \frac{2\ell}{a} , \tag{5.49}$$

where $\ell$ is the number of momentum components with $p_\mu = \pi/a$. In the limit $a \to 0$ the doublers become very heavy and decouple from the theory. If one calculates the corresponding momentum space propagator $\tilde{D}(p)^{-1}$, one finds that the unwanted poles are gone and only the physical pole (5.46) remains.

For the free case, the form of the Wilson term in position space can be found by inverse Fourier transformation of the last term in (5.48). It contains nearest neighbor terms which can be made gauge-invariant by inserting link variables. One obtains $(U_{-\mu}(n) \equiv U_\mu(n - \hat{\mu})^\dagger)$:

$$-a \sum_{\mu=1}^{4} \frac{U_\mu(n)_{ab}\, \delta_{n+\hat{\mu},m} - 2\delta_{ab}\, \delta_{n,m} + U_{-\mu}(n)_{ab}\, \delta_{n-\hat{\mu},m}}{2a^2} . \tag{5.50}$$

The Wilson term is a discretization of $-(a/2)\partial_\mu\partial_\mu$, i.e., proportional to the negative Laplace operator.[2] The pre-factor $a$ shows that the Wilson term vanishes in the naive continuum limit $a \to 0$. We can now combine the Wilson term (5.50) with the naive Dirac operator to obtain Wilson's complete Dirac operator. Using a particularly compact notation it reads:

$$D^{(f)}(n|m)_{\substack{\alpha\beta \\ ab}} = \left(m^{(f)} + \frac{4}{a}\right)\delta_{\alpha\beta}\,\delta_{ab}\,\delta_{n,m} - \frac{1}{2a}\sum_{\mu=\pm1}^{\pm4}(\mathbb{1} - \gamma_\mu)_{\alpha\beta}\, U_\mu(n)_{ab}\,\delta_{n+\hat{\mu},m} , \tag{5.51}$$

where we have defined

$$\gamma_{-\mu} = -\gamma_\mu , \quad \mu = 1, 2, 3, 4 . \tag{5.52}$$

In (5.51) we have now also made explicit the flavor dependence of the mass parameter $m^{(f)}$. Thus the fermion action for QCD with $N_f$ flavors is given by

$$S_F[\psi, \overline{\psi}, U] = \sum_{f=1}^{N_f} a^4 \sum_{n,m \in \Lambda} \overline{\psi}^{(f)}(n)\, D^{(f)}(n|m)\, \psi^{(f)}(m) . \tag{5.53}$$

We stress that the right-hand side depends on the gauge field $U$, via the $U$-dependence of the Dirac operator $D$. For notational simplicity we do not display the argument $U$ of the Dirac operator explicitly.

We now have a working formulation of lattice QCD, Wilson's formulation. For the convenience of the reader we collect the defining formulas for Wilson's

---

[2]Taylor expansion gives $(f(x + \varepsilon) - 2f(x) + f(x - \varepsilon))/\varepsilon^2 = f''(x) + \mathcal{O}(\varepsilon^2)$.

formulation of lattice QCD in Appendix A.4. Although we will further refine our formulation in the next chapters, we strongly recommend that the reader has a brief look at Appendix A.4 now. It summarizes all the conceptual steps we have taken so far in this book and highlights again the red line through the construction.

## 5.3 Fermion lines and hopping expansion

In this section we analyze Wilson fermions in the limit of large quark mass. This will lead us to the so-called *hopping expansion*. Although the hopping expansion is not very powerful as an analytical tool, it will provide an important physical insight: Fermions can be viewed as paths of link variables, so-called *fermion lines* [6–8]. Based on this insight we will justify a step which we used in Chap. 3 when constructing the Wilson loop, namely the replacement of the quark–antiquark pair by a Wilson line (compare (3.57)–(3.47)). We will also find an interpretation for the fermion determinant: It can be seen as a collection of closed fermion lines, so-called *fermion loops*.

### 5.3.1 Hopping expansion of the quark propagator

In Sect. 5.1 we have shown that the fermionic expectation value of a fermion and an anti-fermion field, i.e., a two-point function of fermions, is given by the inverse of the Dirac operator,

$$\left\langle \psi(n)_{\substack{\alpha \\ a}} \overline{\psi}(m)_{\substack{\beta \\ b}} \right\rangle_F = a^{-4} D^{-1}(n|m)_{\substack{\alpha\beta \\ ab}} . \tag{5.54}$$

This formula follows from Wick's theorem (5.36) by setting $M = -a^4 D$, $n = 1$ and replacing the multi-index $i_1$ by $(n, \alpha, a)$ and $j_1$ by $(m, \beta, b)$.

The inverse of the Dirac operator, the quark propagator $D^{-1}$, as well as the fermion determinant will now be expanded for large quark mass $m$. In particular we will use Wilson's Dirac operator given in (5.51). This Dirac operator can be written as (we use matrix/vector notation)

$$D = C\left(\mathbb{1} - \kappa H\right) \quad \text{with} \quad \kappa = \frac{1}{2\left(a\,m + 4\right)} , \quad C = m + \frac{4}{a} , \tag{5.55}$$

$$H(n|m)_{\substack{\alpha\beta \\ ab}} = \sum_{\mu=\pm 1}^{\pm 4} \left(\mathbb{1} - \gamma_\mu\right)_{\alpha\beta} U_\mu(n)_{ab} \, \delta_{n+\hat{\mu},m} . \tag{5.56}$$

The term $H$ collects all nearest neighbor terms in the Dirac operator and thus is referred to as *hopping matrix*. The real number $\kappa$ is the *hopping parameter*. The constant $C$ is irrelevant, since it can be absorbed in a redefinition of the quark fields $\psi \to \sqrt{C}\,\psi$, $\overline{\psi} \to \sqrt{C}\,\overline{\psi}$. Thus we can equally work well with the Dirac operator in the form

$$D = \mathbb{1} - \kappa H , \quad \kappa = \frac{1}{2\,(a\,m + 4)} . \tag{5.57}$$

The essence of the hopping expansion is that $\kappa$ becomes small for large mass. The idea is to expand $D^{-1}$ and $\det[D]$ in powers of $\kappa$. For the quark propagator one can use the well-known geometric series

$$D^{-1} = (\mathbb{1} - \kappa H)^{-1} = \sum_{j=0}^{\infty} \kappa^j H^j . \tag{5.58}$$

The series converges for $\kappa \|H\| < 1$. It is relatively easy to show that the norm[3] of the hopping term obeys $\|H\| \le 8$ and thus the series converges for $\kappa < 1/8$.

Let us look at (5.58) in more detail and display all indices explicitly:

$$D^{-1}(n|m)_{\substack{\alpha\,\beta \\ a\,b}} = \sum_{j=0}^{\infty} \kappa^j\, H^j(n|m)_{\substack{\alpha\,\beta \\ a\,b}} . \tag{5.59}$$

Here $H^j(n|m)_{\substack{\alpha\,\beta \\ a\,b}}$ denotes an entry of the $j$-th power of $H$. Using (5.56) we find for these powers

$$H^0(n|m)_{\substack{\alpha\,\beta \\ a\,b}} = \delta_{\alpha\beta}\,\delta_{ab}\,\delta_{n,m} ,$$

$$H^1(n|m)_{\substack{\alpha\,\beta \\ a\,b}} = \sum_{\mu=\pm 1}^{\pm 4} (\mathbb{1} - \gamma_\mu)_{\alpha\beta}\, U_\mu(n)_{ab}\, \delta_{n+\hat\mu,m} ,$$

$$H^2(n|m)_{\substack{\alpha\,\beta \\ a\,b}} = \sum_{l,\rho,c} H(n,l)_{\substack{\alpha\,\rho \\ a\,c}} H(l,m)_{\substack{\rho\,\beta \\ c\,b}} \tag{5.60}$$

$$= \sum_{\mu,\nu=\pm 1}^{\pm 4} ((\mathbb{1}-\gamma_\mu)(\mathbb{1}-\gamma_\nu))_{\alpha\beta}\, (U_\mu(n)U_\nu(n+\hat\mu))_{ab}\, \delta_{n+\hat\mu+\hat\nu,m} ,$$

$$H^j(n|m)_{\substack{\alpha\,\beta \\ a\,b}} = \sum_{\mu_i=\pm 1}^{\pm 4} \left( \prod_{i=1}^{j} (\mathbb{1}-\gamma_{\mu_i}) \right)_{\alpha\beta} P_{\mu_1\ldots\mu_j}(n)_{ab}\, \delta_{n+\hat\mu_1+\ldots+\hat\mu_j,m} .$$

In the last line we used the abbreviation

$$P_{\mu_1\ldots\mu_j}(n)_{ab} = \left( U_{\mu_1}(n) U_{\mu_2}(n+\hat\mu_1) \ldots U_{\mu_j}(n+\hat\mu_1+\hat\mu_2 \ldots \hat\mu_{j-1}) \right)_{ab} , \tag{5.61}$$

for the emerging products of link variables. The 0-th and first powers of $H$ are trivial. For computing the second power in (5.60) we wrote the matrix multiplication explicitly as a sum over $l, \rho, c$ and then used (5.56). The sum over $l$ vanishes due to the Kronecker deltas. Iterating this step one calculates higher powers of $H$ giving the result in the last line of (5.60).

---

[3] $\|H\| = \max_\phi \sqrt{(H\phi, H\phi)/(\phi,\phi)}$, with $\phi$ a complex vector of length $12\,|\Lambda|$.

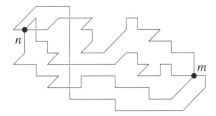

**Fig. 5.1.** The propagator from $n$ to $m$ is a sum over paths of link variables connecting the two points. A path of length $j$ comes with a factor of $\kappa^j$

The terms in (5.60) have a simple interpretation. Due to the Kronecker delta $\delta_{n+\hat{\mu}_1+\ldots+\hat{\mu}_j,m}$ they are nonvanishing only if the positions $n$ and $m$ on the lattice are related by

$$ m = n+\hat{\mu}_1+\ldots+\hat{\mu}_j \text{ for a combination of } \mu_i \in \{\pm 1, \pm 2, \pm 3, \pm 4\} . \quad (5.62) $$

If the condition (5.62) is obeyed, then the two sites $n$ and $m$ are connected by the product of link variables on the path as given in (5.61). In Dirac space one finds products

$$ \prod_{i=1}^{j} (\mathbb{1} - \gamma_{\mu_i}) , \quad (5.63) $$

where the $\mu_i$ are the orientations of the links on the path. It is interesting to note that

$$ (\mathbb{1} - \gamma_\mu)(\mathbb{1} + \gamma_\mu) = 0 . \quad (5.64) $$

This property excludes *back-tracking* paths, i.e., paths containing 180° turns.

To summarize our hopping expansion of the quark propagator, we find that $D^{-1}(n|m)$ can be written as a sum of non-back-tracking paths. A path of length $j$ comes with a power of $\kappa^j$. For given values $n$ and $m$, the leading term is the shortest path (or a sum over the shortest paths) connecting $n$ and $m$. Higher orders in $\kappa$ will contribute longer and longer paths connecting the two points. Along the links of these paths the products of the link variables $U_\mu(n)$ provide the factors in color space and the products (5.63) the factors in Dirac space. Some paths contributing in the hopping expansion are illustrated in Fig. 5.1.

With this result we have also justified the replacement of the product of fermion fields by a Wilson line in our construction of the Wilson loop (Sects. 3.3.1–3.3.4): The Wilson line is simply one of the terms in the hopping expansion. In particular in the limit of infinite quark mass ($\kappa \to 0$), which defines the potential of a static quark, only the shortest possible path will contribute in the hopping expansion, i.e., the straight Wilson line.

### 5.3.2 Hopping expansion for the fermion determinant

Similar to the quark propagator, one can also expand the fermion determinant in $\kappa$. The central equation is

$$\det[D] = \det[\mathbb{1} - \kappa H] = \exp\left(\operatorname{tr}\left[\ln\left(\mathbb{1} - \kappa H\right)\right]\right) = \exp\left(-\sum_{j=1}^{\infty} \frac{1}{j}\,\kappa^j \operatorname{tr}\left[H^j\right]\right).$$
$$(5.65)$$

In the first step we have used (5.57) and in the second a well-known formula for the determinant (see Appendix A.5). In the last step the logarithm was expanded in a power series.

From (5.65) it follows that for the hopping expansion we have to compute traces of powers of the hopping term. These powers have already been calculated in (5.60). Taking the trace means, in addition to the trace in Dirac and color space, that we have to set $n = m$ in (5.60) and to sum over the lattice points $m$. When $n = m$, the paths of links which we have obtained are turned into closed loops, so-called *fermion loops*.[4] Thus the fermion determinant is the exponential of a sum over closed loops of link variables. Since for each loop the trace over color indices is taken, these loops are gauge-invariant, and so is the determinant. A loop of length $j$ comes with a factor $\kappa^j/j$. In Dirac space we have again products of the type (5.63) but now the trace is taken. Again (5.64) excludes non-back-tracking loops.

This concludes our discussion of the hopping expansion. The important message to be learned is that quark propagators are sums over fermion lines, i.e., paths of link variables, and the fermion determinant is a sum over closed fermion loops.

## 5.4 Discrete symmetries of the Wilson action

Let us now discuss some symmetries of the Wilson fermion action. Certainly the lattice discretization breaks many of the symmetries which we have in the continuum, such as translation or rotation invariance. On the lattice we are reduced to discrete translations and rotations, but one can, e.g., demonstrate that in the continuum limit continuous rotational symmetry is recovered.

In this section, however, we concentrate on the discrete symmetries like charge conjugation $\mathcal{C}$ and parity $\mathcal{P}$. These symmetries are important for the construction of hadron interpolators which we will address in Chap. 6.

### 5.4.1 Charge conjugation

We begin our discussion of the discrete symmetries with the charge conjugation $\mathcal{C}$. Charge conjugation transforms particles into antiparticles and acts on the spinor fields via (we omit the flavor index in the following)

---

[4] Note that these are not the loops of weak coupling perturbation theory.

$$\psi(n) \xrightarrow{\mathcal{C}} \psi(n)^{\mathcal{C}} = C^{-1}\overline{\psi}(n)^T ,$$

$$\overline{\psi}(n) \xrightarrow{\mathcal{C}} \overline{\psi}(n)^{\mathcal{C}} = -\psi(n)^T C . \qquad (5.66)$$

In this formula the transposition $T$ is used for notational convenience. It acts on both color and Dirac indices, converting a $3 \times 4$ column spinor to a row-spinor. The charge conjugation matrix $C$ acts only on the Dirac indices and is defined to obey the relations

$$C\gamma_\mu C^{-1} = -\gamma_\mu^T . \qquad (5.67)$$

In Appendix A.2 we give the explicit form of $C$ for the chiral representation of the gamma matrices in (A.24).

The link variables transform under charge conjugation as

$$U_\mu(n) \xrightarrow{\mathcal{C}} U_\mu(n)^{\mathcal{C}} = U_\mu(n)^* = \left(U_\mu(n)^\dagger\right)^T . \qquad (5.68)$$

When writing the link variable as $U_\mu(n) = \exp(iaA_\mu(n))$ (compare (2.39)), we find that charge conjugation (5.68) corresponds to changing $A_\mu(n) \to -A_\mu(n)^T$. This change takes into account that an antiparticle has the opposite charge of the corresponding particle.

Let us now show that the Wilson action is indeed invariant under charge conjugation. For the trivial (mass-type) term we find (sums and constant factors are omitted for notational convenience)

$$\overline{\psi}(n)^{\mathcal{C}} \psi(n)^{\mathcal{C}} = -\psi(n)^T CC^{-1}\overline{\psi}(n)^T = -\psi(n)^T \overline{\psi}(n)^T = \overline{\psi}(n)\psi(n) , \qquad (5.69)$$

where the interchange of the Grassmann variables in the last step removes the overall minus sign.

For the hopping part we find (we use the conventions $\gamma_{-\mu} = -\gamma_\mu$ and $U_{-\mu}(n) = U_\mu(n - \hat{\mu})^\dagger$)

$$a^4 \sum_n \sum_{\mu=\pm 1}^{\pm 4} \overline{\psi}(n)^{\mathcal{C}} \frac{\mathbb{1} - \gamma_\mu}{2a} U_\mu(n)^{\mathcal{C}} \psi(n + \hat{\mu})^{\mathcal{C}}$$

$$= -a^4 \sum_n \sum_{\mu=\pm 1}^{\pm 4} \psi(n)^T C\frac{\mathbb{1} - \gamma_\mu}{2a} C^{-1} U_\mu(n)^* \overline{\psi}(n + \hat{\mu})^T$$

$$= -a^4 \sum_n \sum_{\mu=\pm 1}^{\pm 4} \psi(n)^T \frac{\mathbb{1} + \gamma_\mu^T}{2a} \left(U_\mu(n)^\dagger\right)^T \overline{\psi}(n + \hat{\mu})^T . \qquad (5.70)$$

In the last equation we have a product of transposed matrices and spinors. This can be written as the transpose of the product of the matrices and spinors in reverse order. Reversing the order of the two Grassmann objects produces an overall minus sign. Since the whole product is a scalar, transposition can be dropped altogether in the end. Thus we continue the last equation as

$$a^4 \sum_n \sum_{\mu=\pm 1}^{\pm 4} \overline{\psi}(n + \hat{\mu}) \frac{\mathbb{1} + \gamma_\mu}{2a} U_\mu(n)^\dagger \psi(n)$$

$$= a^4 \sum_{\mu=\pm 1}^{\pm 4} \sum_{m=n+\hat{\mu}} \overline{\psi}(m) \frac{\mathbb{1} + \gamma_\mu}{2a} U_\mu(m - \hat{\mu})^\dagger \psi(m - \hat{\mu})$$

$$= a^4 \sum_m \sum_{\mu=\pm 1}^{\pm 4} \overline{\psi}(m) \frac{\mathbb{1} - \gamma_{-\mu}}{2a} U_{-\mu}(m) \psi(m - \hat{\mu})$$

$$= a^4 \sum_m \sum_{\mu=\pm 1}^{\pm 4} \overline{\psi}(m) \frac{\mathbb{1} - \gamma_\mu}{2a} U_\mu(m) \psi(m + \hat{\mu}) , \tag{5.71}$$

showing that also the hopping part is invariant under charge conjugation.

### 5.4.2 Parity and Euclidean reflections

A parity transformation $\mathcal{P}$ acts on our lattice fields as

$$\psi(\boldsymbol{n}, n_4) \xrightarrow{\mathcal{P}} \psi(\boldsymbol{n}, n_4)^{\mathcal{P}} = \gamma_4 \psi(-\boldsymbol{n}, n_4) ,$$
$$\overline{\psi}(\boldsymbol{n}, n_4) \xrightarrow{\mathcal{P}} \overline{\psi}(\boldsymbol{n}, n_4)^{\mathcal{P}} = \overline{\psi}(-\boldsymbol{n}, n_4) \gamma_4 ,$$
$$U_i(\boldsymbol{n}, n_4) \xrightarrow{\mathcal{P}} U_i(\boldsymbol{n}, n_4)^{\mathcal{P}} = U_i(-\boldsymbol{n} - \hat{i}, n_4)^\dagger , \quad i = 1, 2, 3 ,$$
$$U_4(\boldsymbol{n}, n_4) \xrightarrow{\mathcal{P}} U_4(\boldsymbol{n}, n_4)^{\mathcal{P}} = U_4(-\boldsymbol{n}, n_4) . \tag{5.72}$$

Here a comment on the labeling of the lattice sites is in order: In our definition of the lattice $\Lambda$ (see (2.25) or (A.27)) the spatial lattice sites $n_i$ run from 0 to $N - 1$. This convention is particularly convenient for numerical studies. However, one can equally well label the lattice sites as $n_i = -N/2+1, -N/2+2, \ldots, N/2$. This latter convention, together with the periodic identification $N/2 + 1 \leftrightarrow -N/2 + 1$ is used here when we apply the reflection $\boldsymbol{n} \to -\boldsymbol{n}$.

The diagonal term of the Wilson action is trivially invariant under the parity transformation (5.72), since in the sum over the spatial components, $\boldsymbol{n}$ can always be replaced by a sum over $-\boldsymbol{n}$. For the spatial part of the hopping term we find

$$a^4 \sum_{\boldsymbol{n}, n_4} \sum_{i=\pm 1}^{\pm 3} \overline{\psi}(\boldsymbol{n}, n_4)^{\mathcal{P}} \frac{\mathbb{1} - \gamma_i}{2a} U_i(\boldsymbol{n}, n_4)^{\mathcal{P}} \psi(\boldsymbol{n} + \hat{i}, n_4)^{\mathcal{P}}$$

$$= a^4 \sum_{\boldsymbol{n}, n_4} \sum_{i=\pm 1}^{\pm 3} \overline{\psi}(-\boldsymbol{n}, n_4) \gamma_4 \frac{\mathbb{1} - \gamma_i}{2a} \gamma_4 U_i(-\boldsymbol{n} - \hat{i}, n_4)^\dagger \psi(-\boldsymbol{n} - \hat{i}, n_4)$$

$$= a^4 \sum_{\boldsymbol{m}=-\boldsymbol{n}} \sum_{n_4} \sum_{i=\pm 1}^{\pm 3} \overline{\psi}(\boldsymbol{m}, n_4) \frac{\mathbb{1} - \gamma_{-i}}{2a} U_{-i}(\boldsymbol{m}, n_4) \psi(\boldsymbol{m} - \hat{i}, n_4)$$

$$= a^4 \sum_{\boldsymbol{m}} \sum_{n_4} \sum_{i=\pm 1}^{\pm 3} \overline{\psi}(\boldsymbol{m}, n_4) \frac{\mathbb{1} - \gamma_i}{2a} U_i(\boldsymbol{m}, n_4) \psi(\boldsymbol{m} + \hat{i}, n_4) . \tag{5.73}$$

For the temporal hopping part the equivalent steps are

$$a^4 \sum_{\boldsymbol{n}, n_4} \sum_{\mu=\pm 4} \overline{\psi}(\boldsymbol{n}, n_4)^{\mathcal{P}} \frac{\mathbb{1} - \gamma_\mu}{2a} U_\mu(\boldsymbol{n}, n_4)^{\mathcal{P}} \psi(\boldsymbol{n}, n_4 \pm 1)^{\mathcal{P}}$$

$$= a^4 \sum_{\boldsymbol{n}, n_4} \sum_{\mu=\pm 4} \overline{\psi}(-\boldsymbol{n}, n_4) \frac{\mathbb{1} - \gamma_\mu}{2a} U_\mu(-\boldsymbol{n}, n_4) \psi(-\boldsymbol{n}, n_4 \pm 1)$$

$$= a^4 \sum_{\boldsymbol{m}, n_4} \sum_{\mu=\pm 4} \overline{\psi}(\boldsymbol{m}, n_4) \frac{\mathbb{1} - \gamma_\mu}{2a} U_\mu(\boldsymbol{m}, n_4) \psi(\boldsymbol{m}, n_4 \pm 1) . \qquad (5.74)$$

Thus we have established that the Wilson fermion action is invariant under parity transformations.

It is straightforward to show that also the Wilson gauge action is invariant under $\mathcal{C}$ and $\mathcal{P}$, and we leave this as an exercise to the reader. It has to be stressed that also other lattice actions, which we discuss later, are invariant under parity and charge conjugation. We will, however, not repeat the explicit steps to show that.

In Minkowski space the time component and the spatial components are distinguished by a relative sign in the metric. In Euclidean space there is no such distinction and the Wilson action actually is invariant under the following four, more general transformations $\mathcal{P}_\mu, \mu = 1, 2, 3, 4,$

$$\psi(n) \xrightarrow{\mathcal{P}_\mu} \psi(n)^{\mathcal{P}_\mu} = \gamma_\mu \psi\left(P_\mu(n)\right) ,$$

$$\overline{\psi}(n) \xrightarrow{\mathcal{P}_\mu} \overline{\psi}(n)^{\mathcal{P}_\mu} = \overline{\psi}\left(P_\mu(n)\right) \gamma_\mu ,$$

$$U_\nu(n) \xrightarrow{\mathcal{P}_\mu} U_\nu(n)^{\mathcal{P}_\mu} = U_\nu\left(P_\mu(n) - \hat{\nu}\right)^\dagger , \quad \nu \neq \mu ,$$

$$U_\mu(n) \xrightarrow{\mathcal{P}_\mu} U_\mu(n)^{\mathcal{P}_\mu} = U_\mu\left(P_\mu(n)\right) , \qquad (5.75)$$

where $P_\mu(n)$ is the vector $n$ with the sign of all components reversed, except for $n_\mu$.

The fact that the action is invariant under under all four reflections $\mathcal{P}_\mu$ shows that in the Euclidean formulation none of the four directions is singled out. We remark that the product operation $\mathcal{P}_1 \mathcal{P}_2 \mathcal{P}_3$ corresponds to the Euclidean equivalent of time reflection. This reflection plays an important role for the formal reconstruction of the Hilbert space for the Minkowski theory [9–11]. For this reconstruction it is also necessary to impose anti-periodic temporal boundary conditions for the fermions. For this reason, numerical calculations are often performed with anti-periodic temporal boundary conditions for the fermions, while the gauge fields are usually periodic in all four directions. In the thermodynamic limit, however, the choice of boundary conditions is irrelevant, except for studies of QCD at high temperature which we discuss in Chap. 12.

### 5.4.3 $\gamma_5$-hermiticity

Let us finally discuss a more abstract symmetry of lattice Dirac operators which, however, has important implications. Almost all Dirac operators $D$ (except when a chemical potential, a $\theta$-term or a twisted mass term are introduced, cf. Chaps. 10 and 12) are $\gamma_5$-hermitian, i.e., they obey

$$(\gamma_5\, D)^\dagger = \gamma_5\, D \quad \text{or, equivalently,} \quad D^\dagger = \gamma_5\, D\, \gamma_5 \ . \tag{5.76}$$

It is a straightforward exercise to show this, e.g., for Wilson's lattice Dirac operator (5.51): The diagonal term of (5.51) is real and does not change when multiplied with $\gamma_5$ from both sides. In the hopping term the factor $(\mathbb{1} - \gamma_\mu)$ turns into $(\mathbb{1} + \gamma_\mu)$ due to the properties of the $\gamma$-matrices. Since the sum runs over both signs of $\mu$ we then may write

$$\sum_{\mu=\pm 1}^{\pm 4} \left(\gamma_5\, (\mathbb{1} - \gamma_\mu)\, \gamma_5\right)_{\alpha\beta} U_\mu(n)_{ab}\, \delta_{n+\hat{\mu},m} \tag{5.77}$$

$$= \sum_{\mu=\pm 1}^{\pm 4} (\mathbb{1} + \gamma_\mu)_{\alpha\beta}\, U_\mu(n)_{ab}\, \delta_{n+\hat{\mu},m} = \sum_{\mu=\pm 1}^{\pm 4} (\mathbb{1} - \gamma_\mu)_{\alpha\beta}\, U_{-\mu}(n)_{ab}\, \delta_{n-\hat{\mu},m}$$

$$= \sum_{\mu=\pm 1}^{\pm 4} (\mathbb{1} - \gamma_\mu)_{\alpha\beta}\, U_\mu(n-\hat{\mu})^\dagger_{ab}\, \delta_{n-\hat{\mu},m} = \sum_{\mu=\pm 1}^{\pm 4} (\mathbb{1} - \gamma_\mu)_{\alpha\beta}\, U_\mu(m)^\dagger_{ab}\, \delta_{n,m+\hat{\mu}} \ .$$

In the last step we have used the $\delta$-function to replace the position of the gauge field $n - \hat{\mu}$ by $m$. Comparison with (5.51) shows that $U_\mu$ has been replaced by $U_\mu^\dagger$ and $n$ and $m$ have been exchanged. We have indeed demonstrated the validity of (5.76).

The property is inherited by the inverse operator, the quark propagator, which turns out to be useful in calculations of hadron correlation functions in Sect. 6.2. Another characteristic of $\gamma_5$-hermitian Dirac operators is that their eigenvalues are either real or come in complex conjugate pairs, as we discuss in more detail in Sect. 7.3.1. This property also implies that the fermion determinant is real. The reality of the determinant will turn out to be crucial for Monte Carlo simulations of lattice QCD with fermions, which we discuss in Chap. 8.

## References

1. J. W. Negele and H. Orland: *Quantum Many Particle Systems* (Westview Press – Advanced Book Classics, Boulder 1998)
2. J. Smit: *Introduction to Quantum Fields on a Lattice* (Cambridge University Press, Cambridge 2002)
3. J. Zinn-Justin: *Quantum Field Theory and Critical Phenomena* (Clarendon Press, Oxford 1996)

4. T. Matthews and A. Salam: Nuovo Cim. **12**, 563 (1954)
5. T. Matthews and A. Salam: Nuovo Cim. **2**, 120 (1955)
6. C. B. Lang and H. Nicolai: Nucl. Phys. B **200** [**FS4**], 135 (1982)
7. A. Hasenfratz, P. Hasenfratz, Z. Kunszt, and C. B. Lang: Phys. Lett. B **115**, 289 (1982)
8. A. Hasenfratz, P. Hasenfratz, Z. Kunszt, and C. B. Lang: Phys. Lett. B **117**, 81 (1982)
9. J. Glimm and A. Jaffe: *Quantum Physics. A Functional Integral Point of View*, 2nd ed. (Springer, New York 1987)
10. K. Osterwalder and E. Seiler: Ann. Phys. **110**, 440 (1978)
11. E. Seiler: *Gauge Theories as a Problem of Constructive Quantum Field Theory and Statistical Mechanics*, Lect. Notes Phys. **159**. Springer, Berlin, Heidelberg, New York (1982)

# 6

# Hadron spectroscopy

The simplest quantities involving fermions that one can compute on the lattice are the masses of hadrons. Since there are many combinations of quantum numbers such as spin, parity, flavor content etc. there is a wealth of possible hadrons. Reproducing all their masses correctly is already a powerful test for the correctness of QCD. Over the last 20 years lattice QCD calculations of the mass spectrum have improved continually and, for ground state baryons, have reached impressive agreement with experimental data.

In this chapter we discuss the conceptual background and the actual implementation of a numerical spectroscopy calculation. The hadron correlation functions are not only necessary for computing the spectrum but enter also in the determination of hadronic matrix elements addressed in later chapters.

We first discuss how to construct operators with the correct quantum numbers and their correlation functions and address the quenched approximation. Subsequently we present techniques for quark sources and show how quark propagators can be computed. We continue with the analysis of the resulting hadron propagators and discuss how to obtain the hadron masses.

## 6.1 Hadron interpolators and correlators

In a typical QCD calculation one generates gauge configurations according to the requested distribution and then computes various observables. These include simple expectation values of plaquettes or of more extended operators such as Wilson loops. Also quantities like the chiral condensate $\langle \overline{\psi} \psi \rangle$ have been studied and will be discussed in later chapters.

Here we concentrate on determining hadron masses, i.e., we perform a *hadron spectroscopy calculation*. For this purpose one first computes quark propagators for each gauge configuration. These are then suitably combined to construct the hadron propagators. Finally, averaging over all gauge configurations provides our estimate of the hadron propagators for subsequent analysis.

Gattringer, C., Lang, C.B.: *Hadron Spectroscopy*. Lect. Notes Phys. **788**, 123–156 (2010)
DOI 10.1007/978-3-642-01850-3_6      © Springer-Verlag Berlin Heidelberg 2010

The first step of such a spectroscopy calculation is the identification of *hadron interpolators* $O, \overline{O}$ such that the corresponding Hilbert space operators $\hat{O}, \hat{O}^\dagger$ annihilate and create the particle states we want to analyze. A hadron interpolator is a functional of the lattice fields with the quantum numbers of the state one is interested in.[1] Once the interpolators are identified we consider the Euclidean correlator (1.12) of hadron interpolators $O(n_t), \overline{O}(0)$ located at time slices $n_4 = n_t$ and $n_4 = 0$.

For hadron spectroscopy one studies interpolators constructed out of quarks and gluons. These are by construction gauge-invariant color singlets. Such operators include:

- Local meson operators, like $O_M(n) \equiv \overline{\psi}(n)\Gamma\psi(n)$, consisting of quarks and antiquarks or baryon operators made out of three quarks; these should give information on the mesons and baryons as predicted by QCD.
- Extended interpolators, also contributing to hadron propagators, involving, e.g., for mesons terms like $\overline{\psi}(n)U_\mu(n)\psi(n+\hat{\mu})$ and similar for baryons.
- Pure gauge field interpolators like the plaquette, longer closed loops, or more general gluonic operators. These interpolators couple to gluonic objects such as glueballs, but also to mesons.
- Other, more exotic color-singlet combinations of quarks and antiquarks like $\overline{\psi}\psi\psi\psi$ and $\overline{\psi}\psi\psi\psi\psi$.
- States with $3n$ quarks; these are just the nuclei. For example, the deuterium is a proton–neutron state made out of six valence quarks, three of up-type and three of down-type.

According to (1.21), physically allowed states can be observed in the spectral decomposition of the propagators of these interpolators,

$$\langle O(n_t)\overline{O}(0)\rangle = \sum_k \langle 0|\hat{O}|k\rangle\langle k|\hat{O}^\dagger|0\rangle \, \mathrm{e}^{-n_t a\, E_k}$$

$$= A\,\mathrm{e}^{-n_t a\, E_H}\left(1 + \mathcal{O}\left(\mathrm{e}^{-n_t a\, \Delta E}\right)\right), \qquad (6.1)$$

where $A$ is a constant, $E_H$ is the energy of the lowest state $|H\rangle$ with $\langle 0|\hat{O}|H\rangle \neq 0$, and $\Delta E$ is the energy difference to the first excited state. From the leading exponential decay we thus can extract the energy $E_H$ of the hadron.

### 6.1.1 Meson interpolators

The discrete symmetries which we have discussed in Sect. 5.4 play an important role in the classification of hadrons and are needed for the construction of hadron interpolators. To illustrate their use we discuss as an example the construction of pion interpolators in more detail. Other meson interpolators can then be constructed along the same lines with the help of Table 6.1, which lists the quantum numbers of fermion bilinears.

---

[1] According to common practice, we will often use "operator" and "interpolator" synonymously in the following.

To be specific, we consider QCD with only two or three flavors (up, down, strange). Instead of using a flavor index $f$ attached to spinors denoted by $\psi, \overline{\psi}$, we often use a different symbol for each flavor. Thus $u, \overline{u}, d, \overline{d}$ and $s, \overline{s}$ denote the spinors for the three light flavors.

The pions, which we now consider in our example, are made from $u$ and $d$ quarks only. An up-quark $u$ has isospin $I = 1/2$, $I_z = +1/2$ and charge $Q = 2/3\,e$, where in our convention the charge of the electron is $-e$. For the down-quark $d$ we have $I = 1/2$, $I_z = -1/2$ and $Q = -1/3\,e$. The pseudoscalar combinations can be classified with respect to their isospin and are grouped into an iso-triplet ($I = 1$) containing the particles $\pi^+$, $\pi_0$, $\pi^-$ with isospin components $I_z = 1, 0, -1$ and an iso-singlet ($I = 0$) containing the $\eta$-meson.[2]

The charged pions $\pi^+$ and $\pi^-$ have a mass of 140 MeV. They have zero spin ($J = 0$), negative parity ($P = -1$), isospin $I = 1$, $I_z = \pm 1$, and electric charge $Q = \pm e$. Thus, for obtaining the correct charge and isospin, $\pi^+$ must be a $\overline{d}$–$u$ combination and $\pi^-$ is of the type $\overline{u}$–$d$. We also expect them to be color singlets. More explicitly we can use the pseudoscalar interpolators (summation convention for all indices)

$$O_{\pi^+}(n) = \overline{d}(n)\,\gamma_5\,u(n) = \overline{d}(n)_{\underset{c}{\alpha}}\,(\gamma_5)_{\alpha\beta}\,u(n)_{\underset{c}{\beta}}\,,$$
$$O_{\pi^-}(n) = \overline{u}(n)\,\gamma_5\,d(n) = \overline{u}(n)_{\underset{c}{\alpha}}\,(\gamma_5)_{\alpha\beta}\,d(n)_{\underset{c}{\beta}}\,. \qquad (6.2)$$

For their transformation under parity (5.72) we find

$$
\begin{aligned}
O_{\pi^+}(\boldsymbol{n}, n_4) &= \overline{d}(\boldsymbol{n}, n_4)\,\gamma_5\,u(\boldsymbol{n}, n_4) \\
&\xrightarrow{P} \overline{d}(-\boldsymbol{n}, n_4)\,\gamma_4\gamma_5\gamma_4\,u(-\boldsymbol{n}, n_4) = -\overline{d}(-\boldsymbol{n}, n_4)\,\gamma_5\,u(-\boldsymbol{n}, n_4) \\
&= -O_{\pi^+}(-\boldsymbol{n}, n_4)\,, \qquad (6.3)
\end{aligned}
$$

showing that our interpolator $O_{\pi^+}$ indeed has negative parity (the same holds for $O_{\pi^-}$). The fact that the parity transformation changes the spatial vector $\boldsymbol{n}$ into $-\boldsymbol{n}$ is irrelevant for interpolators where we project to zero momentum, i.e., where we sum over all $\boldsymbol{n}$ (see Sect. 6.1.4). Applying charge conjugation (5.66) gives

$$
\begin{aligned}
O_{\pi^+}(n) = \overline{d}(n)\,\gamma_5\,u(n) &\xrightarrow{C} -d(n)^T\,C\gamma_5 C^{-1}\,\overline{u}(n)^T = -d(n)^T\,\gamma_5^T\,\overline{u}(n)^T \\
&= \overline{u}(n)\,\gamma_5\,d(n) = O_{\pi^-}(n)\,, \qquad (6.4)
\end{aligned}
$$

where we used $C\gamma_5 C^{-1} = \gamma_5^T$, which follows from (5.67). The interchange of the Grassmann variables cancels the overall minus sign. Equation (6.4) establishes that charge conjugation transforms $O_{\pi^+}$ into $O_{\pi^-}$ and vice versa. The interpolator for the $I_z = 0$ component of the iso-triplet, the $\pi^0$, is

$$O_{\pi^0}(n) = \frac{1}{\sqrt{2}}\left(\overline{u}(n)\,\gamma_5\,u(n) - \overline{d}(n)\,\gamma_5\,d(n)\right)\,. \qquad (6.5)$$

---

[2]We remark that the physical $\eta$ also has an admixture of $\overline{s}s$. See, e.g., [1].

**Table 6.1.** Quantum numbers of the most commonly used meson interpolators according to the general form (6.9). We remark that the classification with respect to $C$ is for the flavor neutral interpolators only

| State | $J^{PC}$ | $\Gamma$ | Particles |
|---|---|---|---|
| Scalar | $0^{++}$ | $\mathbb{1}$ , $\gamma_4$ | $f_0, a_0, K_0^*, \ldots$ |
| Pseudoscalar | $0^{-+}$ | $\gamma_5$ , $\gamma_4\gamma_5$ | $\pi^\pm, \pi^0, \eta, K^\pm, K^0, \ldots$ |
| Vector | $1^{--}$ | $\gamma_i$ , $\gamma_4\gamma_i$ | $\rho^\pm, \rho^0, \omega, K^*, \phi, \ldots$ |
| Axial vector | $1^{++}$ | $\gamma_i\gamma_5$ | $a_1, f_1, \ldots$ |
| Tensor | $1^{+-}$ | $\gamma_i\gamma_j$ | $h_1, b_1, \ldots$ |

The iso-singlet state ($I = 0$), the $\eta$-meson, is described by the interpolator

$$O_\eta(n) = \frac{1}{\sqrt{2}} \left( \overline{u}(n)\, \gamma_5\, u(n) + \overline{d}(n)\, \gamma_5\, d(n) \right) . \tag{6.6}$$

The properties of $O_{\pi^0}$ and $O_\eta$ under the parity transformation are the same as for the interpolators $O_{\pi^\pm}$, i.e., they have $P = -1$. Concerning charge conjugation, $O_{\pi^0}$ and $O_\eta$ are eigenstates with $C = +1$.

Other mesons differ from the examples we have discussed so far by their flavor content as well as by spin and parity. We can, e.g., obtain the interpolator for the strange $K^+$ meson from the interpolator for $\pi^+$, by replacing the $d$ quark with an $s$ quark,

$$O_{K^+}(n) = \overline{s}(n)\, \gamma_5\, u(n) . \tag{6.7}$$

Different spin and parity correspond to different gamma matrices in the fermionic bilinears. For example we obtain from the interpolator for $\pi^+$ the interpolator for the $\rho^+$ vector meson ($I = 1, I_z = +1, Q = +e, J = 1, P = -1$) by replacing $\gamma_5$ with $\gamma_i$, $i = 1, 2, 3$:

$$O_{\rho^+}(n)_i = \overline{d}(n)\, \gamma_i\, u(n) , \quad i = 1, 2, 3 . \tag{6.8}$$

A general local meson interpolator has the form

$$O_M(n) = \overline{\psi}^{(f_1)}(n)\, \Gamma\, \psi^{(f_2)}(n) , \tag{6.9}$$

where $\Gamma$ is a monomial of gamma matrices, and we have for this equation switched back to the notation with upper flavor indices $f_i$. In the case of degenerate flavors ($f_1 = f_2$) combinations of (6.9) are formed to obtain the desired flavor symmetries, as was done in our interpolators for $\pi^0$ and $\eta$ in (6.5) and (6.6). In Table 6.1 we list the matrices $\Gamma$ for the most commonly used interpolators together with the corresponding quantum numbers. We remark that in addition to the local interpolators gauge-invariant combinations of quarks and antiquarks at different lattice sites may also be used. (See Sects. 6.2.2 and 6.2.3.)

### 6.1.2 Meson correlators

So far we have constructed the interpolator $O_M$ that corresponds to the operator $\hat{O}_M$ acting in the physical Hilbert space. For the Euclidean correlator we also need to find the interpolator corresponding to $\hat{O}_M^\dagger$ which generates the meson state from the vacuum. In order to identify this operator we formally conjugate the interpolator (6.9) (we drop the space–time argument $n$ here),

$$\left(\overline{\psi}^{(f_1)}\,\Gamma\,\psi^{(f_2)}\right)^\dagger = -\psi^{(f_2)\,\dagger}\,\Gamma^\dagger\overline{\psi}^{(f_1)\,\dagger} \equiv -\overline{\psi}^{(f_2)}\gamma_4\Gamma^\dagger\gamma_4\psi^{(f_1)} = \pm\overline{\psi}^{(f_2)}\,\Gamma\,\psi^{(f_1)}\,. \tag{6.10}$$

The minus sign in the first step comes from the interchange of the Grassmann variables. In the second step we have used the relation $\overline{\psi} = \psi^\dagger\gamma_4$ which holds for the operators in Hilbert space and performed the equivalent replacement for the Grassmann spinors in our interpolator. The last step in (6.10) reflects the simple algebraic property $\gamma_4\Gamma^\dagger\gamma_4 = \pm\Gamma$. Thus, up to a possible overall sign, the interpolator that corresponds to $\hat{O}_M^\dagger$ is obtained by interchanging $\psi$'s and $\overline{\psi}$'s and ordering the $\overline{\psi}$ to the left. The sign is irrelevant for the exponential decay in (6.1). Thus, for mesons we use interpolators

$$O_M(n) = \overline{\psi}^{(f_1)}(n)\,\Gamma\,\psi^{(f_2)}(n)\,, \quad \overline{O}_M(m) = \overline{\psi}^{(f_2)}(m)\,\Gamma\,\psi^{(f_1)}(m)\,, \tag{6.11}$$

combined in correlators $\langle O_M(n)\,\overline{O}_M(m)\rangle$. At the moment the space–time arguments $m, n$ are arbitrary and only later we will set the time arguments to $m_4 = 0$ and $n_4 = n_t$ to match the form of the hadron correlator (6.1). In Sect. 6.1.4 we will discuss what to do with the spatial components $\boldsymbol{m}, \boldsymbol{n}$.

For evaluating the correlators we have to compute the Grassmann integrals that appear when calculating the fermionic part $\langle\ldots\rangle_F$ of the expectation value (compare (5.2)). In this step there is an important difference between the correlators for iso-triplet operators such as (6.2) and (6.5) and correlators for an iso-singlet operator such as (6.6). For an iso-triplet operator of the form $O_T = \overline{d}\,\Gamma\,u$ we find (summation convention is used for all indices)

$$\begin{aligned}
\left\langle O_T(n)\,\overline{O}_T(m)\right\rangle_F &= \left\langle \overline{d}(n)\Gamma u(n)\,\overline{u}(m)\Gamma d(m)\right\rangle_F \\
&= \Gamma_{\alpha_1\beta_1}\Gamma_{\alpha_2\beta_2}\left\langle \overline{d}(n)_{\underset{c_1}{\alpha_1}}\,u(n)_{\underset{c_1}{\beta_1}}\,\overline{u}(m)_{\underset{c_2}{\alpha_2}}\,d(m)_{\underset{c_2}{\beta_2}}\right\rangle_F \\
&= -\Gamma_{\alpha_1\beta_1}\Gamma_{\alpha_2\beta_2}\left\langle u(n)_{\underset{c_1}{\beta_1}}\,\overline{u}(m)_{\underset{c_2}{\alpha_2}}\right\rangle_u\left\langle d(m)_{\underset{c_2}{\beta_2}}\,\overline{d}(n)_{\underset{c_1}{\alpha_1}}\right\rangle_d \\
&= -\Gamma_{\alpha_1\beta_1}\Gamma_{\alpha_2\beta_2}\,D_u^{-1}(n|m)_{\underset{c_1 c_2}{\beta_1\alpha_2}}\,D_d^{-1}(m|n)_{\underset{c_2 c_1}{\beta_2\alpha_1}} \\
&= -\,\mathrm{tr}\left[\Gamma D_u^{-1}(n|m)\,\Gamma D_d^{-1}(m|n)\right]\,. \tag{6.12}
\end{aligned}$$

In the third line of this equation we have reordered the Grassmann variables and then used the fact that the fermionic expectation value factorizes with respect to the flavors: $\langle\ldots\rangle_F = \langle\ldots\rangle_u\langle\ldots\rangle_d$. Subsequently we applied Wick's

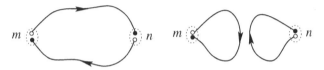

**Fig. 6.1.** Connected (*left-hand side plot*) and disconnected (*right-hand side plot*) pieces of a meson correlator

theorem (5.36) for each of the two flavors (compare also (5.54)). This step is often referred to as *fermion contraction*.

The Dirac operators $D_u$, $D_d$ for $u$ and $d$ quark differ only by the value of the mass parameter (compare (5.51)). Often the small difference between the $u$ and the $d$ quark masses is ignored and one uses $D_u = D_d$, i.e., exact isospin symmetry. It is, however, important to keep in mind that also in this case, only Grassmann variables with equal flavor can be contracted with each other.

The result in the last line of (6.12) has a simple interpretation: The propagator $D_u^{-1}(n|m)$ propagates a $u$ quark from space–time point $m$ to the point $n$, while the propagator $D_d^{-1}(m|n)$ transports a $d$ quark in the opposite direction. Such a contribution is referred to as *connected piece* and is depicted in the left-hand side plot of Fig. 6.1. We remark that each of the individual lines in this figure symbolizes a collection of fermion lines (cf. Fig. 5.1).

In the correlator of an iso-singlet operator $O_S = (\overline{u}\Gamma u + \overline{d}\Gamma d)/\sqrt{2}$, such as (6.6), also another type of contribution appears. The fermion contractions for this correlator are obtained by following the same steps as in (6.12),

$$
\begin{aligned}
\langle O_S(n)\overline{O}_S(m)\rangle_F = &-\frac{1}{2}\,\text{tr}\left[\Gamma D_u^{-1}(n|m)\Gamma D_u^{-1}(m|n)\right]\\
&+\frac{1}{2}\,\text{tr}\left[\Gamma D_u^{-1}(n|n)\right]\,\text{tr}\left[\Gamma D_u^{-1}(m|m)\right] \qquad (6.13)\\
&+\frac{1}{2}\,\text{tr}\left[\Gamma D_u^{-1}(n|n)\right]\,\text{tr}\left[\Gamma D_d^{-1}(m|m)\right]\,+\,u\leftrightarrow d\,.
\end{aligned}
$$

The first type of contribution are the connected pieces we have already discussed. However, one also gets propagators $D_u^{-1}(n|n)$, $D_u^{-1}(m|m)$ which transport a $u$ quark from a space time-point back to the same point. Such terms are called *disconnected pieces* and are depicted in the right-hand side plot of Fig. 6.1. Numerically these contributions need more computational effort and higher statistics than the connected parts and many studies avoid considering such mesons or drop the disconnected pieces.

We remark that the interpolator $O_{T,I_z=0} = (\overline{u}\Gamma u - \overline{d}\Gamma d)/\sqrt{2}$ for the $I_z = 0$ component of the iso-triplet differs from the singlet interpolator only by a relative minus sign between the $u$ and the $d$ terms (compare (6.5) and (6.6)). The corresponding correlator is like in (6.13), but with a minus sign in the third term. In the case of exact isospin symmetry, $D_u = D_d$, the disconnected pieces cancel. The resulting correlator is the same as for the other members of the triplet (compare (6.12)), implying that the masses of all triplet states are degenerate. In nature this degeneracy is quite accurate, since $m_{\pi^\pm} = 140$ MeV and $m_{\pi^0} = 135$ MeV. The small discrepancy is due to slightly different masses for $u$ and $d$ quarks and electroweak corrections.

### 6.1.3 Interpolators and correlators for baryons

Baryons are objects made out of three valence quarks. Baryon interpolators $O$ with well-defined quantum numbers have to be constructed such that the corresponding Hilbert space operator $\hat{O}$ in (6.1) projects onto the state we are interested in. As for the case of mesons we first discuss an example in detail, the construction of nucleon interpolators, and then address other baryons. A review on baryon interpolators, going beyond our introductory presentation, can be found in [2].

The proton $p$ and the neutron $n$ are the $I_z = +1/2$ and $I_z = -1/2$ components of an iso-doublet ($I = 1/2$). Their almost degenerate masses of $m_p = 938$ MeV and $m_n = 940$ MeV demonstrate again that isospin is a good symmetry, as we already stressed for the pion system. The proton has electric charge $Q = e$, while the neutron has vanishing charge. Consequently the proton is a $uud$ state and the neutron is of the type $ddu$. Due to the facts that the electric charge is irrelevant in QCD and $p$ and $n$ are strongly related by isospin symmetry, usually one does not distinguish between the two and refers to them simply as *nucleons*. We discuss only the $uud$-type interpolator ($I_z = 1/2$). The $I_z = -1/2$ component is obtained by interchanging $u$ and $d$.

The simplest interpolator for the nucleon $N$ is given by

$$O_N(n) = \epsilon_{abc}\, u(n)_a\, \left(u(n)_b^T\, C\gamma_5\, d(n)_c\right) . \tag{6.14}$$

In this equation we show explicitly only the color indices $a, b, c$, while for the Dirac indices we use vector/matrix notation. The transposition $T$ acts on the Dirac indices and turns the column 4-spinor $u(n)_b$ into a row spinor $u(n)_b^T$. Summing the color indices with the epsilon tensor makes the interpolator a color singlet and gauge-invariant, as follows from the action of a gauge transformation on the fermions (2.30) and (3.31).

The term in the parentheses combines a $u$ and a $d$ quark into a so-called *diquark*, using the charge conjugation matrix $C$ and $\gamma_5$. The diquark has isospin $I = 0$ and spin $J = 0$. The notion diquark here serves to discuss the quantum numbers and no dynamical meaning is implied. Thus the full interpolator $O_N$ has $I = 1/2, I_z = +1/2$ and $J = 1/2$ as needed for the nucleon.

The last quantum number we have to discuss is parity, which is $P = +1$ for proton and neutron. Under the parity transformation $\mathcal{P}$ (see (5.72)) our nucleon interpolator $O_N(n)$ transforms as

$$O_N^{\mathcal{P}}(\boldsymbol{n}, n_4) = \epsilon_{abc}\, \gamma_4\, u(-\boldsymbol{n}, n_4)_a\, u(-\boldsymbol{n}, n_4)_b^T\, \gamma_4^T C\gamma_5\gamma_4\, d(-\boldsymbol{n}, n_4)_c$$
$$= \epsilon_{abc}\, \gamma_4 u(-\boldsymbol{n}, n_4)_a\, u(-\boldsymbol{n}, n_4)_b^T\, C\gamma_5\, d(-\boldsymbol{n}, n_4)_c = \gamma_4\, O_N(-\boldsymbol{n}, n_4) , \tag{6.15}$$

where in the second step we used $\gamma_\mu^T C = -C\gamma_\mu$ (compare (A.23)). As for mesons, the change of the spatial vector $\boldsymbol{n}$ into $-\boldsymbol{n}$ is irrelevant when we project to zero momentum, i.e., when we sum over all $\boldsymbol{n}$ (see Sect. 6.1.4). The nontrivial transformation of the spinor indices with $\gamma_4$ can be taken into account by considering the combinations

$$O_{N_\pm}(n) = \frac{1}{2}\left(O_N(n) \pm O_N^P(n)\right) = \epsilon_{abc}\, P_\pm\, u(n)_a \left(u(n)_b^T\, C\gamma_5\, d(n)_c\right) , \quad (6.16)$$

where the parity projectors $P_\pm$ are defined as

$$P_\pm = \tfrac{1}{2}\left(\mathbb{1} \pm \gamma_4\right) . \tag{6.17}$$

After projection to zero momentum, the interpolators in (6.16) have definite parity $P = \pm 1$. The interpolator $O_{N_+}$ describes the positive parity nucleon, while $O_{N_-}$ couples to its negative parity partner $N(1535)$ with a mass of 1535 MeV.

Similar to the steps performed for the meson correlator in (6.10), one finds that the interpolator for the corresponding creation operator is given by (up to an overall sign)

$$\overline{O}_{N_\pm}(n) = \epsilon_{abc}\left(\overline{u}(n)_a\, C\gamma_5\, \overline{d}(n)_b^T\right) \overline{u}(n)_c\, P_\pm . \tag{6.18}$$

Interpolators for other spin-$1/2$ baryons are obtained by changing the flavor content. Furthermore, different diquark structures can be considered and we obtain more general spin-$1/2$ interpolators:

$$O_{N_\pm} = \epsilon_{abc}\, P_\pm\, \Gamma^A u_a \left(u_b^T\, \Gamma^B\, d_c\right) , \quad \overline{O}_{N_\pm} = \epsilon_{abc}\left(\overline{u}_a\, \Gamma^B\, \overline{d}_b^T\right) \overline{u}_c\, \Gamma^A\, P_\pm ,$$

$$O_{\Sigma_\pm} = \epsilon_{abc}\, P_\pm\, \Gamma^A u_a \left(u_b^T\, \Gamma^B\, s_c\right) , \quad \overline{O}_{\Sigma_\pm} = \epsilon_{abc}\left(\overline{u}_a\, \Gamma^B\, \overline{s}_b^T\right) \overline{u}_c\, \Gamma^A\, P_\pm ,$$

$$O_{\Xi_\pm} = \epsilon_{abc}\, P_\pm\, \Gamma^A s_a \left(s_b^T\, \Gamma^B\, u_c\right) , \quad \overline{O}_{\Xi_\pm} = \epsilon_{abc}\left(\overline{s}_a\, \Gamma^B\, \overline{u}_b^T\right) \overline{s}_c\, \Gamma^A\, P_\pm ,$$

$$O_{\Lambda_\pm} = \epsilon_{abc}\, P_\pm \Gamma^A \left(2s_a\left(u_b^T\, \Gamma^B\, d_c\right) + d_a\left(u_b^T\, \Gamma^B\, s_c\right) - u_a\left(d_b^T\, \Gamma^B\, s_c\right)\right) ,$$

$$\overline{O}_{\Lambda_\pm} = \epsilon_{abc}\left(2\left(\overline{u}_a^T\, \Gamma^B\, \overline{d}_b\right)\overline{s}_c + \left(\overline{u}_a^T\, \Gamma^B\, \overline{s}_b\right)\overline{d}_c - \left(\overline{d}_a^T\, \Gamma^B\, \overline{s}_b\right)\overline{u}_c\right) \Gamma^A\, P_\pm, (6.19)$$

where we list the interpolators for nucleon, for the strange $\Sigma$ ($I = 1, S = -1$), $\Xi$ ($I = 1/2, S = -2$), and $\Lambda$ ($I = 0, S = -1$) baryons of both parities. Three possible choices for the matrices $\Gamma^A$ and $\Gamma^B$, all giving rise to $J^P = 1/2^+$, are $(\Gamma^A, \Gamma^B) = (\mathbb{1}, C\gamma_5)$, $(\gamma_5, C)$, or $(\mathbb{1}, i\gamma_4 C\gamma_5)$.

Baryons with spin $J = 3/2$ can be obtained by using $(\Gamma^A, \Gamma^B) = (\mathbb{1}, C\gamma_j)$. The spatial gamma matrix $\gamma_j$ gives rise to a diquark with spin $J = 1$ and together with the quark outside the diquark the resulting interpolator has spin $J = 3/2$ contributions, but also an admixture of spin $J = 1/2$. For the details of the necessary projection to definite spin we refer to [2].

Note that all our baryon interpolators $O_{B_\pm}$ and $\overline{O}_{B_\pm}$ still have an open Dirac index – they describe fermions after all. Usually this index is summed over and one considers correlators of the type ($\alpha$ is summed)

$$\left\langle O_{B_\pm}(n)_\alpha \, \overline{O}_{B_\pm}(m)_\alpha \right\rangle . \tag{6.20}$$

We conclude with displaying as an example the fermion contractions for the nucleon interpolator (6.14) (we use $P_\pm^2 = P_\pm$ and vector/matrix notation),

$$
\begin{aligned}
\left\langle O_{N_\pm}(n)_\alpha \, \overline{O}_{N_\pm}(m)_\alpha \right\rangle_F &= - \left\langle \overline{O}_{N_\pm}(m)_\alpha \, O_{N_\pm}(n)_\alpha \right\rangle_F \\
&= - \left\langle \epsilon_{abc} \, \epsilon_{a'b'c'} \left( \overline{u}(m)_a \, C\gamma_5 \, \overline{d}(m)_b^T \right) \overline{u}(m)_c \, P_\pm \, u(n)_{c'} \left( u(n)_{a'}^T \, C\gamma_5 \, d(n)_{b'} \right) \right\rangle_F \\
&= \epsilon_{abc} \, \epsilon_{a'b'c'} \, (C\gamma_5)_{\alpha'\beta'} \, (C\gamma_5)_{\alpha\beta} \, (P_\pm)_{\gamma\gamma'} \, \underset{b'b}{D_d^{-1}(n|m)_{\beta'\beta}} \times
\end{aligned}
\tag{6.21}
$$

$$
\left( \underset{a'a}{D_u^{-1}(n|m)_{\alpha'\alpha}} \, \underset{c'c}{D_u^{-1}(n|m)_{\gamma'\gamma}} - \underset{a'c}{D_u^{-1}(n|m)_{\alpha'\gamma}} \, \underset{c'a}{D_u^{-1}(n|m)_{\gamma'\alpha}} \right) ,
$$

demonstrating that for baryons only connected pieces occur.

### 6.1.4 Momentum projection

There is a final step to be implemented for our hadron interpolators: We want our hadron states to be states of definite spatial momentum $\boldsymbol{p}$. Thus we define

$$\widetilde{O}(\boldsymbol{p}, n_t) = \frac{1}{\sqrt{|\Lambda_3|}} \sum_{\boldsymbol{n} \in \Lambda_3} O(\boldsymbol{n}, n_t) \, e^{-i \, a \, \boldsymbol{n} \, \boldsymbol{p}} . \tag{6.22}$$

Note that this Fourier transformation runs only over the spatial components $\boldsymbol{n}$ in the spatial lattice $\Lambda_3 = \{\boldsymbol{n} = (n_1, n_2, n_3) \mid n_i = 0, 1, \ldots, N{-}1\}$. Accordingly, the momenta $\boldsymbol{p}$ are spatial momenta with components $p_i = 2\pi k_i/(aN), k_i = -N/2{+}1, \ldots, N/2$. Thus, the interpolator $\widetilde{O}(\boldsymbol{p}, n_t)$ is projected to definite spatial momentum and is located on a single time slice $n_t$.

It is sufficient to project only one of the two interpolators of a correlation function to definite momentum, typically the operator $O(\boldsymbol{n}, n_t)$ at the sink. The source operator $\overline{O}(\boldsymbol{0}, 0)$ can remain in real space and usually is placed at the origin. That this is sufficient can be seen by writing the real space interpolator $\overline{O}$ as a sum of its Fourier components (compare (A.37)) and using the fact that states with different momenta are orthogonal to each other such that only the zero momentum term survives. Note, however, that this is strictly correct only for the exact expectation values, whereas it is only approximate for a sum over a finite number of configurations.

Thus, according to (1.21), our final formula for Euclidean hadron correlators with definite momentum $\boldsymbol{p}$ reads

$$
\begin{aligned}
\left\langle \widetilde{O}(\boldsymbol{p}, n_t) \, \overline{O}(\boldsymbol{0}, 0) \right\rangle &= \frac{1}{\sqrt{|\Lambda_3|}} \sum_{\boldsymbol{n} \in \Lambda_3} e^{-i a \boldsymbol{n} \boldsymbol{p}} \left\langle O(\boldsymbol{n}, n_t) \, \overline{O}(\boldsymbol{0}, 0) \right\rangle \\
&= A \, e^{-a n_t E(\boldsymbol{p})} \left( 1 + \mathcal{O}\left( e^{-a n_t \Delta E} \right) \right) ,
\end{aligned}
\tag{6.23}
$$

where the energy $E(\boldsymbol{p})$ is related to the hadron mass $m_H$ through the relativistic dispersion relation (in our units the speed of light is $c = 1$)

$$E(\boldsymbol{p}) = \sqrt{m_H^2 + \boldsymbol{p}^2} \, (1 + \mathcal{O}(ap)) \; , \qquad (6.24)$$

and for zero momentum we find $E(\boldsymbol{0}) = m_H$.

Determining the hadron propagators for non-zero momenta allows one to study the spectral relation between energy and momentum as realized on the lattice. Since the asymptotic states describe free particles in the lattice world, the adequate spectral relation is that of free lattice particles. It may be obtained, e.g., from the free boson (or fermion) lattice propagator in momentum space. For the simplest, nearest neighbor discretization one identifies the pole at momentum $(iE(\boldsymbol{p}), \boldsymbol{p})$ and finds the relation $(m_H = E(\boldsymbol{0}))$

$$\cosh\left(a \, E(\boldsymbol{p})\right) = \cosh(a \, m_H) + \sum_{k=1}^{3} (1 - \cos(a \, p_k)) \; , \qquad (6.25)$$

which approaches the continuum relation for vanishing lattice spacing $a$, cf. (6.24). Careful analysis of hadron propagators along these lines allows one to check for the recovery of full Euclidean invariance.

### 6.1.5 Final formula for hadron correlators

Central for obtaining results to be compared with experiment are the correlation functions of products of fields. For any product of operators, denoted by $A$, one has to evaluate

$$\langle A \rangle = \langle\langle A \rangle_F \rangle_G = \frac{1}{Z} \int \mathcal{D}[U] \, e^{-S_G[U]} \, \mathcal{D}[\psi, \overline{\psi}] \, e^{-S_F[\psi, \overline{\psi}, U]} \, A[\psi, \overline{\psi}, U] \; . \quad (6.26)$$

The integral over the Grassmann variables can be computed in closed form (see Chap. 5) and the integral (5.4) over the gauge field remains.

For definiteness let us discuss the example of a propagator for a meson built out of two quark flavors as in (6.12). After the Grassmann integration we end up with a ratio of purely bosonic integrals over the gauge fields,

$$\langle O_T(n) \overline{O}_T(m) \rangle = -\frac{1}{Z} \int \mathcal{D}[U] \, e^{-S_G[U]} \, \det[D_u] \det[D_d]$$
$$\times \operatorname{tr} \left[ \Gamma D_u^{-1}(n|m) \, \Gamma D_d^{-1}(m|n) \right] \; ,$$

$$Z = \int \mathcal{D}[U] \, e^{-S_G[U]} \, \det[D_u] \, \det[D_d] \; . \qquad (6.27)$$

We stress again that the fermion determinants depend on $U$ and one has to perform a Monte Carlo simulation using $Z^{-1} \exp(-S_G[U]) \det[D_u] \det[D_d]$ as the distribution weight for the gauge fields. This seems to be a straightforward task at first sight. However, computing the determinant of the Dirac operator matrix $D$ is highly nontrivial. Remember that this matrix has $12 |\Lambda|$ rows and columns, where $|\Lambda|$ is the number of lattice sites. This dimension therefore is typically of the order of millions and a brute force calculation of the determinant is prohibitively expensive.

## 6.1.6 The quenched approximation

Following the first applications of the Monte Carlo method to lattice QCD, most of the results in the 1980s and 1990s were obtained in the *quenched approximation*.[3] In particular the successful determination of the ground state spectrum of hadrons with light quarks is a milestone of this approach.

Before discussing the quenched approximation, let us briefly remind ourselves of the physical interpretation of the determinant factor. From the hopping expansion discussed in Chap. 5, we found that it may be written as an effective action, which is a sum over closed loops of gauge field link variables. In essence it describes the fermionic vacuum where virtual pairs of quarks and antiquarks are created and annihilated. These quarks are usually called *sea quarks*, since they are related to Dirac's picture of a fermion sea, where the holes denote anti-fermions. Of course sea quarks are identical to generic quarks and we only call them such for clarification of the subsequent approximation.

Putting the fermion determinants to unity by hand corresponds to neglecting the effect of the vacuum quark loops (Sect. 5.3.2). Actually, since in the expansion of the determinant the leading term of the effective action is a sum over plaquettes, part of the effective action is intrinsically included in the gauge action by redefining the gauge coupling. This is no problem, since the gauge coupling itself is a bare parameter and only physical quantities are of relevance.

Omitting the determinants defines the quenched approximation. Instead of using the full expression for, e.g., the meson correlator (6.27), one computes

$$\langle O_T(n)\,\overline{O}_T(m)\rangle_{\text{quenched}} = -\frac{1}{Z}\int \mathcal{D}[U]\,e^{-S_G[U]}\,\text{tr}\left[\Gamma D_u^{-1}(n|m)\,\Gamma D_d^{-1}(m|n)\right]\,,$$

$$Z = \int \mathcal{D}[U]\,e^{-S_G[U]}\,. \tag{6.28}$$

In this approximation one constructs the Markov chain of gauge configurations as for pure gauge theory (compare Chap. 3) and evaluates the quark propagators on those configurations. Combinations of the quark propagators build up the hadron correlators as discussed in the last section. In order to distinguish these propagating quarks from the sea quarks, and since they are responsible for the quantum numbers of the hadron, they are called *valence quarks*. The quenched approximation is therefore also called *valence approximation*.

The quenched approximation may be considered as the limit, in which the sea quarks become infinitely heavy and therefore cannot be generated from the vacuum as particle–antiparticle pairs. This can be seen from the hopping expansion for the fermion determinant (5.65). Since the expansion parameter $\kappa$ defined in (5.55) vanishes for $m \to \infty$, the determinant becomes 1 in this limit.

---

[3]The terms come from quenching – rapid cooling – of steel, where the carbon atoms are frozen to their random positions.

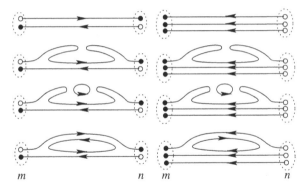

**Fig. 6.2.** Sample of quark lines contributing in hadron propagation (mesons l.h.s., baryons r.h.s.). In the quenched approximation only contributions in the upper two rows, i.e., without sea quark loops (= closed loops) contribute. The diagrams in the second row appear in the quenched approximation and resemble the dynamical diagrams in the bottom rows. They are called hairpin diagrams and give rise to logarithmic singularities in the quenched approximation

A simulation including the determinant and therefore allowing for the full dynamical vacuum structure of fermions is called a simulation with *dynamical quarks*, in contrast to the quenched simulation. Figure 6.2 gives examples for contributing quark lines in the quenched and the fully dynamical simulation.

In full QCD there is no difference between sea quarks and valence quarks. However, in our computer laboratory we may give them different masses. Such intermediate stages, where the masses of the sea quarks and the valence quarks differ, are called *partial quenching*, and are sometimes useful for a better control of the various systematic uncertainties.

We remark that neither the quenched nor the partially quenched approximation are proper quantum field theories with a valid Hilbert space construction and correct positivity properties for the fermions. This leads to some problems in analyzing in particular the results toward small quark masses.

Forbidding dynamical fermion loops also prevents one from studying resonances. However, these provide a conceptual problem anyhow. They are not asymptotic states of the field theory and one has to introduce intricate tools for their identification and the determination of scattering amplitudes in general (cf. Chap. 11). We expect, however, that resonances should in the quenched case to leading order be observed as bound states at nearby energy values.

Since dynamical simulations will be treated in Chap. 8, in this chapter we will show exclusively results for the hadron spectrum in the quenched situation. Obviously these are of much higher statistical precision than those obtained from simulations with dynamical quarks, albeit with the intrinsic uncertainty of the effect of quenching. Special features due to quenching (like occurrence of extra singularities and ghosts) will be discussed in Sect. 6.4.3.

We stress, however, that most of the formulas and techniques presented in this chapter apply to both quenched and dynamical simulations.

## 6.2 Strategy of the calculation

In this section we discuss techniques for computing and storing the quark propagators needed for the hadron correlators presented in the last section. We describe the construction of realistic quark sources which improve the overlap of the lattice hadron interpolators with the physical states. The problem of exceptional configurations and its amelioration through smoothing of the gauge field are addressed.

### 6.2.1 The need for quark sources

A complete quark propagator is a matrix as large as the fermion matrix itself. Depending on the lattice action, the Dirac operator may be sparse, with many vanishing entries, but the propagator $D^{-1}$ is not and consists of $\mathcal{O}(10^{12})$ complex numbers. Each entry $D^{-1}(n|m)^{\beta\alpha}_{ba}$ connects a source point $(m, \alpha, a)$ with a sink point $(n, \beta, b)$. Here the notion "point" relates to site- as well as Dirac and color indices. Since the matrix provides the information for propagation in one particular gauge field configuration, its entries will be highly correlated and this correlation reduces the information content. So for two reasons one does not try to store the complete propagator matrix: It is wasteful to calculate all entries and it would need too much computer memory.

However, if we consider instead just the propagator from a fixed site $m_0$, a fixed Dirac index $\alpha_0$, and a fixed color index $a_0$, to any site of the lattice, this is just one column of the inverse Dirac operator,

$$D^{-1}(n|m_0)^{\beta\alpha_0}_{ba_0} = \sum_{m,\alpha,a} D^{-1}(n|m)^{\beta\alpha}_{ba}\, S_0^{(m_0,\alpha_0,a_0)}(m)_{\substack{\alpha \\ a}} , \qquad (6.29)$$

where we have introduced the so-called *point sources*

$$S_0^{(m_0,\alpha_0,a_0)}(m)_{\substack{\alpha \\ a}} = \delta(m - m_0)\, \delta_{\alpha\alpha_0}\, \delta_{aa_0} . \qquad (6.30)$$

The expression (6.29) must be evaluated for 12 sources, one for each combination $\alpha_0, a_0$ of Dirac and color indices. The point sources are placed at the position $m_0$ of the source interpolator $\overline{O}_H(m_0)$ and the column $D^{-1}(n|m_0)$ is the propagator to the position $n$ of the sink operator $O_H(n)$.

Since mesons contain a backward running antiquark (compare Fig. 6.1), one also needs the propagator in the other direction, i.e., from $n$ to $m_0$. Naively one would expect that this requires placing sources also at $n$. However, we can utilize $\gamma_5$-hermiticity, as discussed in Sect. 5.4.3, in order to obtain the backward running propagator for free.

Using this symmetry property we may write (repeated indices are summed)

$$\gamma_5 \, D^{-1} \, \gamma_5 = D^{-1\dagger} \quad \Longleftrightarrow \quad (\gamma_5)_{\alpha\alpha'} \, D^{-1}(m_0|n)_{\substack{\alpha'\beta' \\ c\,d}} \, (\gamma_5)_{\beta'\beta} = D^{-1}(n|m_0)^{*}_{\substack{\beta\alpha \\ dc}} \,,$$
$$(6.31)$$

and so obtain the backward running propagator. This can, e.g., be used to write the iso-triplet pion propagator (6.12) as

$$\langle O_T(n) \, \overline{O}_T(m_0) \rangle_F = - \mathrm{tr}\left[ D^{-1}(n|m_0)\gamma_5 D^{-1}(m_0|n)\gamma_5 \right] = - \sum_{\alpha,\beta,c,d} |D^{-1}(n|m_0)_{\substack{\alpha\beta \\ cd}}|^2.$$
$$(6.32)$$

This example shows explicitly that for the pion, and similarly for other mesons, only the forward propagator $D^{-1}(n|m_0)$ needs to be calculated. For baryons we only need forward propagating quarks anyhow (compare (6.21)).

### 6.2.2 Point source or extended source?

In order to get clear and strong correlation signals, which allow a reliable analysis, we have to optimize the interpolating fields. Although any operator with the correct quantum numbers contributes to the physical state, some may be more important than others. The overlap can be improved considerably by providing more realistic spatial wave functions. We present the idea using a meson interpolator as an example, but the generalization to baryons is straightforward.

We consider a meson interpolator on a single fixed time slice $n_t$, located at $n_0 = (\boldsymbol{n}_0, n_t)$. A more general wave function may be introduced by writing

$$O_M(n_0) = \sum_{\boldsymbol{n}_1, \boldsymbol{n}_2} F(n_0; \boldsymbol{n}_1, \boldsymbol{n}_2)_{\substack{\alpha_1 \alpha_2 \\ a_1 a_2}} \overline{\psi}^{(f)}(\boldsymbol{n}_1, n_t)_{\substack{\alpha_1 \\ a_1}} \psi^{(f')}(\boldsymbol{n}_2, n_t)_{\substack{\alpha_2 \\ a_2}} , \quad (6.33)$$

where we have introduced a distribution function $F(n_0; \boldsymbol{n}_1, \boldsymbol{n}_2)$ which combines field values $\overline{\psi}(\boldsymbol{n}_1, n_t)$ and $\psi(\boldsymbol{n}_2, n_t)$ at spatial positions $\boldsymbol{n}_1, \boldsymbol{n}_2$ in the vicinity of $\boldsymbol{n}_0$. In the simplest case, where the complete interpolator is localized on the single site $n_0 = (\boldsymbol{n}_0, n_t)$, $F$ reduces to

$$F(n_0; \boldsymbol{n}_1, \boldsymbol{n}_2)_{\substack{\alpha_1 \alpha_2 \\ a_1 a_2}} = \delta(\boldsymbol{n}_0 - \boldsymbol{n}_1) \, \delta_{\alpha_0 \alpha_1} \, \delta_{a_0 a_1} \, \Gamma_{\alpha_0 \beta_0} \, \delta(\boldsymbol{n}_0 - \boldsymbol{n}_2) \, \delta_{\beta_0 \alpha_2} \, \delta_{a_0 a_2}$$
$$\Rightarrow \quad O_M(n_0) = \overline{\psi}^{(f)}(\boldsymbol{n}_0, n_t)_{\substack{\alpha_0 \\ a_0}} \Gamma_{\alpha_0 \beta_0} \psi^{(f')}(\boldsymbol{n}_0, n_t)_{\substack{\beta_0 \\ a_0}} , \quad (6.34)$$

where $\Gamma$ is an element of the Clifford algebra.

However, a more realistic wave function can be obtained by choosing a less trivial function $F$. In order to be able to work with only one set of 12 sources one usually uses factorizable functions ($\alpha_0, \beta_0, a_0$ are summed)

$$F(n_0; \boldsymbol{n}_1, \boldsymbol{n}_2)_{\substack{\alpha_1 \alpha_2 \\ a_1 a_2}} = S_i^{(n_0, \alpha_0, a_0)}(\boldsymbol{n}_1)_{\substack{\alpha_1 \\ a_1}} \Gamma_{\alpha_0 \beta_0} S_k^{(n_0, \beta_0, a_0)}(\boldsymbol{n}_2)^{*}_{\substack{\alpha_2 \\ a_2}} . \quad (6.35)$$

We have introduced the subscripts $i$ and $k$ to indicate that the source functions for $\overline{\psi}$ and $\psi$ may differ. Using the form (6.35) one can rewrite (6.33) in terms of so-called *smeared fermions*. To indicate the type of source we mark the smeared fermions with a lower index $i$ or $k$, respectively:

$$\psi_k^{(f')}(\boldsymbol{n}_0, n_t)_{\substack{\alpha_0 \\ a_0}} \equiv \sum_{n_2} S_k^{(n_0,\alpha_0,a_0)}(\boldsymbol{n}_2)^*_{\substack{\alpha_2 \\ a_2}} \psi^{(f')}(\boldsymbol{n}_2, n_t)_{\substack{\alpha_2 \\ a_2}} ,$$

$$\overline{\psi}_i^{(f)}(\boldsymbol{n}_0, n_t)_{\substack{\alpha_0 \\ a_0}} \equiv \sum_{n_1} S_i^{(n_0,\alpha_0,a_0)}(\boldsymbol{n}_1)_{\substack{\alpha_1 \\ a_1}} \overline{\psi}^{(f)}(\boldsymbol{n}_1, n_t)_{\substack{\alpha_1 \\ a_1}} . \qquad (6.36)$$

This form makes it obvious that the smeared field at $\boldsymbol{n}_0$ is combined from the original fields at positions $\boldsymbol{n}_i$. The meson operator in terms of the smeared fermions becomes simply (this is (6.34), but now for smeared fields):

$$O_M(\boldsymbol{n}_0) = \overline{\psi}_i^{(f)}(\boldsymbol{n}_0, n_t)_{\substack{\alpha_0 \\ a_0}} \Gamma_{\alpha_0 \beta_0} \psi_k^{(f')}(\boldsymbol{n}_0, n_t)_{\substack{\beta_0 \\ a_0}} . \qquad (6.37)$$

Introducing these smearing functions for sources $S_i$, located at $m_0$, and sinks $S_k$, located at $n_0$, leads to the smeared quark propagator

$$G_{ki}(n_0|m_0)_{\substack{\beta_0\alpha_0 \\ b_0a_0}} \equiv \left\langle \psi_k^{(f)}(n_0)_{\substack{\beta_0 \\ b_0}} \overline{\psi}_i^{(f)}(m_0)_{\substack{\alpha_0 \\ a_0}} \right\rangle_F \qquad (6.38)$$

$$= \left\langle \sum_{m,n} S_k^{(n_0,\beta_0,b_0)}(\boldsymbol{n})^*_{\substack{\beta \\ b}} \psi^{(f)}(n)_{\substack{\beta \\ b}} S_i^{(m_0,\alpha_0,a_0)}(\boldsymbol{m})_{\substack{\alpha \\ a}} \overline{\psi}^{(f)}(m)_{\substack{\alpha \\ a}} \right\rangle_F$$

$$= \sum_{n,m} S_k^{(n_0,\beta_0,b_0)}(\boldsymbol{n})^*_{\substack{\beta \\ b}} \left( D_f^{-1} \right)(n|m)_{\substack{\beta\alpha \\ ba}} S_i^{(m_0,\alpha_0,a_0)}(\boldsymbol{m})_{\substack{\alpha \\ a}} .$$

In the construction of the meson and baryon correlators, as discussed in Sect. 6.1, one now simply replaces the original quark propagator $D_f^{-1}$ by the smeared propagator $G_{ki}$. In an actual numerical calculation one computes the propagator from a smeared source $S_i$ to a point-like sink according to

$$\sum_{m,\alpha,a} D^{-1}(n|m)_{\substack{\beta\alpha \\ ba}} S_i^{(m_0,\alpha_0,a_0)}(\boldsymbol{m})_{\substack{\alpha \\ a}} . \qquad (6.39)$$

The smearing of the sink can then be done in the subsequent construction of the hadron propagators by applying the smearing operator presented in the next subsection.

### 6.2.3 Extended sources

For smearing functions $S$ that are not gauge-covariant, one has to fix the gauge in the time slice using the methods from Sect. 3.2. Various source shapes have been used in that context. The extreme among those is a source constant in a time slice, the so-called *wall-source*. Attractive are Gaussian shapes motivated by intuition. The final incentive always is the optimization of the signal.

A gauge-covariant source, with a shape similar to a Gaussian, is obtained by *Jacobi smearing* [3, 4]. One acts with a smearing operator $M$ on the point source $S_0$ of (6.30) in time slice $n_t$ to obtain the smeared source:

$$S^{(n_0,\alpha_0,a_0)} = M\, S_0^{(n_0,\alpha_0,a_0)} \,, \quad M = \sum_{n=0}^{N} \kappa^n\, H^n \,. \tag{6.40}$$

The operator $H$ is essentially the spatial part of the Wilson term (5.50) without the constant piece ($n_t$ is fixed to the time slice of the source),

$$H(\boldsymbol{n},\boldsymbol{m}) = \sum_{j=1}^{3} \Big( U_j(\boldsymbol{n},n_t)\,\delta(\boldsymbol{n}+\hat{\jmath},\boldsymbol{m}) + U_j(\boldsymbol{n}-\hat{\jmath},n_t)^\dagger\,\delta(\boldsymbol{n}-\hat{\jmath},\boldsymbol{m}) \Big) \,. \tag{6.41}$$

Note that $H$ and therefore also $M$ are hermitian and act only on the color indices, but are trivial in Dirac space. The operation connects different sites of the time slice to the central site with gauge transporters. Jacobi smearing has two free parameters: the number of smearing steps $N$ and the positive real parameter $\kappa$. These two parameters can be used to adjust the profile (width) of the source. Also combining quark sources of different widths has proved to be efficient for optimizing the hadron propagation signals (cf. Sect. 6.3.3).

### 6.2.4 Calculation of the quark propagator

Having prepared our sources $S$, we need to compute the propagator $D^{-1}$ acting on the source according to (6.39), i.e., we need $G = D^{-1}S$. Put differently, one has to solve the system of equations

$$DG = S \,, \tag{6.42}$$

where $D$ is the Dirac matrix operator, $S$ the source vector, and $G$ the unknown propagator vector. This can be done using iterative methods. In order to get acquainted with that approach let us briefly discuss the grandmother of those algorithms: As shown in the discussion of the hopping expansion (compare (5.57)), after rescaling the fermions we may split $D$ into a constant part and a nontrivial term,

$$D \equiv \mathbb{1} - Q \,. \tag{6.43}$$

Then a series expansion

$$G = (\mathbb{1} - Q)^{-1}\, S = (\mathbb{1} + Q + Q^2 + Q^3 + \ldots)\, S \tag{6.44}$$

leads to the iteration prescription

$$G^{(0)} = S \,, \quad G^{(i+1)} = S + Q\, G^{(i)} \,. \tag{6.45}$$

This is the simple Jacobi iteration which converges when the largest eigenvalue of $Q$ is smaller than 1 in magnitude. In each iteration step one only has to apply the matrix $Q$ to the vector $G^{(i)}$ and add $S$ until convergence is obtained. Relaxation methods like the Gauss–Seidel method interpolate between subsequent steps and may improve convergence. Both methods belong to the class of stationary methods which solve for each of the variables locally, by keeping the current values of the other variables fixed. Other variants derived from the Gauss–Seidel method, but converging an order of magnitude faster, are the SOR (successive overrelaxation) and SSOR (symmetric SOR) methods, which introduce extrapolation parameters [5].

Most calculations, however, use a variant of the *conjugate gradient* (CG) method [5, 6]. This is a so-called non-stationary method and the concept is to iteratively minimize a quadratic functional, equivalent to finding the solution of the system of equations. For real symmetric, positive definite $N \times N$ matrices $A$ the function

$$Q(\boldsymbol{x}) = \frac{1}{2}\,\boldsymbol{x}^T A\,\boldsymbol{x} - \boldsymbol{x}^T \boldsymbol{b} \qquad (6.46)$$

is (up to a constant) a positive definite quadratic form assuming its minimum at $\boldsymbol{x}^*$, where its gradient vanishes,

$$\partial Q(\boldsymbol{x})|_{\boldsymbol{x}^*} = A\,\boldsymbol{x}^* - \boldsymbol{b} = 0 \ . \qquad (6.47)$$

Thus $\boldsymbol{x}^*$ is the solution of the system $A\boldsymbol{x} = \boldsymbol{b}$. The CG method works by iteratively constructing search direction vectors $\boldsymbol{p}^{(i)}$ and iterates $\boldsymbol{x}^{(i)}$ such that in each step $Q(\boldsymbol{x}^{(i)} + \alpha_i\,\boldsymbol{p}^{(i)})$ is minimized as a function of the real parameter $\alpha_i$, leading to the next iterate $\boldsymbol{x}^{(i+1)} = \boldsymbol{x}^{(i)} + \alpha_i\,\boldsymbol{p}^{(i)}$. The vectors $\boldsymbol{p}^{(i)}$ are orthogonal to $A\,\boldsymbol{p}^{(i-1)}$ (and all previous $A\,\boldsymbol{p}^{(j)}$) sequentially build up a vector space $\mathbb{K}^{(i)} = \mathrm{span}(\boldsymbol{p}^{(0)}, \dots, \boldsymbol{p}^{(i)})$, which is called a *Krylov subspace*. All variants of the CG method are based on such an implicit construction of an orthogonal basis for the Krylov subspace. Methods like the Lanczos algorithm for finding the eigensystem of a matrix work with this concept as well and the corresponding iteration may be related to the CG iteration [5].

The iterates $\boldsymbol{x}^{(i)}$ minimize $Q$ in the spaces $\mathbb{K}^{(i)}$ and approach the solution vector $\boldsymbol{x}^*$ in at most $N$ steps. However, often much fewer steps give sufficiently accurate results. This may be checked by computing the norm $\|A\,\boldsymbol{x}^{(i)} - \boldsymbol{b}\|$, and the process is terminated once the norm is smaller than the requested accuracy $\varepsilon$. Only scalar products between vectors and matrix–vector multiplications have to be computed and only vectors have to be kept in storage. Both are important advantages for the large, but often sparse, matrices associated with the Dirac operators.

In its original version CG works only for positive definite symmetric matrices $A$. However, for non-symmetric matrices the residual vectors cannot be made orthogonal within these limitations. The Dirac operator $D$ is not positive definite hermitian; it has complex eigenvalues. For non-symmetric, general matrices a variant called *Bi-Conjugate Gradient* (Bi-CGR) does the job. There two sequences of search directions and residual vectors are generated, which

**Table 6.2.** Pseudocode [5] for the Bi-CGStab algorithm with preconditioning matrix $M$. For no preconditioning set $M = 1$ and simplify the code

---

$r^{(0)} = b - A\,x^{(0)}$

$\tilde{r} = r^{(0)}$ (for example)

**for** $i = 1, 2 \ldots$ **iterate** :

    $\rho_{i-1} = \tilde{r}^\dagger \cdot r^{(i-1)}$

    **if** $\rho_{i-1}$ **equals 0, the method fails.**

    **if** $i = 1$ **then**

        $p^{(1)} = r^{(0)}$

    **else**

        $\beta_{i-1} = \alpha_{i-1}\,\rho_{i-1} / \left( \omega_{i-1}\,\rho_{i-2} \right)$

        $p^{(i)} = r^{(i-1)} + \beta_{i-1} \left( p^{(i-1)} - \omega_{i-1}\,v^{(i-1)} \right)$

    **end if**

    **solve** $M\,\hat{p} = p^{(i)}$

    $v^{(i)} = A\,\hat{p}$

    $\alpha_i = \rho_{i-1} / \left( \tilde{r}^\dagger \cdot v^{(i)} \right)$

    $s = r^{(i-1)} - \alpha_i\,v^{(i)}$

    **if the norm of** $s$ **is small enough, set** $x^{(i)} = x^{(i-1)} + \alpha_i\,\hat{p}$ **and stop.**

    **solve** $M\,\hat{s} = s$

    $t = A\,\hat{s}$

    $\omega_i = t^\dagger \cdot s / \left( t^\dagger \cdot t \right)$

    $r^{(i)} = s - \omega_i\,t$

    $x^{(i)} = x^{(i-1)} + \alpha_i\,\hat{p} + \omega_i\,\hat{s}$

**repeat until convergence** . . .

---

obey a bi-orthogonality relation. Even better, improving convergence properties, the so-called Bi-CGR Stabilized (Bi-CGStab) method [7] and another improvement, Bi-CGStab(2) by [8], have been introduced. For details we refer to the original papers and to [5].

We remark that one could utilize $D^{-1} = D^\dagger (D\,D^\dagger)^{-1}$ and apply an algorithm for hermitian matrices to compute $(D\,D^\dagger)^{-1}$. Furthermore, for $\gamma_5$-hermitian Dirac operators one could also invert the hermitian matrix $D\gamma_5$ and use $D^{-1} = \gamma_5 (D\gamma_5)^{-1}$. In practice, however, it turned out that the strategy of using an algorithm for hermitian matrices does not necessarily lead to faster convergence.

Here we discuss only the Bi-CGStab algorithm along the lines of [5]. In order to solve the matrix equation (6.47) for the unknown vector $x^*$, one starts with a guessed vector $x^{(0)}$. The pseudo-code for the iteration with preconditioning (see below) then may be written as in Table 6.2. For the application of Bi-CGStab to our problem (6.42), one sets $A = D$, $b = S$, and obtains the desired propagator vector $G = D^{-1}S$ as $G = \lim_{i \to \infty} x^{(i)}$.

We remark that $\tilde{r}$ remains constant and one may also choose other $\tilde{r}$, as long as $\rho_0 \neq 0$. For repeating the iteration one needs $\omega_i \neq 0$ (the

Bi-CGStab(2) algorithm helps in this respect [8]). One stops the iterations as soon as $r^{(i)\dagger} \cdot r^{(i)} < \varepsilon$ for some requested accuracy $\varepsilon$. The algorithm needs two matrix–vector multiplications with $A$, two inversions of the preconditioner $M$, and four inner products for each iteration step.

For the simple CG algorithm convergence is guaranteed for positive definite symmetric matrices, at least after $N$ steps for matrices of dimension $N$, but usually significantly faster. For Bi-CGStab convergence may be faster than that of other CG methods, but not necessarily, and in some cases it even breaks down. Other starting values or other methods have to be chosen in such situations.

So-called preconditioning methods sometimes allow one to accelerate the convergence, depending on the Dirac operator. A *preconditioner* is a suitable matrix $M$ which we use to transform the system to

$$M^{-1} A \boldsymbol{x} = M^{-1} \boldsymbol{b} \,. \tag{6.48}$$

Let us assume that we have a matrix $M$, which is numerically cheap to invert, i.e., to solve $M \hat{\boldsymbol{s}} = \boldsymbol{s}$ for $\hat{\boldsymbol{s}}$, and which approximates $A$ in some way. Then the spectral properties of $M^{-1} A$ may be more favorable, allowing faster convergence of the iterative solution. This may be true in particular, if the small eigenvalues of $M$ agree with those of $A$. For a more complete discussion of such methods cf. [5, 9].

When inverting the Dirac operators for different quark masses, the matrices differ just by a constant in the diagonal. One then can simultaneously solve the set of equations for a set of quark masses with little extra cost. The rate of convergence is determined by the convergence for the smallest mass. Such a multi-mass algorithm has been discussed in [10, 11].

Since the determination of the quark propagators is numerically quite expensive, one often stores these for each gauge configuration sampled and computes various hadron observables from them in the later analysis.

### 6.2.5 Exceptional configurations

So far we have not taken into account the Dirac operator matrix $D[U]$ which depends on the gauge field $U$. This implies that the properties of $D[U]$ will change as one changes $U$. In particular, certain fluctuations of the gauge field, so-called *exceptional configurations*, lead to small eigenvalues of $D[U]$ which make the numerical inversion of $D[U]$ problematic.

To understand this phenomenon in more detail let us consider the eigenvalues of $D[U]$. For quarks with bare mass parameter $m$ they are given by

$$m + \lambda_i[U] \,, \tag{6.49}$$

where $\lambda_i[U]$ are the eigenvalues of the massless Dirac operator. In general these are complex numbers, but also real eigenvalues $\lambda_i[U] = r_i[U]$ are possible. If such a real eigenvalue becomes negative and in magnitude similar to $m$, then the Dirac operator has a very small eigenvalue. In this case the numerical inversion of $D[U]$ breaks down, and only increasing the quark mass $m$ resolves the problem. Thus the exceptional configurations limit the quark masses one can reach. For example in quenched simulations with Wilson fermions it turns out that in a typical simulation[4] (e.g., for lattice size $32^3 \times 64$) the quark mass has to be chosen large enough such that the pion mass is at least 300 MeV instead of the physical 135 MeV.

We remark that for chiral fermions, where the Dirac operator obeys the Ginsparg–Wilson equation, a concept which we discuss in Chap. 7, the real eigenvalues do not fluctuate and the problem with exceptional configurations does not occur; therefore realistic pion masses can be reached. Also introducing a twisted mass term (see Chap. 10) solves the problem with exceptional configurations, since it expels the spectrum from a strip along the real axis.

In practical simulations one finds that the fluctuations of the eigenvalues are coupled to the fluctuations of the gauge field. These fluctuations can be damped by either using improved gauge field actions (see Chap. 9), larger lattices, or by smoothing techniques which we discuss in the next subsection.

### 6.2.6 Smoothing of gauge configurations

When aiming for correlation functions one is mainly interested in the long distance behavior. On the other hand, typical for gauge theories are violent short distance fluctuations of the gauge field. One can considerably improve the correlation signal by *smoothing* or *smearing* the gauge field, either only in the time slices or even in space and time. This also helps with the exceptional configurations discussed in the preceding subsection.

When smoothing or smearing one typically replaces the link variables by local averages over short paths connecting the link's endpoints. This is a gauge covariant procedure and one does not have to fix the gauge. The operators and propagators are then constructed on the smeared configurations. As long as the smearing is local enough, i.e., the smearing operator combines only a fixed number of links, the long distance correlation signals should not be affected in the continuum limit.

Smearing algorithms are all averaging products of links along certain paths connecting the endpoints of a given link. For SU(2) this average is proportional to a group element, but for SU(3) this is not the case. Therefore one has to project the average to an SU(3) matrix. We briefly mention here only three variants of such smearing algorithms.

**APE smearing [12]:** In that case the average is over the original link $U_\mu$ and over the six perpendicular *staples* (visualize!) connecting its endpoints,

---

[4]Theoretical arguments show that only for $\kappa < 1/8$ one can be sure that there are no exceptional configurations.

$$V_\mu(n) = (1 - \alpha) U_\mu + \frac{\alpha}{6} \sum_{\nu \neq \mu} C_{\mu\nu}(n) ,$$

$$C_{\mu\nu}(n) = U_\nu(n) U_\mu(n + \hat{\nu}) U_\nu(n + \hat{\mu})^\dagger$$
$$+ U_\nu(n - \hat{\nu})^\dagger U_\mu(n - \hat{\nu}) U_\nu(n - \hat{\nu} + \hat{\mu}) , \quad (6.50)$$

where the real parameter $\alpha$ may be adjusted depending on the gauge coupling. The projection of the sum to SU(3) is usually done by maximizing $\mathrm{Re\,tr}\left[X V_\mu(n)^\dagger\right]$ for $X \in$ SU(3) and using $X$ as new link variable $U'_\mu(n)$.

**HYP smearing** [13]: In that approach the average is over paths lying within the hypercubes containing the link variable. Again the sum has to be projected to SU(3).

**Stout smearing** [14]: This method uses a particular way of projection, defining the new link after a smearing step by

$$U'_\mu(n) = \mathrm{e}^{\mathrm{i}\, Q_\mu(n)} U_\mu(n) . \quad (6.51)$$

$Q_\mu(n)$ is a traceless, hermitian matrix constructed from staples,

$$Q_\mu(n) = \frac{\mathrm{i}}{2} \left( \Omega_\mu(n)^\dagger - \Omega_\mu(n) - \frac{1}{3} \mathrm{tr} \left[ \Omega_\mu(n)^\dagger - \Omega_\mu(n) \right] \right) ,$$

$$\Omega_\mu(n) = \left( \sum_{\nu \neq \mu} \rho_{\mu\nu} C_{\mu\nu}(n) \right) U_\mu(n)^\dagger . \quad (6.52)$$

The $C_{\mu\nu}$ are the same as in (6.50) and the real weight factors $\rho_{\mu\nu}$ are tunable parameters. Common choices are taking them constant, $\rho_{\mu\nu} = \rho$, or smearing only the spatial links: $\rho_{\mu 4} = \rho_{4\mu} = 0$, $\rho_{nm} = \rho$. The new links have gauge transformation properties like the original ones. This approach of getting an SU(3) element from a sum of matrices may be used for other combinations of paths too, of course. The advantage of stout smearing is that $U'_\mu(n)$ is differentiable with respect to the link variables. This is a prime benefit in applications like the hybrid Monte Carlo method for dynamical fermions discussed in Chap. 8.

Such link smearing methods lead to what is often called a *fat link* for obvious reasons. There is a variety of other such suggestions, among them the FLIC link [15] or differentiable variants of the HYP smearing [16, 17].

All such smearing steps can be iterated. However, the smearing then affects not just neighboring variables but extends over increasingly larger distances. Eventually the asymptotic behavior of the propagators will be affected rendering such results problematic. Anyway, most of the positive effect of smearing is achieved already after the first smearing step(s).

## 6.3 Extracting hadron masses

Once the quark propagators are computed as described in the last section, they can be combined to the hadron correlators discussed in Sect. 6.1. From these correlators one can extract the corresponding hadron masses and in this section we present the techniques necessary for this step.

In spectroscopy calculations we analyze hadron correlators of the form (6.1). In these formulas energies $E$, masses $m$, and momenta $p$ always come with a factor of the lattice constant $a$, making the products $aE$, $am$, and $ap$ dimensionless. It is convenient to drop the factor of the lattice constant and use so-called *lattice units*. In lattice units (6.1) assumes the form (6.53) below. For going back to physical units, an energy $E$ has to be replaced by $aE$, and equivalently for masses and momenta (see Sect. 6.4.2). Unless stated otherwise, from now on we use lattice units for masses and energies.

### 6.3.1 Effective mass curves

In the analysis of the mass spectrum one studies the correlation function of an operator $\widehat{O}$ with the quantum numbers of a particular hadron or set of hadrons. However, these operators usually do not create eigenstates of the Hamiltonian. They have contributions from several eigenstates and the correlation function is the spectral sum (1.21), discussed in Chap. 1. If the sink operator is projected to zero momentum (by summation over the respective time slices) we find

$$C(n_t) \equiv \left\langle \widetilde{O}(\mathbf{0}, n_t)\, \overline{O}(\mathbf{0}, 0) \right\rangle = \sum_k \langle 0|\widehat{O}|k\rangle\langle k|\widehat{O}^\dagger|0\rangle\, \mathrm{e}^{-n_t\, E_k} . \qquad (6.53)$$

For finite lattice size the values $E_k$ are discrete. If the operators couple to single particle intermediate states, the masses $m_k$ of those particles will correspond to the low-lying energy values, i.e., $E_k = m_k$. If the operators couple to two- or more-particle intermediate states, the energy values $E_k$ are related to masses and relative momenta of the involved particles.

Due to the exponential form of the terms in the sum (6.53), the relative contribution of higher lying energy states varies with $n_t$. Only for large $n_t$ the lowest energy state dominates. For smaller $n_t$ the mixing with other states blurs the picture and the sum (6.53) has many relevant contributions,

$$C(n_t) = A_0\, \mathrm{e}^{-n_t\, E_0} + A_1\, \mathrm{e}^{-n_t\, E_1} \;\ldots\, . \qquad (6.54)$$

In Sect. 6.3.3 we will discuss methods to analyze the case when more than one energy level has to be determined. For the moment let us consider the simplest situation where one is interested only in the lowest energy $E_0$ corresponding to the ground state.

The left-hand side plot of Fig. 6.3 shows an example for the correlation function of a mesonic operator with the quantum numbers of the pion in a sample calculation. The linear regions in the log-plot correspond to the range of $n_t$-values where a single exponential behavior $\sim \exp(-n_t\, E_0)$ dominates.

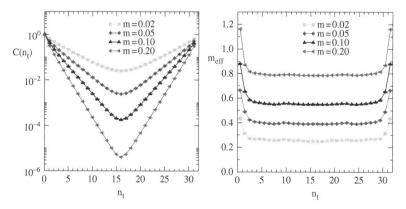

**Fig. 6.3.** Result of a Monte Carlo simulation on a $16^3 \times 32$ lattice at lattice spacing $a \approx 0.15$ fm. L.h.s.: log-plot for the pion correlation function; r.h.s.: effective mass plot (in lattice units). The different sets correspond to different values of the quark mass in lattice units. The points are connected to guide the eye

For mesons propagation in $n_t$ and $(N_T - n_t)$ is identical up to a possible relative minus sign and, when only the ground state is considered, the correlator shows (depending on source and sink interpolators) a cosh- or sinh-dependence on $n_t$:

$$A_0\, e^{-n_t\, E_0} \pm A_0\, e^{-(N_T - n_t)\, E_0} = \begin{cases} 2\, A_0\, e^{-N_T\, E_0/2}\, \cosh\left((N_T/2 - n_t)E_0\right) \\ 2\, A_0\, e^{-N_T\, E_0/2}\, \sinh\left((N_T/2 - n_t)E_0\right) \end{cases}.$$
$$(6.55)$$

In order to analyze in which range of $n_t$ the contribution of the sub-leading exponentials in (6.54) can be neglected, one defines an *effective mass* as

$$m_{\text{eff}}\left(n_t + \tfrac{1}{2}\right) = \ln \frac{C(n_t)}{C(n_t + 1)}.$$
$$(6.56)$$

Once the correlator $C(n_t)$ is dominated by the ground state energy, $m_{\text{eff}}$ becomes constant and forms an *effective mass plateau* at $m_{\text{eff}} = E_0$.

If one wants to respect periodicity in $n_t$ according to (6.55), one sets

$$\frac{C(n_t)}{C(n_t + 1)} = \frac{\cosh\left(m_{\text{eff}}(n_t - N_T/2)\right)}{\cosh\left(m_{\text{eff}}(n_t + 1 - N_T/2)\right)},$$
$$(6.57)$$

(and equivalently for the case of sinh) and solves for $m_{\text{eff}}$ at each $n_t$. This was done in the right-hand side plot in Fig. 6.3. Depending on the mass, such modifications due to periodicity effects may not be necessary if one stays away from the region $n_t \simeq N_T/2$.

In the right-hand side plot of Fig. 6.3 the effective mass plateaus are well pronounced for $4 < n_t < 28$. In this range the contribution of excited states can be neglected and a two-parameter fit to the simple form $A_0 \exp(-n_t\, E_0)$

can be performed. If the correlator is symmetric, as in our example, then one fits to $A_0 \cosh((n_t - N_T/2) E_0)$.

A generalization of effective mass techniques is combining correlation function values from different time slices subtracting the leading exponential decay. This may be extended to a tool for extraction excited hadron masses [18].

We stress that the symmetry of correlators, i.e., equal propagation in $n_t$ and $N_T - n_t$ up to a sign, is not a general feature. In particular, after projection to definite parity (see (6.16)) a baryon propagates in $n_t$, while its parity partner propagates in $N_T - n_t$. Since the two masses are different, the propagator is not symmetric, and instead of cosh or sinh, the simple exponential form has to be used for effective masses and the fit function.

### 6.3.2 Fitting the correlators

Although plotting $m_{\text{eff}}$ gives a first estimate of the ground state mass, a correlated fit in a range of $n_t$ values should be performed. For the correlated analysis one fits the data in a range $n_{\min} \leq n_t, n'_t \leq n_{\max}$ by minimizing

$$\chi^2 = \sum_{n_t, n'_t = n_{\min}}^{n_{\max}} (C(n_t) - f(n_t)) \, w(n_t, n'_t) \, (C(n'_t) - f(n'_t)) \,, \qquad (6.58)$$

with regard to the parameters $A_0, E_0$ of the hypothesis function

$$f(n_t) = A_0 \, e^{-E_0 \, n_t} \ \text{or} \ f(n_t) = A_0 \cosh ((n_t - N_T/2) E_0) \quad \text{(or sinh)} .$$
$$(6.59)$$

The weight $w$ ideally should be the inverse of the exact covariance matrix, i.e., $w(n_t, n'_t) = \text{Cov}^{-1}(n_t, n'_t)$. In actual Monte Carlo runs one only knows an estimator, i.e., the measured covariance matrix of the data,

$$\text{Cov}_N(n_t, n'_t) = \frac{1}{N-1} \, \langle (C(n_t) - \langle C(n_t) \rangle_N) \, (C(n'_t) - \langle C(n'_t) \rangle_N) \rangle_N \,, \ (6.60)$$

where $\langle \ldots \rangle_N$ denotes the statistical average over $N$ gauge configurations. In (6.58) one thus uses $w(n_t, n'_t) = \text{Cov}_N^{-1}(n_t, n'_t)$.

However, this estimator is often too badly determined and due to statistical fluctuations there may be accidental small eigenvalues destabilizing the fit. In such a case often either some smooth approximation or just the diagonal part of the covariance matrix (neglecting the correlation between different data points) are used. The latter gives rise to $w(n_t, n'_t) = \delta_{n_t, n'_t} / \sigma(n_t)^2$ and the $\chi^2$-functional of (6.58) reduces to the form familiar from uncorrelated fits. For a critical account of correlated fits, see [19].

One method to determine the fit range $[n_{\min}, n_{\max}]$ in (6.58) is the analysis of effective mass plateaus discussed above. Another criterion is to require that the $\chi^2$ per degree of freedom of the fit does not change by more than $\mathcal{O}(1)$ when the region is extended. Although a generally valid way to determine the fit interval is desirable, often – in particular for data with only moderate statistics – it is affected by individual prejudice. Any such decision involves some arbitrariness and should be checked by visual inspection of the fits.

A simple method to estimate the statistical error of the fit result for the mass is to use the statistical bootstrap or jackknife methods. One repeats the analysis for large subsets of the data and determines the final error of the mass a posteriori, as discussed in Chap. 4.

### 6.3.3 The calculation of excited states

In particular at smaller time distances one observes the contribution of states with energies higher than that of the ground state (see the right-hand side plot of Fig. 6.3). Obviously it is desirable to get an estimate of these energy levels corresponding to excited states. At the same time, disentangling ground states from higher states provides cleaner signals for both.

A direct fit of the correlation function to a hypothesis function $f(n_t)$ which is a sum of $K$ exponentials,

$$f(n_t) = \sum_{k=1}^{K} c_k \, e^{-n_t \, E_k} \, , \tag{6.61}$$

would formally work, if the number of parameters were smaller than the number of data points and if the data $C(n_t)$ were exact. In reality the data result from Monte Carlo simulations and do have errors. It can be easily demonstrated that already a sum of three exponentials may lead to unreliable results in such fits. Several alternative approaches have been discussed in the literature.

**Bayesian analysis (conditionally biased fits):** In order to stabilize the fit to a hypothesis function of the type (6.61), one does not use the $\chi^2$ functional (6.58), but instead one minimizes

$$F = \chi^2 + \lambda \phi \, , \tag{6.62}$$

where $\phi$ is some stabilizing function of the fit parameters and $\lambda$ a positive real multiplier. This technique is well known for so-called ill-posed problems like analytic continuation of functions given only by data points [20].

In [21] it has been proposed to use as the stabilizing function

$$\phi = \sum_{k=1}^{K} \left( a_k \, (E_k - \widehat{E}_k)^2 + b_k \, (c_k - \widehat{c}_k)^2 \right) \, , \tag{6.63}$$

where $\widehat{E}_k$ and $\widehat{c}_k$ are prejudices for the expected energy values and coefficients. Adjusting the parameters $a_k$ and $b_k$ determines the relative weight for the corresponding bias. The fit is thus stabilized, even if the number of fit parameters exceeds the number of data, leading to unique results depending on the multiplier $\lambda$, which defines the "amount of prejudice". One tries to find regions for the resulting fit parameters that depend only weakly on $\lambda$.

One technique based on such ideas [22] has been to start with simple one-exponential fits at large-$n_t$ values, then continue with two-exponential fits including the result of the earlier fit as a bias along the lines discussed. Experience shows that one usually needs good statistics (small statistical errors of the data) in order to successfully employ that method.

**Maximum entropy method:** For this approach one writes the correlation function as the Laplace transform of a spectral density,

$$C(n_t) = \int_0^\infty dE \, \rho(E) \, e^{-n_t E} \,. \tag{6.64}$$

In the continuum the resulting spectral density $\rho(E)$ should exhibit peaks near the energy values dominating the correlation function. When computing $\rho(E)$ from lattice data, one discretizes the energy in small steps, $E_n = n\,\Delta E$, and uses the values $\rho(E_n)$ of the density at the discrete energies $E_n$ as fit parameters. The technical challenge is to recover the spectral density from the (few) values of $C(n_t)$ by minimizing a suitable distance functional (6.62). Various approaches to define the stabilizing functional $\phi$ have been used.

In the maximum entropy method the functional $\phi$ in (6.62) is chosen based on probabilistic arguments. We normalize the spectral density according to

$$C(0) = \int_0^\infty dE \, \rho(E) \equiv 1 \,. \tag{6.65}$$

Then $\rho(E)$ should be distributed such that each "quantum" has the same a priori chance to contribute to the energy value $E$. This leads to the stabilizer

$$\phi[\rho] = \int_0^\infty dE \, \rho(E) \ln(\rho(E)) \,. \tag{6.66}$$

Minimizing the joint probability leads to a smooth solution. A drawback is that in this method the spectral density has to be positive. Also, the functional is non-linear and dedicated numerical methods have to be used to solve the minimization problem.

This approach is heavily used in statistical physics. It also has been applied to nucleon propagators [23, 24]. From experience with statistical physics it is expected, however, that one needs data for many values of $n_t$ and with very small errors for a reliable result. Sample calculations show significant ambiguities when one tries to extract more energy levels from moderately precise correlation data.

**Variational analysis:** The situation concerning excited states can be improved by considering not only a single correlator, but by computing a matrix of *cross correlators*

$$C_{ij}(n_t) \equiv \left\langle \widetilde{O}_i(\mathbf{0}, n_t) \overline{O}_j(\mathbf{0}, 0) \right\rangle = \sum_k \langle 0|\widehat{O}_i|k\rangle\langle k|\widehat{O}_j^\dagger|0\rangle \, e^{-n_t E_k} \,, \tag{6.67}$$

for a set of $N$ *basis interpolators* $O_i$, $i = 1, \ldots, N$, all with the quantum numbers of the state one is interested in. Different interpolators can, e.g., be constructed by using different Dirac structures as discussed in Sect. 6.1, or with quark propagators acting on different quark sources.

Utilizing the special form of the spectral decomposition on the right-hand side of (6.67) it has been shown [25–27] that diagonalizing the cross-correlation matrix $C(n_t)$ allows to disentangle the physical states to some amount. Indeed the eigenvalues $\lambda^{(k)}(n_t)$ of the correlation matrix can be shown to behave as

$$\lambda^{(k)}(n_t) \propto e^{-n_t\, E_k}\left(1 + \mathcal{O}(e^{-n_t\, \Delta E_k})\right) , \tag{6.68}$$

where $\Delta E_k$ is the distance of $E_k$ to nearby energy levels. Even better results can be obtained when using the generalized eigenvalue equation

$$C(n_t)\, \boldsymbol{v} = \lambda(n_t)\, C(n_0)\, \boldsymbol{v} , \tag{6.69}$$

where the eigenvalues behave again as in (6.68), but the amplitude of the correction term is typically smaller. The generalized eigenvalue problem (6.69) may be rewritten as

$$Q(n_0)^{-1}\, C(n_t)\, Q^\dagger(n_0)^{-1}\, \boldsymbol{u} = \lambda(n_t)\, \boldsymbol{u} \text{ with } Q(n_0)\, Q^\dagger(n_0) = C(n_0) . \tag{6.70}$$

Numerically this form is expected to be more stable than diagonalizing $C(n_0)^{-1} C(n_t)$, which is another way of rewriting the generalized eigenvalue problem as a regular eigenvalue problem.

In the generalized eigenvalue problem the normalization at some time slice $n_0 < n_t$ ought to improve the signal by suppressing the contributions from higher excited states. Indeed that effect can be observed in sample calculations using model data: Both approaches, the straightforward diagonalization, as well as the generalized eigenvalue problem, give the correct answers asymptotically, but the generalized eigenvalue approach does so already at smaller distances $n_t$.

The method improves when the number of basis interpolators is increased. These should be independent and have good overlap with the eigenstates of the problem. On the other hand, in realistic calculations, including more interpolators enhances the statistical noise and thus affects the diagonalization. It always pays off to choose lattice operators which are close to the expected physical content of the eigenstates.

The eigenvectors of the problem give the overlap of the eigenstates, supposedly the states close to the physical states, with the basis interpolators used in the correlation matrix (6.67). Thus, one may derive information on the wave function of the physical modes. In [28–30] lattice operators have been obtained by combining Jacobi smeared quark sources of different widths thus allowing for nodes in the spatial wave function. A different approach is followed in [31, 32] where extended lattice interpolators are constructed using irreducible representations of the symmetry group of a cubic lattice.

## 6.4 Finalizing the results for the hadron masses

Now we are at a stage where the hadron masses have been extracted from the corresponding correlators and are ready for analysis. We discuss the necessary steps for finalizing the spectroscopy calculation and illustrate them using data from quenched calculations.

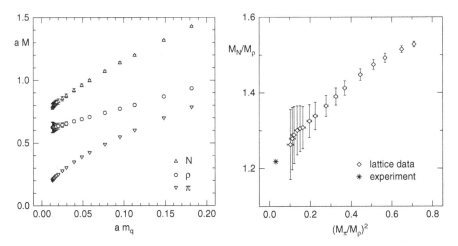

**Fig. 6.4.** Raw data for hadron masses. The *left-hand side plot* shows pion, $\rho$, and nucleon masses in lattice units as a function of the bare quark mass. The *right-hand side plot* is a so-called *APE plot* where dimensionless ratios are compared to the experimental number. The data are taken from the quenched calculation [33]

### 6.4.1 Discussion of some raw data

Following the steps of the last sections we have computed hadron masses in lattice units, i.e., we have obtained numbers for the dimensionless products $aM$. Usually, the hadron masses are evaluated for a range of light quark masses $m_q \equiv m_u = m_d$. The left-hand side plot in Fig. 6.4 shows such raw data for pion, $\rho$, and nucleon masses from a quenched simulation [33].

Converting masses in lattice units to numbers, that can be compared to experimental data, can be done by considering ratios of masses where the lattice constant $a$ cancels. The right-hand side plot in Fig. 6.4 shows $M_N/M_\rho$, the ratio of nucleon to $\rho$ mass, as a function of $(M_\pi/M_\rho)^2$, where $M_\pi$ is the pion mass. Such a plot is called *APE plot* [34]. An alternative is the so-called *Edinburgh plot* [35], where one uses $M_\pi/M_\rho$ on the horizontal axis.

In both, APE and Edinburgh plots, the data for $M_N/M_\rho$ can be compared to the physical point, $M_N/M_\rho \approx 1.209$ at $M_\pi/M_\rho \approx 0.180$. Although we do not perform an extrapolation, it is obvious that the data shown in the APE plot of Fig. 6.4 clearly extrapolate to the physical point already in the quenched approximation.

### 6.4.2 Setting the scale and the quark mass parameters

The results of a lattice calculation are always dimensionless numbers like, e.g., mass ratios. In order to compare with the experiment, one has to relate a mass or length unit to an experimental value. Consequently, the data have to be converted to physical units such as MeV. In other words, one has to determine the physical value of the lattice constant $a$ in order to obtain a mass $M$ from the calculated number for $aM$. There are two main techniques for this step called *scale setting*.

**Setting the scale with the Sommer parameter:** For this approach one uses the Sommer parameter to determine the lattice constant $a$ from the static potential as discussed in Sect. 3.5. The result has the form $a = x$ fm, where $x$ is a dimensionless number.

From our hadron correlators we have computed the dimensionless number $X = a\,M$ for some hadron mass $M$. Consequently the physical mass $M = X/a$ is given in inverse fm. This mass unit is a consequence of using natural units $\hbar = 1$, $c = 1$. For converting this result to MeV one uses

$$1 = \hbar c = 197.327\,\text{MeV fm} \quad \Rightarrow \quad 1\,\text{fm}^{-1} = 197.327\,\text{MeV} . \tag{6.71}$$

For the hadron mass in MeV we then obtain

$$M = \frac{X}{a} = \frac{X}{x}\,\text{fm}^{-1} = \frac{X}{x}\,197.327\,\text{MeV} . \tag{6.72}$$

**Setting the scale with a hadron mass:** In this approach one uses one of the hadron masses $M_0$ for setting the scale. Popular candidates are the $\rho$ meson or the nucleon. For that hadron, e.g., the $\rho$-meson, the mass in lattice units, $aM_0(m_q)$, has to be known at several values of the bare quark mass $m_q$. One then extrapolates the data to the value $m_q^*$ of the mass parameter, where the dimensionless pion to $\rho$ mass ratio assumes its physical value, $M_\pi(m_q^*)/M_\rho(m_q^*) = M_\pi^{\text{exp}}/M_\rho^{\text{exp}} = 0.180$. Alternatively one may also extrapolate $aM_0(m_q)$ to the chiral limit, i.e., to $m_q = 0$. This is only an approximation for the true physical point, but on the other hand this choice avoids the statistical error due to the measured pion to $\rho$ mass ratio. For both extrapolations the result is a number $X = aM_0(m_q^*)$ (or $X = aM_0(0)$), and we identify $M_0(m_q^*) \equiv M_0^{\text{exp}}$ (or $M_0(0) \equiv M_0^{\text{exp}}$), where $M_0^{\text{exp}}$ is the experimental mass of our hadron in MeV. One obtains the lattice constant in inverse MeV: $a = X/M_0^{\text{exp}}$. This scale $a$ can be directly used to obtain other masses $M$ in MeV from the values $aM$ in lattice units.

For simulations with three quark flavors, in addition to the scale $a$, also the light quark mass $m_q = m_u = m_d$ and the strange quark mass $m_s$ have to be determined. For the light quark mass one either identifies the physical point by setting $m_q = m_q^*$, with $m_q^*$ determined as discussed above, or alternatively one assumes that the light u and d quarks are massless. For setting the strange quark mass, a similar strategy is applied: One uses a hadron containing the strange quark, typically the $K^\pm$ pseudoscalar meson, the vector meson $\phi$, or baryons like $\Sigma$ or $\Omega$. The $K^\pm$ has quark content u, s and thus its mass is a function of the u and s quark masses, $M_{K^\pm}(m_q, m_s)$ with $m_q = m_u$. In a first step the data are extrapolated either to the chiral limit, $m_q = 0$, or to $m_q = m_q^*$. This gives the $K^\pm$ mass $M_{K^\pm}(m_s)$ at different values of the strange quark mass (the scale is set with one of the methods discussed above). This is done for several values of $m_s$ and the data for $M_{K^\pm}(m_s)$ are interpolated. The value $m_s^*$, where the interpolated data coincide with the experimental value, $M_{K^\pm}(m_s^*) = M_{K^\pm}^{\text{exp}} = 494$ MeV, gives the bare strange quark mass parameter $m_s = m_s^*$. When using the $\phi$, the situation is even simpler, since $\phi$ is predominantly a strange–antistrange state and no chiral extrapolation is needed.

### 6.4.3 Various extrapolations

Analyzing lattice data one faces the notorious, three-fold group of problems:

- Finite size effects and infinite volume limit $V \to \infty$.
- Scaling in the continuum limit $a \to 0$.
- Chiral extrapolation $m \to 0$.

**Finite size:** All numerical lattice results are obtained in finite volumes. At fixed other parameters one therefore has to study the dependence on the lattice size and, if possible, extrapolate to infinite spatial volume. For the extrapolation one may either use a naive parameterization, e.g., in powers of $1/L$ for spatial extent $L$, or, if possible, a model based on theory.

Actually, like in the analysis of the spin models of statistical physics, finite size effects may even be useful to learn about infinite volume results. An example is chiral perturbation theory which provides a systematic expansion with the infinite volume quantities appearing as parameters [36, 37]. We apply such formulas in the next chapter when we discuss a lattice calculation of the chiral condensate on finite lattices.

For most spectroscopy studies one uses $N^3 \times N_T$ lattices which are much larger in the time direction than in the spatial directions and $L = a N$. Indeed, neglecting the finite time extent (and thus finite temperature effects) the leading finite size effects are due to the spatial volume. Effects from interaction around the spatial torus lead to exponential corrections to the mass of $\mathcal{O}(\exp(-\alpha L))$ (cf. [38, 39] and the discussion in Chap. 11). The leading contribution comes from the smallest mass hadron, the pion, and is of $\mathcal{O}(\exp(-L M_\pi))$. As a rule of thumb, for $L M_\pi > 4$ finite size effects from this mechanism can be ignored.

For small volumes numerical studies of the volume dependence of hadron masses seem to indicate a $1/L^n$ behavior, with $n \approx 2$–$3$. This would be compatible with the intuitive picture that the wave function is squeezed at too small spatial volume and that the observed shift in energy is in leading order due to this squeezing effect (cf. [40, 41] and references therein). Most analyses agree on the observation that ground state hadron masses obtained for linear lattice sizes larger than 3 fm show no noticeable volume dependence. For excited states with wave functions of larger extent, the squeezing effect may be more serious.

**Continuum limit and scaling:** A measured mass $M$ (physical units) has $a$-dependent corrections,

$$M(a) = M_{\text{phys}} \left(1 + \mathcal{O}\left(a^\alpha\right)\right) , \qquad (6.73)$$

which may be different for different lattice actions. Whereas with the Wilson fermion action a quadratic contribution to the mass was identified [42], one finds that improving the fermion action reduces this dependence [33].

Eventually we want to obtain results for continuous space–time. This corresponds to the limit $a \to 0$, or, equivalently, $\beta \to \infty$. However, decreasing the lattice spacing for a fixed number of lattice points shrinks the physical volume. Thus, as we decrease $a$ (increase $\beta$), we have to increase the number of lattice points, $N^3 \times N_T$, which in turn drives up the numerical cost of the simulation. Typical simulations try to afford three or four values of $a$ and perform an extrapolation to $a = 0$.

**Chiral extrapolation:** Often numerical results are obtained for unphysically large quark masses. Eventually we want to approach the small quark masses corresponding to the physical pion. From a conceptual point of view one even might be interested in the massless case, the chiral limit $m \to 0$. Massless 2-flavor QCD (i.e., assuming vanishing $m_u$ and $m_d$) is an interesting and nontrivial, one-parameter theory, where hadron masses are expected to be close to their experimental values. One could even call this a zero-parameter theory, since the gauge coupling is only used to fix the scale. Unfortunately this limit is hard to achieve for several reasons. Toward smaller quark masses a Pandora's box of problems opens up.

For fermion actions that do not preserve chiral symmetry (or do not introduce a regulator like the twisted mass action, cf. Chaps. 7 and 10), there are zero modes occurring even for nonvanishing quark mass, due to the exceptional configurations discussed in Sect. 6.2.5. This limits the smallest pion masses that can be reached.

Another class of problems at small $m_q$ turns up when using the quenched approximation. Omitting the fermion determinant has several effects:

- The absence of the fermion determinant alters the physics by omitting sea quark loops (compare Fig. 6.2). Thus, hadrons cannot decay in the quenched approximation.
- The contribution of the anomaly (see next chapter) is removed and the $\eta'$-meson becomes light – an additional Goldstone mode appears.
- States with negative coefficients in their correlation function appear, so-called ghosts. The quenched $\eta'$ combined with a hadron provides an example for such a state [43–45], cf. the hairpin diagrams in Fig. 6.2.
- Small eigenvalues of the Dirac operator are not suppressed by the determinant (which is the product of all eigenvalues). Consequently, fluctuations leading to exceptional configurations may occur more often.

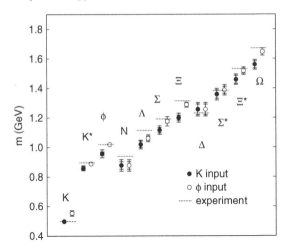

**Fig. 6.5.** Quenched light hadron spectrum compared with experiments; statistical and systematic errors are indicated. The plot is from a high statistics simulation of the CP–PACS collaboration using Wilson valence quarks [42]. (Reprinted figure with permission from S. Aoki et al., Phys. Rev. D 67, 034503 (2003). Copyright (2003) by the American Physical Society)

For extrapolations in the quark mass, *chiral perturbation theory* provides the suitable conceptional background (for introductory accounts, see [46–49]). Chiral perturbation theory is an expansion in the pion mass around the chiral limit taking into account the symmetries of QCD. The resulting extrapolation formulas are typically expressed in terms of the pion mass, and as an example we show the result for the extrapolation of the nucleon [50, 51]

$$M_{\rm N} = c_0 + c_2 M_\pi^2 + c_3 M_\pi^3 + c_4 M_\pi^4 \ln(M_\pi) + \mathcal{O}\left(M_\pi^4\right), \qquad (6.74)$$

where the expansion coefficients $c_i$ are related to low energy constants such as the pion decay constant in the chiral limit.

Chiral perturbation theory can be modified to take into account quenching [52–54]. Formulas such as (6.74) obtain corrections which may become singular in the chiral limit. This is due to the additional Goldstone boson already addressed.

### 6.4.4 Some quenched results

High precision determinations of hadron masses involve measurements on several hundred gauge field configurations on lattices with size up to $64^3 \times 112$ for quenched Wilson fermions (see e.g. [42, 55]; simulations with other formulations will be addressed in later chapters). The lattice spacing in these simulations was as small as 0.05 fm at a linear physical lattice size of $\mathcal{O}(3.2 \text{ fm})$.

To illustrate what has been achieved in quenched lattice spectroscopy, in Fig. 6.5 we show the results of a high statistics study from the CP–PACS collaboration [42]. The mass values have been obtained after extrapolation to infinite volume, vanishing lattice spacing and the chiral limit. Results from setting the strange quark mass with either $K$ or $\phi$ are compared. Although the calculation was done in the quenched approximation, a good agreement with experimental data is found for a large range of hadrons. In Chap. 8 we will compare Fig. 6.5 to a more recent spectroscopy result with fully dynamical quarks.

# References

1. C. Amsler et al. [Particle Data Group]: Phys. Lett. B **667**, 1 (2008)
2. D.B. Leinweber et al.: in Lect. Notes Phys. **663**, 71. Springer, Berlin, Heidelberg (2005)
3. S. Güsken et al.: Phys. Lett. B **227**, 266 (1989)
4. C. Best et al.: Phys. Rev. D **56**, 2743 (1997)
5. R. Barrett et al.: *Templates for the Solution of Linear Systems: Building Blocks for Iterative Methods* (SIAM, Philadelphia 1994)
6. W. H. Press, S. A. Teukolsky, W. T. Vetterling, and B. P. Flannery: *Numerical Recipes in C*, 2nd ed. (Cambridge University Press, Cambridge, New York 1999)
7. H. A. van der Vorst: SIAM J. Sci. Statist. Comput. **13**, 631 (1992)
8. M. H. Gutknecht: SIAM J. Sci. Statist. Comput. **14**, 1020 (1993)
9. I. Montvay and G. Münster: *Quantum Fields on a Lattice* (Cambridge University Press, Cambridge, New York 1994)
10. A. Frommer et al.: Int. J. Mod. Phys. C **6**, 627 (1995)
11. B. Jegerlehner: Nucl. Phys. B (Proc. Suppl.) **63**, 958 (1998)
12. M. Albanese et al.: Phys. Lett. B **192**, 163 (1987)
13. A. Hasenfratz and F. Knechtli: Phys. Rev. D **64**, 034504 (2001)
14. C. Morningstar and M. Peardon: Phys. Rev. D **69**, 054501 (2004)
15. W. Kamleh, D. Adams, D. B. Leinweber, and A. G. Williams: Phys. Rev. D **66**, 014501 (2002)
16. A. Hasenfratz, R. Hoffmann, and S. Schaefer: JHEP **0705**, 029 (2007)
17. S. Dürr: Comput. Phys. Commun. **180**, 1338 (2009)
18. D. Guadagnoli, M. Papinutto, and S. Simula: Phys. Lett. B **604**, 74 (2004)
19. C. Michael: Phys. Rev. D **49**, 2616 (1994)
20. S. Ciulli, C. Pomponiu, and I. Sabba-Stefanescu: Phys. Rep. **17**, 133 (1975)
21. G. P. Lepage et al.: Nucl. Phys. (Proc. Suppl.) **106**, 12 (2002)
22. Y. Chen et al. (unpublished) arXiv:hep-lat/0405001 (2004)
23. M. Asakawa, T. Hatsuda, and Y. Nakahara: Prog. Part. Nucl. Phys. **46**, 459 (2001)
24. K. Sasaki, S. Sasaki, T. Hatsuda, and M. Asakawa: Nucl. Phys. B (Proc. Suppl.) **129**, 212 (2004)
25. C. Michael: Nucl. Phys. B **259**, 58 (1985)
26. M. Lüscher and U. Wolff: Nucl. Phys. B **339**, 222 (1990)
27. B. Blossier et al.: JHEP **0904**, 094 (2009)
28. T. Burch et al.: Phys. Rev. D **70**, 054502 (2004)

29. T. Burch et al.: Phys. Rev. D **73**, 094505 (2006)
30. T. Burch et al.: Phys. Rev. D **74**, 014504 (2006)
31. S. Basak et al.: Phys. Rev. D **72**, 094506 (2005)
32. S. Basak et al.: Phys. Rev. D **72**, 074501 (2005)
33. C. Gattringer et al. [BGR (Bern-Graz-Regensburg) collaboration]: Nucl. Phys. B **677**, 3 (2004)
34. P. Bacilieri et al.: Phys. Lett. B **214**, 115 (1988)
35. K. C. Bowlers et al.: Phys. Lett. B **162**, 354 (1985)
36. J. Gasser and H. Leutwyler: Ann. Phys. **158**, 142 (1984)
37. P. Hasenfratz and H. Leutwyler: Nucl. Phys. B **343**, 241 (1990)
38. M. Lüscher: in *Progress in Gauge Field Theory, Carg´ese, 1983*, edited by G. 't Hooft et al. (Plenum, New York 1984)
39. M. Lüscher: Commun. Math. Phys. **104**, 177 (1986)
40. M. Fukugita et al.: Phys. Lett. B **294**, 380 (1992)
41. K. Sasaki and S. Sasaki: Phys. Rev. D **72**, 034502 (2005)
42. S. Aoki et al. [CP-PACS collaboration]: Phys. Rev. D **67**, 034503 (2003)
43. S. R. Sharpe: Phys. Rev. D **56**, 7052 (1997)
44. S. R. Sharpe: Phys. Rev. D **62**, 099901 (2000)
45. J. Heitger, R. Sommer, and H. Wittig: Nucl. Phys. B **588**, 377 (2000)
46. H. Leutwyler: in *Festschrift in Honor of B.L. Ioffe: At the Frontier of Particle Physics*, edited by M. Shifman, Vol. 1, p. 271 (World Scientific, Singapore 2001)
47. U. Meißner: in *Festschrift in Honor of B.L. Ioffe: At the Frontier of Particle Physics*, edited by M. Shifman, Vol. 1, p. 417 (World Scientific, Singapore 2001)
48. U. Meißner: PoS **LAT2005**, 009 (2006)
49. V. Bernard and U. Meißner: Ann. Rev. Nucl. Part. Sci. **57**, 33 (2007)
50. T. Becher and H. Leutwyler: Eur. Phys. J. C **9**, 643 (1999)
51. M. Procura, T. R. Hemmert, and W. Weise: Phys. Rev. D **69**, 034505 (2004)
52. S. R. Sharpe: Phys. Rev. D **46**, 3146 (1992)
53. C. W. Bernard and M. F. L. Golterman: Phys. Rev. D **46**, 853 (1992)
54. C. W. Bernard and M. F. L. Golterman: Phys. Rev. D **49**, 486 (1994)
55. K. C. Bowlers et al. [UKQCD collaboration]: Phys. Rev. D **62**, 054506 (2000)

# 7

---

# Chiral symmetry on the lattice

Chiral symmetry and its spontaneous breaking are central properties of QCD with important phenomenological implications. They explain why pions have unexpectedly small masses and why we do not see degenerate masses for parity partners in the baryon sector. This important role of chiral symmetry and the mechanism of its breaking require that a reasonable lattice version of QCD has to implement chiral symmetry properly.

However, putting chiral symmetry onto the lattice turned out to be a formidable challenge. The underlying reason is the doubling problem. In Chap. 5 we have added the Wilson term to the naive fermion action in order to remove the doublers. This term, however, breaks chiral symmetry explicitly. Only more than 20 years after Wilson's introduction of lattice gauge theory the problem of chiral symmetry on the lattice was solved with a generalization of chiral symmetry through the so-called "Ginsparg–Wilson equation" for the lattice Dirac operator. With this new concept a clean implementation of chiral symmetry on the lattice has been achieved.

This chapter is devoted to chiral symmetry on the lattice. After a brief review of chiral symmetry in continuum QCD we discuss why Wilson fermions fail to show chiral symmetry. Subsequently we present the Ginsparg–Wilson equation and explore its consequences concerning the axial anomaly and the index theorem. Finally we discuss an explicit solution of the Ginsparg–Wilson equation, the so-called *overlap operator*.

## 7.1 Chiral symmetry in continuum QCD

### 7.1.1 Chiral symmetry for a single flavor

To warm up, let us first discuss chiral symmetry for the case of only a single flavor in the continuum. The action for a massless fermion reads (for notational convenience we drop the space–time argument $x$ in the following)

Gattringer, C., Lang, C.B.: *Chiral Symmetry on the Lattice*. Lect. Notes Phys. **788**, 157–184 (2010)
DOI 10.1007/978-3-642-01850-3_7        © Springer-Verlag Berlin Heidelberg 2010

$$S_F[\psi, \overline{\psi}, A] = \int d^4 x \, L\left(\psi, \overline{\psi}, A\right) , \tag{7.1}$$

$$L\left(\psi, \overline{\psi}, A\right) = \overline{\psi} \, \gamma_\mu \left(\partial_\mu + i \, A_\mu\right) \psi = \overline{\psi} \, D \psi ,$$

where $D$ denotes the massless Dirac operator.

We now perform a *chiral rotation* of the fermion fields

$$\psi \rightarrow \psi' = e^{i\alpha\gamma_5} \psi , \quad \overline{\psi} \rightarrow \overline{\psi}' = \overline{\psi} \, e^{i\alpha\gamma_5} , \tag{7.2}$$

where $\gamma_5$ is the chirality matrix acting in Dirac space and $\alpha$ is a constant, real parameter. The Lagrangian density is invariant under the chiral rotation,

$$\begin{aligned} L\left(\psi', \overline{\psi}', A\right) &= \overline{\psi}' \, \gamma_\mu \left(\partial_\mu + i \, A_\mu\right) \psi' = \overline{\psi} \, e^{i\alpha\gamma_5} \gamma_\mu \left(\partial_\mu + i \, A_\mu\right) e^{i\alpha\gamma_5} \psi \\ &= \overline{\psi} \, e^{i\alpha\gamma_5} e^{-i\alpha\gamma_5} \gamma_\mu \left(\partial_\mu + i \, A_\mu\right) \psi = L\left(\psi, \overline{\psi}, A\right) , \end{aligned} \tag{7.3}$$

where in the second step we have used $\gamma_\mu \gamma_5 = -\gamma_5 \gamma_\mu$. A mass term breaks this invariance since it transforms nontrivially:

$$m \, \overline{\psi}' \psi' = m \, \overline{\psi} \, e^{i2\alpha\gamma_5} \, \psi . \tag{7.4}$$

The chiral symmetry decouples the action of left- and right-handed massless fermions. To see this we introduce the right- and left-handed projectors

$$P_R = \frac{\mathbb{1} + \gamma_5}{2} , \quad P_L = \frac{\mathbb{1} - \gamma_5}{2} , \tag{7.5}$$

which obey

$$\begin{aligned} P_R^2 = P_R , \quad P_L^2 = P_L , \quad P_R \, P_L = P_L \, P_R = 0 , \quad P_R + P_L = \mathbb{1} , \\ \gamma_\mu P_L = P_R \gamma_\mu , \quad \gamma_\mu P_R = P_L \gamma_\mu . \end{aligned} \tag{7.6}$$

With these we can define right- and left-handed fermion fields

$$\psi_R = P_R \psi , \quad \psi_L = P_L \psi , \quad \overline{\psi}_R = \overline{\psi} \, P_L , \quad \overline{\psi}_L = \overline{\psi} \, P_R . \tag{7.7}$$

After a few lines of algebra the decoupling of left- and right-handed components follows

$$L\left(\psi, \overline{\psi}, A\right) = \overline{\psi}_L \, D \, \psi_L + \overline{\psi}_R \, D \, \psi_R . \tag{7.8}$$

Thus we find that left- and right-handed components "do not talk to each other." A mass term, however, mixes the components:

$$m \, \overline{\psi} \, \psi = m \left(\overline{\psi}_R \, \psi_L + \overline{\psi}_L \, \psi_R\right) . \tag{7.9}$$

Since chiral symmetry of the action holds only for massless quarks, the limit of vanishing quark mass is often referred to as the *chiral limit*.

One can summarize the essence of chiral symmetry in the simple equation

$$D \, \gamma_5 + \gamma_5 \, D = 0 . \tag{7.10}$$

It expresses the fact that the massless Dirac operator $D = \gamma_\mu(\partial_\mu + i A_\mu)$ anticommutes with $\gamma_5$.

### 7.1.2 Several flavors

Let us now delve somewhat deeper into the symmetry properties of the QCD action and consider the theory with $N_f$ flavors of quarks. Now the fermion fields $\psi, \overline{\psi}$ carry also a flavor index. However, for notational convenience we use vector notation for the flavor index, too. In this notation the fermion action reads

$$S_F\left[\psi, \overline{\psi}, A\right] = \int \mathrm{d}^4 x \, \overline{\psi} \, \left(\gamma_\mu \left(\partial_\mu + \mathrm{i}\, A_\mu\right) + M\right) \psi \,, \qquad (7.11)$$

where we introduce a mass matrix

$$M = \mathrm{diag}\left(m_1, m_2, \ldots, m_{N_f}\right) \qquad (7.12)$$

acting in flavor space (in the real world $m_1 = m_u$, $m_2 = m_d$, etc.). For generalizing the chiral transformations of the last paragraph we use matrices that mix the different flavors. In particular we denote the generators of SU($N_f$) by $T_i, i = 1, 2, \ldots, N_f^2 - 1$ (see Appendix A.1).

Before we discuss the chiral rotations, let us first note that the action for massless quarks is also invariant under the $N_f^2$ *vector transformations*[1]

$$\psi' = \mathrm{e}^{\mathrm{i}\,\alpha T_i}\,\psi \,, \qquad \overline{\psi}' = \overline{\psi}\,\mathrm{e}^{-\mathrm{i}\,\alpha T_i} \,, \qquad (7.13)$$

$$\psi' = \mathrm{e}^{\mathrm{i}\,\alpha \mathbb{1}}\,\psi \,, \qquad \overline{\psi}' = \overline{\psi}\,\mathrm{e}^{-\mathrm{i}\,\alpha \mathbb{1}} \,, \qquad (7.14)$$

where $\mathbb{1}$ denotes the $N_f \times N_f$ unit matrix. The invariance of (7.11) for $M = 0$ under transformations (7.13) and (7.14) is evident. However, the invariance under (7.13) extends also to the case of degenerate masses $M = \mathrm{diag}(m, m, \ldots, m)$. This symmetry is the isospin symmetry, generalized to $N_f$ flavors. Symmetry (7.14) even holds for arbitrary masses and the corresponding conserved quantity is baryon number.

The *chiral* or *axial vector* rotations are defined as

$$\psi' = \mathrm{e}^{\mathrm{i}\,\alpha \gamma_5 T_i}\,\psi \,, \qquad \overline{\psi}' = \overline{\psi}\,\mathrm{e}^{\mathrm{i}\,\alpha \gamma_5 T_i} \,, \qquad (7.15)$$

$$\psi' = \mathrm{e}^{\mathrm{i}\,\alpha \gamma_5 \mathbb{1}}\,\psi \,, \qquad \overline{\psi}' = \overline{\psi}\,\mathrm{e}^{\mathrm{i}\,\alpha \gamma_5 \mathbb{1}} \,, \qquad (7.16)$$

where now also the left- and right-handed components of the different flavors mix. Like for the case of a single flavor one finds that the action (7.11) is invariant under (7.15) and (7.16) for the case of $M = 0$.

Taken altogether, the massless action has the symmetry

$$\mathrm{SU}(N_f)_L \times \mathrm{SU}(N_f)_R \times \mathrm{U}(1)_V \times \mathrm{U}(1)_A \,. \qquad (7.17)$$

The notation used for the two SU($N_f$) pieces stresses that the terms for the left- and right-handed components are symmetric under independent SU($N_f$) rotations.

---

[1] The name "vector transformations" comes from the fact that the corresponding Noether currents are vector currents.

However, when one considers the fully quantized theory one finds that the fermion determinant is not invariant under (7.16) and the corresponding so-called $U(1)_A$ *axial symmetry* is broken explicitly by a noninvariance of the fermion integration measure (we will show this in Sect. 7.3). Taking into account this so-called *axial anomaly* [1, 2] we find that for the quantized, massless theory the symmetry is broken explicitly to the remaining symmetry

$$SU(N_f)_L \times SU(N_f)_R \times U(1)_V . \qquad (7.18)$$

Introducing nonvanishing, degenerate masses, $M = \text{diag}(m, m \ldots m)$, breaks the symmetry $SU(N_f)_L \times SU(N_f)_R$ explicitly to its subgroup $SU(N_f)_V$,

$$SU(N_f)_V \times U(1)_V , \qquad (7.19)$$

and allowing for nondegenerate masses reduces the symmetry further to

$$U(1)_V \times U(1)_V \times \ldots \times U(1)_V \qquad (N_f \text{ factors}) . \qquad (7.20)$$

It is interesting to have a second look at the explicit breaking due to the quark masses. The lightest quarks have masses that are relatively small compared to the typical QCD scale of 1 GeV given by, e.g., the mass of the proton ($\approx 940$ MeV). In particular one has (in the $\overline{MS}$ renormalization scheme)

$$m_u \approx m_d \approx 5 \,\text{MeV} \;,\;\; m_s \approx 100 \,\text{MeV} , \qquad (7.21)$$

such that the explicit breaking of the chiral symmetry is very small for $u$ and $d$ quarks and a 10% effect for the $s$ quark. Thus we would expect that (7.18) is a good approximate symmetry for $N_f = 2$ and in part also for $N_f = 3$.

### 7.1.3 Spontaneous breaking of chiral symmetry

If $u$ and $d$ quarks were massless, $SU(2)_L \times SU(2)_R \times U(1)_V$ would be an exact symmetry of QCD (for $N_f = 2$) at least on the Lagrangian level. Due to the small mass of the $u$ and $d$ quarks the explicit breaking is very small and traces of the symmetry should be visible. One of the consequences is that one expects degenerate masses for the nucleon and its partner with negative parity, the so-called $N^\star$.

We briefly sketch the argument for such a degeneracy of the masses: In the last chapter we have shown that the masses of the nucleon and its negative parity partner are obtained from the exponential decay of the zero momentum correlators

$$C_\pm(t) = -\int d^3x \left\langle \epsilon_{abc} \left( \overline{u}(0)_a C\gamma_5 \overline{d}(0)_b^T \right) \overline{u}(0)_c \frac{\mathbb{1} \pm \gamma_4}{2} \epsilon_{efg} u_e(x) \left( u(x)_f^T C\gamma_5 d(x)_g \right) \right\rangle ,$$
$$(7.22)$$

where $x = (\boldsymbol{x}, t)$. This equation gives the continuum version of the lattice correlator (6.21) after projection to zero momentum. The exponential decay of $C_+(t)$ gives the mass of the nucleon (for forward running $t$) and $C_-(t)$ corresponds to the $N^\star$. Now consider the transformation

$$\psi \rightarrow \exp\left(\mathrm{i}\,\alpha\,\gamma_5\,T_3\right)\psi\,, \quad \overline{\psi} \rightarrow \overline{\psi}\exp\left(\mathrm{i}\,\alpha\,\gamma_5\,T_3\right)\,, \tag{7.23}$$

where $\psi$ and $\overline{\psi}$ are vectors in flavor space: $\psi = (u,d)^T$, $\overline{\psi} = (\overline{u},\overline{d})$. Using $T_3 = \sigma_3/2 = \mathrm{diag}(1/2,-1/2)$ we find that the individual flavors transform as

$$u \rightarrow \exp\left(+\mathrm{i}\frac{\alpha}{2}\gamma_5\right)u\,, \quad \overline{u} \rightarrow \overline{u}\exp\left(+\mathrm{i}\frac{\alpha}{2}\gamma_5\right)\,,$$
$$d \rightarrow \exp\left(-\mathrm{i}\frac{\alpha}{2}\gamma_5\right)d\,, \quad \overline{d} \rightarrow \overline{d}\exp\left(-\mathrm{i}\frac{\alpha}{2}\gamma_5\right)\,. \tag{7.24}$$

When setting $\alpha = \pi$ this simplifies to

$$u \rightarrow \mathrm{i}\gamma_5\,u\,, \quad \overline{u} \rightarrow \mathrm{i}\overline{u}\,\gamma_5\,, \quad d \rightarrow -\mathrm{i}\gamma_5\,d\,, \quad \overline{d} \rightarrow -\mathrm{i}\overline{d}\,\gamma_5\,. \tag{7.25}$$

Inserting this transformation into (7.22) one finds that the correlators transform under (7.25) as

$$C_\pm(t) \rightarrow -C_\mp(t)\,. \tag{7.26}$$

Thus, up to an overall minus sign, the correlator $C_+(t)$ for the nucleon and the correlator $C_-(t)$ for the $N^\star$ are transformed into each other. Since an overall minus sign does not alter the exponential decay properties of the correlator, the invariance under (7.23) implies that the nucleon and its parity partner $N^\star$ have equal masses.

However, no such degeneracy of the nucleon and $N^\star$ masses is observed. The nucleon (proton or neutron) has a mass of 940 MeV, while the $N^\star$ is found at 1535 MeV. The mass difference of almost 600 MeV is far too large to be explained by the small explicit breaking of chiral symmetry due to the $u$ and $d$ quark masses. So another mechanism which breaks the symmetry under (7.23) must be at work. Since the action itself is invariant (up to the small explicit breaking by the quark masses), the strong breaking effect observed in nature must come from a *spontaneous breaking of chiral symmetry*, except for $U(1)_A$ which is broken explicitly by the anomaly.

The spontaneous breaking of symmetries is a concept which is well known from, e.g., spin systems describing ferromagnets. While the Hamiltonian of the system is invariant under a global rotation of the spins, in the ground state of the system all spins point in the same direction, thus leading to a macroscopic magnetization of the system.

A similar effect is taking place in QCD. While the action is invariant under chiral rotations, the ground state is not. An order parameter for chiral symmetry breaking is

$$\langle \overline{u}(x)u(x)\rangle\,, \tag{7.27}$$

the so-called *chiral condensate*. The chiral condensate transforms like a mass term and is not invariant under chiral rotations. Consequently, when the theory has a nonvanishing chiral condensate this implies that chiral symmetry is broken spontaneously. In Sect. 7.3 we will discuss the calculation of the chiral condensate on the lattice and show that it has a nonvanishing value.

Another important consequence of a spontaneously broken continuous symmetry is the appearance of so-called *Goldstone modes* or *Goldstone bosons* (see standard text books on quantum field theory such as [3, 4]). Goldstone modes are massless bosonic excitations. For the example of the spin system with continuous symmetry group these are spin waves. In the case of QCD the pions are interpreted as the "would-be" Goldstone bosons of chiral symmetry breaking. For massless quarks a spontaneous symmetry breaking mechanism of QCD would explain massless pions. Their relatively small masses of about 140 MeV can be understood as resulting from the explicit breaking by the $u$ and $d$ quark masses.

To summarize, we find that for the light quark flavors the action is essentially invariant under chiral transformations – up to the chiral U(1) transformation symmetry, which is broken explicitly by the axial anomaly. However, the experimental findings do not reflect the chiral symmetry and we must conclude that chiral symmetry is broken spontaneously due to the dynamics of QCD. Chiral symmetry at the classical level and its spontaneous breaking provide an explanation for several phenomena, such as the nondegenerate masses of baryonic parity partners or the small masses of the pions.

## 7.2 Chiral symmetry and the lattice

Having convinced ourselves of the crucial importance of chiral symmetry and its breaking for QCD phenomenology, let us now return to the lattice. We start with analyzing the chiral properties of Wilson fermions.

### 7.2.1 Wilson fermions and the Nielsen–Ninomiya theorem

Upon inspecting the Wilson Dirac operator (5.51) we find that (7.10) is violated also for vanishing quark mass. The culprit is the Wilson term (5.50) which we had to add to the naive action in order to remove the doublers. This term comes with the unit matrix $\mathbb{1}$ in Dirac space which does not anticommute with $\gamma_5$ and therefore violates (7.10). Thus, even for massless quarks we break chiral symmetry explicitly. Consequently we cannot expect to capture such a subtle effect as spontaneous chiral symmetry breaking with Wilson fermions.

This situation cannot be simply overcome by adding a term for removing the doublers in a different way. There is a fundamental theorem by Nielsen and Ninomiya [5–7] which states that on the lattice one cannot implement chiral symmetry in the form of (7.10) and at the same time have a theory free of doublers. More explicitly [8] consider a Euclidean action for free fermions in the form

$$S = \sum_{n,m,\mu} \overline{\psi}(n)\, \mathrm{i}\, \gamma_\mu\, F_\mu(n|m)\, P_R\, \psi(m) \,. \tag{7.28}$$

The kernel $F(n|m)$ is assumed to be translationally invariant, $F_\mu(n|m) = F_\mu(n-m)$, and should give rise to a hermitian Hamilton operator which implies $F_\mu(-n) = F_\mu(n)^*$. Furthermore, $F_\mu(n)$ must be local, i.e., decrease sufficiently fast, such that its Fourier transform $\widetilde{F}_\mu(p)$ exists and all its derivatives are continuous. Due to the right-handed projector $P_R$ one would naively expect that action (7.28) describes only right-handed fermions. However, the arguments in [8], based on the expansion of $\widetilde{F}_\mu(p)$ around its zeros (poles in the propagator), show that an equal number of left- and right-handed fermions is described by an action of the form (7.28).

For many years the Nielsen–Ninomiya theorem seemed to define the ultimate limitation for a further development of lattice QCD. A deep understanding of chiral symmetry on the lattice became possible only after the rediscovery of the Ginsparg–Wilson equation.

### 7.2.2 The Ginsparg–Wilson equation

In their seminal paper [9], Ginsparg and Wilson formulated the essential equation for chiral symmetry on the lattice. Based on renormalization group transformations (which we will discuss in Chap. 9) they proposed to replace the continuum expression (7.10) by

$$D\,\gamma_5 + \gamma_5\, D = a\, D\, \gamma_5\, D \,. \tag{7.29}$$

In this equation the anti-commutator (7.10) is augmented with a nonvanishing right-hand side. For dimensional reasons – the Dirac operator $D$ has dimension $1/\text{length}$ – an extra factor of the lattice constant $a$ appears for the term quadratic in $D$. Thus the right-hand side vanishes for $a \to 0$ and the continuum form of chiral symmetry is recovered in a naive continuum limit. However, (7.29) allows to define chiral symmetry on the lattice also for finite $a$. We will discuss this construction and the consequences of the Ginsparg–Wilson equation in great detail below.

It is interesting to analyze the effect of the extra term for the quark propagator $D^{-1}$. Assuming that $D$ has no zero modes and is invertible, we may multiply (7.29) with $D^{-1}$ from both sides and obtain

$$\gamma_5\, D^{-1}(n|m) + D^{-1}(n|m)\, \gamma_5 = a\, \gamma_5\, \delta(n-m) \,, \tag{7.30}$$

where we write explicitly the space–time arguments $n, m$ of the propagators. The term introduced by Ginsparg and Wilson is simply a contact term, i.e., the anti-commutator of the propagator with $\gamma_5$ is modified only for $n = m$.

Before we explore the consequences of the Ginsparg–Wilson equation, we should remark that the original paper [9] from 1982 had no immediate consequences and was "forgotten," since no solution of the nonlinear equation

(7.29) was imagined for the interacting case (nontrivial gauge links) at that time. Only in 1997 and 1998 it was realized [10, 11] that two independent approaches to chiral symmetry on the lattice, *overlap* and *fixed point fermions* introduced earlier, give rise to Dirac operators $D$ obeying (7.29). We discuss these in Sect. 7.4 and Chap. 9.

### 7.2.3 Chiral symmetry on the lattice

We have already announced that, based on the Ginsparg–Wilson equation, it is possible to define a modified chiral rotation which leaves the lattice action for massless quarks invariant [12]. Let $D$ be a lattice Dirac operator obeying the Ginsparg–Wilson equation (7.29). Using $D$ we can define a chiral rotation, which for $a \to 0$ reduces to the continuum transformation (7.2),

$$\psi' = \exp\left(i\alpha\,\gamma_5\left(\mathbb{1} - \frac{a}{2}D\right)\right)\psi \;, \quad \overline{\psi}' = \overline{\psi}\exp\left(i\alpha\left(\mathbb{1} - \frac{a}{2}D\right)\gamma_5\right)\;. \quad (7.31)$$

In this equation the action of the Dirac operator is understood in the sense

$$(D\psi)\,(n)\,\underset{a}{_\alpha} = \sum_{m,\beta,b} D(n|m)\,\underset{ab}{_{\alpha\beta}}\,\psi(m)\,\underset{b}{_\beta} \;, \quad (\overline{\psi}D)\,(m)\,\underset{b}{_\beta} = \sum_{n,\alpha,a} \overline{\psi}(n)\,\underset{a}{_\alpha}\,D(n|m)\,\underset{ab}{_{\alpha\beta}} \;.$$
$$(7.32)$$

In the subsequent equations we refer to this convention, but use matrix/vector notation for all indices: space–time, Dirac, and color.

The Lagrangian density for massless fermions for such a Dirac operator is invariant under the transformation (7.31):

$$\begin{aligned} L\left(\psi',\overline{\psi}'\right) &= \overline{\psi}'\,D\,\psi' \\ &= \overline{\psi}\,\exp\left(i\alpha\left(\mathbb{1} - \frac{a}{2}D\right)\gamma_5\right)D\exp\left(i\alpha\,\gamma_5\left(\mathbb{1} - \frac{a}{2}D\right)\right)\psi \\ &= \overline{\psi}\,\exp\left(i\alpha\left(\mathbb{1} - \frac{a}{2}D\right)\gamma_5\right)\exp\left(-i\alpha\left(\mathbb{1} - \frac{a}{2}D\right)\gamma_5\right)D\psi \\ &= \overline{\psi}\,D\,\psi = L\left(\psi,\overline{\psi}\right)\;. \end{aligned} \quad (7.33)$$

In the step from the second to the third line we have used

$$D\gamma_5\left(\mathbb{1} - \frac{a}{2}D\right) + \left(\mathbb{1} - \frac{a}{2}D\right)\gamma_5 D = 0\;, \quad (7.34)$$

which is just another way of writing the Ginsparg–Wilson equation (7.29).

Decomposition (7.7) into left- and right-handed components can be generalized to the lattice. We define a new kind of projectors

$$\widehat{P}_R = \frac{1 + \widehat{\gamma}_5}{2}\;, \quad \widehat{P}_L = \frac{1 - \widehat{\gamma}_5}{2}\;, \quad \widehat{\gamma}_5 = \gamma_5\left(\mathbb{1} - a\,D\right)\;. \quad (7.35)$$

Due to the Ginsparg–Wilson equation (7.29) one finds $\widehat{\gamma}_5^2 = \mathbb{1}$, implying

$$\widehat{P}_R^2 = \widehat{P}_R \,, \quad \widehat{P}_L^2 = \widehat{P}_L \,, \quad \widehat{P}_R \widehat{P}_L = \widehat{P}_L \widehat{P}_R = 0 \,, \quad \widehat{P}_R + \widehat{P}_L = \mathbb{1} \,. \quad (7.36)$$

Invoking again the Ginsparg–Wilson equation one shows

$$D \widehat{P}_R = P_L D \,, \quad D \widehat{P}_L = P_R D \,. \quad (7.37)$$

Based on this last equation a sensible definition of left- and right-handed components, which generalizes the continuum form (7.7), is

$$\psi_R = \widehat{P}_R \psi \,, \quad \psi_L = \widehat{P}_L \psi \,, \quad \overline{\psi}_R = \overline{\psi} \, P_L \,, \quad \overline{\psi}_L = \overline{\psi} \, P_R \,. \quad (7.38)$$

This definition, together with (7.37), implies the vanishing of terms like $\overline{\psi}_L D \psi_R$ and $\overline{\psi}_R D \psi_L$. The action thus can be decomposed like the continuum action (7.8) into left- and right-handed parts:

$$\overline{\psi} D \psi = \overline{\psi}_L D \psi_L + \overline{\psi}_R D \psi_R \,. \quad (7.39)$$

We thus may identify the continuum fields $\psi_L$, $\psi_R$ with the lattice quantities (7.38). The symmetry breaking mass term (7.9) is identified as the lattice term

$$m \left( \overline{\psi}_R \psi_L + \overline{\psi}_L \psi_R \right) = m \overline{\psi} \left( P_L \widehat{P}_L + P_R \widehat{P}_R \right) \psi = m \overline{\psi} \left( \mathbb{1} - \frac{a}{2} D \right) \psi \,. \quad (7.40)$$

Thus Ginsparg–Wilson fermions with mass are described by the operator [13]

$$D_m = D + m \left( \mathbb{1} - \frac{a}{2} D \right) = \omega D + m \mathbb{1} \,, \quad (7.41)$$

with the abbreviation

$$\omega \equiv 1 - \frac{a m}{2} \,. \quad (7.42)$$

   With chiral rotation (7.31) and decomposition (7.39) we have successfully implemented the corresponding continuum structures (7.2) and (7.8) on the lattice. The generalization to several flavors proceeds along the same steps as for the continuum case.

   Let us stress the most important difference between the concept of chirality in the continuum and on the lattice. In the continuum chirality is a strictly local concept independent of the gauge field. Rotation (7.2), as well as projection (7.7), only involves the spinors at a given space–time point $x$, i.e., only $\psi(x)$ or $\overline{\psi}(x)$ is needed. This is different on the lattice. Both chiral rotation (7.31) and projection (7.38) require the application of the lattice Dirac operator $D$. Thus the chiral rotation and the decomposition into components involve neighboring sites and depend on the gauge field. Consequently the chirality of a lattice fermion is determined using information from the gauge field and from neighboring lattice sites.

## 7.3 Consequences of the Ginsparg–Wilson equation

In this section we discuss some of the most important consequences of the Ginsparg–Wilson equation (7.29) without specifying a particular solution. Actual solutions will be presented in Sect. 7.4, the so-called overlap action, and in Sects. 9.2 and 9.3 fixed point fermions will be discussed. We remark that chiral symmetry will also play a major role when we discuss Ward identities in Chap. 11. This subject is not addressed here.

### 7.3.1 Spectrum of the Dirac operator

We begin our discussion of the consequences of the Ginsparg–Wilson equation by analyzing the eigenvalue spectrum of the Dirac operator on a finite lattice. The corresponding eigenvalue equation is denoted by

$$D\, v_\lambda = \lambda\, v_\lambda \ . \tag{7.43}$$

At the moment we do not invoke the Ginsparg–Wilson equation and only require the Dirac operator $D$ to be $\gamma_5$-*hermitian* (compare Sect. 5.4.3),

$$\gamma_5\, D\, \gamma_5 = D^\dagger \ . \tag{7.44}$$

This property is obeyed by most commonly used Dirac operators.[2] For the example of the Wilson Dirac operator this can be shown in a few lines of algebra, and the overlap operator, which we will discuss below, inherits $\gamma_5$-hermiticity from the kernel Dirac operator used for its construction.

The requirement of $\gamma_5$-hermiticity alone already has interesting consequences for the eigensystem. For the characteristic polynomial $P(\lambda)$ of $D$ one finds (the asterisk denotes complex conjugation)

$$\begin{aligned}
P(\lambda) &= \det[D - \lambda\mathbb{1}] = \det[\gamma_5^2(D - \lambda\mathbb{1})] = \det[\gamma_5(D - \lambda\mathbb{1})\gamma_5] \\
&= \det[D^\dagger - \lambda\mathbb{1}] = \det[D - \lambda^*\mathbb{1}]^* = P(\lambda^*)^* \ ,
\end{aligned} \tag{7.45}$$

where we have used $\gamma_5^2 = \mathbb{1}$ and (7.44). The eigenvalues $\lambda$ are the zeros of $P(\lambda)$ and (7.45) implies that if $\lambda$ is a zero, so is $\lambda^*$. Thus for a $\gamma_5$-hermitian Dirac operator the eigenvalues are either real or come in complex conjugate pairs.

Equation (7.44) also has an interesting consequence for the $\gamma_5$ matrix element of the eigenvectors. Writing the inner product of two vectors $u, v$ as $u^\dagger v = (u, v)$ we find

$$\lambda(v_\lambda, \gamma_5\, v_\lambda) = (v_\lambda, \gamma_5 D\, v_\lambda) = (v_\lambda, D^\dagger \gamma_5\, v_\lambda) = (D\, v_\lambda, \gamma_5\, v_\lambda) = \lambda^*(v_\lambda, \gamma_5\, v_\lambda) \ , \tag{7.46}$$

and thus $(\mathrm{Im}\,\lambda)\,(v_\lambda, \gamma_5\, v_\lambda) = 0$. This implies

$$(v_\lambda, \gamma_5\, v_\lambda) = 0 \ , \quad \text{unless} \quad \lambda \in \mathbb{R} \ . \tag{7.47}$$

---

[2] Unless a chemical potential, a $\theta$-angle or a twisted mass term is introduced.

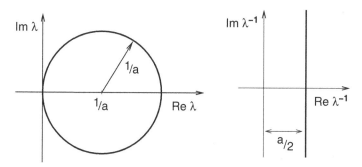

**Fig. 7.1.** Allowed regions for the eigenvalues $\lambda$ of a Ginsparg–Wilson Dirac operator in the complex plane (*left-hand side plot*) and for the eigenvalues $\lambda^{-1}$ of the corresponding propagator (*right-hand side plot*)

Hence only eigenvectors $v_r$ with real eigenvalues $r$ can have nonvanishing *chirality*, i.e., $(v_r, \gamma_5 v_r) \neq 0$.

In addition to being $\gamma_5$-hermitian, we now also require the Dirac operator $D$ to obey the Ginsparg–Wilson equation (7.29). Multiplying this equation with $\gamma_5$ from either left or right and using (7.44) we obtain

$$D^\dagger + D = a\, D^\dagger D , \quad D + D^\dagger = a\, D\, D^\dagger . \tag{7.48}$$

A first consequence of these two equations is that a $\gamma_5$-hermitian Ginsparg–Wilson Dirac operator $D$ is a *normal operator*, i.e., $D$ and $D^\dagger$ commute, since the left-hand sides of the two Equations (7.48) are equal. Normality implies that the eigenvectors form an orthogonal basis (see Appendix A.5). Furthermore, multiplying the first equation in (7.48) with a normalized eigenvector $v_\lambda$ from the right and with $v_\lambda^\dagger$ from the left, one finds

$$\lambda^* + \lambda = a\,\lambda^*\lambda . \tag{7.49}$$

Writing the eigenvalue as $\lambda = x + i\,y$ this equation turns into

$$\left(x - \frac{1}{a}\right)^2 + y^2 = \frac{1}{a^2} , \tag{7.50}$$

which shows that the eigenvalues of a $\gamma_5$-hermitian Ginsparg–Wilson Dirac operator are restricted to a circle in the complex plane (compare the left-hand side plot of Fig. 7.1). This so-called *Ginsparg–Wilson circle* has its center at $1/a$ on the real axis and a radius of $1/a$. We remark that the doubler modes, i.e., those modes where in the free case at least one of the momentum components equals to $\pi/a$, end up near $2/a$ in the complex plane and decouple as $a \to 0$.

A convenient parametrization of the eigenvalues $\lambda$ of a Ginsparg–Wilson Dirac operator is given by

$$\lambda = \frac{1}{a}\left(1 - e^{i\varphi}\right) , \quad \varphi \in (-\pi, \pi] . \tag{7.51}$$

Using this parametrization one finds that the eigenvalues $1/\lambda$ of the quark propagator $D^{-1}$ fall on a line parallel to the imaginary axis (see Fig. 7.1),

$$\frac{1}{\lambda} = \frac{a}{2} + i\frac{a}{2}\frac{\sin(\varphi)}{1 - \cos(\varphi)} \; . \tag{7.52}$$

Since the Ginsparg–Wilson circle touches the origin of the complex plane, $D$ may have exact zero eigenvalues. Let $v_0$ be a zero mode, then

$$D v_0 = 0 \quad \Rightarrow \quad \gamma_5 D v_0 = 0 \quad \Rightarrow \quad D \gamma_5 v_0 = 0 \; , \tag{7.53}$$

using the Ginsparg–Wilson equation in the last step. Thus, on the subspace of zero modes, the Dirac operator commutes with $\gamma_5$ and the zero modes can be chosen as eigenstates of $\gamma_5$. Since $\gamma_5^2 = \mathbb{1}$, the eigenvalues of $\gamma_5$ are $+1$ and $-1$ and we conclude

$$\gamma_5 v_0 = \pm v_0 \; , \tag{7.54}$$

implying that the zero modes are chiral. A zero mode with positive chirality is referred to as *right handed*, while *left handed* is used for negative chirality.

From the left-hand side plot in Fig. 7.1 it is obvious that a Ginsparg–Wilson Dirac operator can have real eigenvalues also at $2/a$. Following the same steps as for the zero modes, one finds that also the eigenmodes with real eigenvalue $2/a$ are chiral. These eigenmodes are the doubler partners of the zero modes shifted to the other side of the Ginsparg–Wilson circle, where they decouple in the limit $a \to 0$.

### 7.3.2 Index theorem

In the continuum the celebrated *Atiyah–Singer index theorem* [14] relates the numbers of left- and right-handed zero modes of the massless Dirac operator to a quantity $Q_{\text{top}}$ which is a property of the gauge fields, the so-called *topological charge*. To derive the lattice equivalent of the index theorem [15] we consider

$$Q_{\text{top}} \equiv \frac{a}{2}\,\text{tr}\,[\gamma_5 D] = -\frac{1}{2}\,\text{tr}\,[\gamma_5(2 - a D)] = -\frac{1}{2}\sum_\lambda (v_\lambda, \gamma_5(2 - a D)\,v_\lambda)$$

$$= -\frac{1}{2}\sum_\lambda (2 - a\lambda)\,(v_\lambda, \gamma_5 v_\lambda) = n_- - n_+ \; , \tag{7.55}$$

where $n_-, n_+$ denote the numbers of left- and right-handed zero modes. In the second step we have used $\text{tr}[\gamma_5] = 0$, and in the third step we expressed the trace as a sum over the eigenvectors $v_\lambda$ of $D$. As discussed above, $D$ is normal and thus its normalized eigenvectors form an orthonormal basis suitable for computing the trace. In the last step we have used (7.47), implying that only eigenvectors with real eigenvalues have nonvanishing chirality and that the factor $(2 - a\lambda)$ cancels the contributions from the doubler modes with $\lambda = 2/a$. The $\gamma_5$ matrix elements of the surviving zero modes are $\pm 1$ due to (7.54).

If one leaves out the second step in (7.55), where the extra $\gamma_5$ was sneaked in, one obtains $Q_{\text{top}} = n'_+ - n'_-$, where $n'_\pm$ are the numbers of left- and right-handed eigenvectors with eigenvalue $2/a$. This gives the same result, since the doublers enter with opposite chirality.

The topological charge $Q_{\text{top}}$ can be written as the space–time sum of the so-called *topological charge density* $q(n)$. Doing so we obtain

$$Q_{\text{top}} = a^4 \sum_{n \in \Lambda} q(n) \,, \quad q(n) = \frac{1}{2\,a^3}\,\text{tr}_{CD}[\gamma_5 D(n|n)] \,, \quad Q_{\text{top}} = n_- - n_+ \,,$$

(7.56)

where $\text{tr}_{CD}$ denotes the trace over color and Dirac indices.

The lattice index theorem (7.56) is a truly remarkable result: The Dirac operator depends on the gauge field variables $U_\mu(n)$. Thus the topological charge $Q_{\text{top}}$ is a functional of the gauge field, and it is a highly nontrivial statement that this functional is an integer number. As the name "topological charge" already indicates the underlying reason is of topological nature.

At this point we remark that in actual lattice calculations one never finds zero modes of both chiralities for a single configuration, i.e., $n_+$ and $n_-$ are not both nonzero simultaneously. It may be argued that the underlying reason is that only a very subtle arrangement of the gauge field would give rise to $n_+ \neq 0$ and $n_- \neq 0$ at the same time. Such an arrangement has zero measure in the path integral, and the fact that at least one of the numbers $n_+, n_-$ vanishes implies $Q_{\text{top}} = n_-$ or $Q_{\text{top}} = -n_+$. This fact is sometimes referred to as *absence of fine tuning*.

In the continuum the index theorem takes the form $Q_{\text{top}}^{\text{cont}} = n_- - n_+$, with the topological charge given by

$$Q_{\text{top}}^{\text{cont}} = \int \mathrm{d}^4 x \; q(x)^{\text{cont}} \,, \quad q(x)^{\text{cont}} = \frac{1}{32\pi^2}\,\epsilon_{\mu\nu\rho\sigma}\,\text{tr}_C\left[F_{\mu\nu}(x)F_{\rho\sigma}(x)\right] \,,$$

(7.57)

where $\epsilon_{\mu\nu\rho\sigma}$ is the completely anti-symmetric rank 4 Levi–Civita tensor and $F_{\mu\nu}$ is the field strength tensor (2.15). The functional $Q_{\text{top}}^{\text{cont}}$ is an expression for the so-called *Pontryagin index*.

An instance of a continuum gauge field configuration with nonvanishing $Q_{\text{top}}^{\text{cont}}$ is so-called *instantons* [16]. There it was also shown that instantons are local minima of the gauge action. Thus they give a contribution in a saddle point evaluation of the path integral. This insight led to a wealth of papers analyzing the role of topological objects in QCD (for reviews see [17, 18] and references therein). For an introductory text on topological field configurations and an elementary proof that $Q_{\text{top}}^{\text{cont}}$ of (7.57) is indeed integer valued we recommend [19].

On the lattice the topological charge (7.55) is an integer due to its definition through the number of zero modes of a chiral Dirac operator. Thus the remaining question is whether the lattice definition (7.56) of the topological charge density $q(n)$ approaches its continuum counterpart $q(x)^{\text{cont}}$ of (7.57) in the limit $a \to 0$. It can be shown that under certain (smoothness) conditions for the gauge fields one has $q(n) = q(x)^{\text{cont}} + \mathcal{O}(a^2)$ as $a \to 0$. Several derivations of this fact have been given for various settings and conditions on the gauge fields, the first results being [15, 20, 21].

The QCD path integral is symmetric with respect to configurations of positive and negative topological charges and thus the expectation value $\langle Q_{\text{top}} \rangle$ vanishes. However, one can consider the *topological susceptibility* defined as

$$\chi_{\text{top}} = \frac{1}{V} \langle Q_{\text{top}}^2 \rangle = \frac{1}{a^4 |\Lambda|} a^8 \sum_{m,n} \langle q(m)\, q(n) \rangle = a^4 \sum_n \langle q(0)\, q(n) \rangle , \quad (7.58)$$

where in the last step we have used translational invariance of the correlator to get rid of the summation over $m$. We remark that $\chi_{\text{top}}$ is volume dependent. The topological susceptibility considered in the infinite volume limit provides information about the distribution of topological charge as a function of $N_f$ and $m$ [22]. Using the fermionic definition (7.55) of the topological charge it has been determined in several quenched calculations [23–25] (and references therein) with a result of $\chi_{\text{top}} \approx (190 \text{ MeV})^4$. Since in the quenched case the fermion determinant is not taken into account, this result is independent of $m$. For the dynamical case the simulations are not so far advanced and the dependence of $\chi_{\text{top}}$ on $N_f$ and $m$ is not finally settled. Phenomenologically, the topological susceptibility plays an interesting role in the so-called *Witten–Veneziano formula* which we briefly address at the end of the next section.

### 7.3.3 The axial anomaly

Having implemented chiral symmetry on the lattice, we are now ready to give a (lattice) derivation [12] of the axial anomaly which we have already addressed in Sect. 7.1. There it was stated that the integration measure for the fermions is not invariant under the flavor diagonal rotation (7.16).

To derive the anomaly on the lattice we consider lattice QCD with $N_f$ massless flavors with a Dirac operator obeying the Ginsparg–Wilson equation. The spinors $\psi, \overline{\psi}$, describing our $N_f$ flavors, are transformed with an infinitesimal chiral rotation of the form

$$\psi' = \left( \mathbb{1} + i\varepsilon\, M\, \gamma_5 \left( \mathbb{1} - \frac{a}{2} D \right) \right) \psi , \quad \overline{\psi}' = \overline{\psi} \left( \mathbb{1} + i\varepsilon\, M \left( \mathbb{1} - \frac{a}{2} D \right) \gamma_5 \right) . \tag{7.59}$$

This transformation is the infinitesimal version of (7.31) obtained by setting $\alpha = \varepsilon$ in (7.31) and keeping only the leading terms in the expansion of the exponential in $\varepsilon$. In addition we consider several flavors by allowing for a mixing matrix $M$ which can either be the identity, $M = \mathbb{1}_{N_f}$, or one of the generators of $SU(N_f)$, $M = T_i$. According to (5.26) for the transformation of the integration measure of Grassmann variables we obtain

$$\mathcal{D}[\psi, \overline{\psi}] = \mathcal{D}[\psi', \overline{\psi}'] \det\left[ \mathbb{1} + i\varepsilon\, M\, \gamma_5 \left( \mathbb{1} - \frac{a}{2} D \right) \right] \det\left[ \mathbb{1} + i\varepsilon\, M \left( \mathbb{1} - \frac{a}{2} D \right) \gamma_5 \right]$$

$$= \mathcal{D}[\psi', \overline{\psi}'] \det\left[ \mathbb{1} + i\varepsilon\, M\, \gamma_5 \left( \mathbb{1} - \frac{a}{2} D \right) \right]^2 , \tag{7.60}$$

where in the second step we used the invariance of the determinant under cyclic permutation of $\gamma_5$. Expansion in $\varepsilon$ gives

$$\det\left[\mathbb{1} + \mathrm{i}\,\varepsilon\, M\,\gamma_5\left(\mathbb{1} - \frac{a}{2}\,D\right)\right]^2 = \exp\left(2\,\mathrm{tr}\left[\ln\left(\mathbb{1} + \mathrm{i}\,\varepsilon\, M\,\gamma_5\left(\mathbb{1} - \frac{a}{2}\,D\right)\right)\right]\right)$$

$$= \exp\left(-2\sum_{j=1}^{\infty}\frac{(-\mathrm{i}\,\varepsilon)^j}{j}\,\mathrm{tr}\left[\left(M\,\gamma_5\left(\mathbb{1} - \frac{a}{2}\,D\right)\right)^j\right]\right)$$

$$= 1 + 2\,\mathrm{i}\,\varepsilon\,\mathrm{tr}\left[M\,\gamma_5\left(\mathbb{1} - \frac{a}{2}\,D\right)\right] + \mathcal{O}\left(\varepsilon^2\right) \qquad (7.61)$$

$$= 1 + 2\,\mathrm{i}\,\varepsilon\,\mathrm{tr}_F[M]\sum_{n\in\Lambda}\left(3\,\mathrm{tr}_D[\gamma_5] - \frac{a}{2}\,\mathrm{tr}_{CD}\left[\gamma_5\,D(n|n)\right]\right) + \mathcal{O}(\varepsilon^2)\,.$$

In the first line of this equation we have used a well-known formula for the determinant (see Appendix A.5) and in the second line the logarithm was expanded in a power series. In the last line we have split the trace into its components, with $\mathrm{tr}_F$, $\mathrm{tr}_C$, and $\mathrm{tr}_D$ denoting the traces over flavor, color, and Dirac indices, respectively. The sum runs over all sites of the lattice $\Lambda$. Since $\gamma_5$ is traceless, the first of the $\mathcal{O}(\varepsilon)$ terms vanishes.

If the flavor matrix $M$ is chosen to be one of the $SU(N_f)$ generators, the anomaly is canceled due to the vanishing trace over the flavor indices. Only for the flavor singlet choice $M = \mathbb{1}_{N_f}$ this trace is nontrivial, $\mathrm{tr}_F[M] = N_f$. Combining (7.60) and (7.61) we obtain for this case

$$\mathcal{D}\left[\psi, \overline{\psi}\right] = \mathcal{D}\left[\psi', \overline{\psi}'\right]\left(1 - 2\,\mathrm{i}\,\varepsilon\, N_f\, Q_{\mathrm{top}} + \mathcal{O}\left(\varepsilon^2\right)\right)\,, \qquad (7.62)$$

where we have used definition (7.56) of the topological charge $Q_{\mathrm{top}}$.

We conclude that indeed a chiral, flavor singlet rotation is not a symmetry of QCD, although the action itself is invariant (for massless quarks). The non-invariance comes from the measure and the symmetry breaking term is the topological charge. This result underlines again the strong connection between chiral symmetry and topological gauge field configurations which we have already addressed when we discussed the index theorem.

The axial anomaly has important physical consequences. Most importantly it implies that the chiral flavor singlet symmetry cannot be broken spontaneously and no flavor singlet Goldstone particle exists. This explains why the $\eta$-meson is much heavier than the pions, when one considers approximate chiral symmetry for $N_f = 2$. For approximate chiral symmetry with $N_f = 3$, this role is taken over by the $\eta'$-meson. The aforementioned Witten–Veneziano formula [26–28] relates the mass of the $\eta'$ to the topological susceptibility $\chi_{\mathrm{top}}$, thus revealing a deep connection of QCD vacuum properties and the mass spectrum. The mathematical subtleties of the Witten–Veneziano formula have been addressed in several recent articles [29–33], partly based on the rigorous framework provided by Ginsparg–Wilson fermions.

### 7.3.4 The chiral condensate

In Sect. 7.1 we have discussed that in the continuum the vacuum expectation value $\langle \overline{u}(x)\, u(x) \rangle$ is not invariant under any of the chiral rotations (7.15) and (7.16). Thus it can serve as an order parameter for spontaneous chiral symmetry breaking, with a nonvanishing condensate signaling the spontaneous breaking of chiral symmetry. We emphasize again that this spontaneous breaking occurs only for $N_f \geq 2$, since for $N_f = 1$ the single axial symmetry is not broken spontaneously, but explicitly by the anomaly.

When discussing chiral symmetry we have identified the correct mass term on the lattice in (7.40), i.e., the scalar bilinear which maximally mixes left- and right-handed components. Accordingly, for lattice QCD with Ginsparg–Wilson fermions we consider the scalar expectation value

$$\Sigma^{\text{lat}}\left(a, m, |\Lambda|\right) \equiv - \left\langle \overline{u}(n)\left(\mathbb{1} - \frac{a}{2}\, D\right) u(n) \right\rangle . \tag{7.63}$$

The expectation value on the right-hand side is evaluated on a lattice $\Lambda$ with $|\Lambda|$ sites and a lattice spacing $a$. We consider $N_f$ flavors of fermions, all with mass $m$, and the corresponding Dirac operator $D_m$ is given by (7.41). The term $a\,D/2$ in (7.63) cancels the real part of the eigenvalues (7.52) of the massless propagator, thus shifting its spectrum onto the imaginary axis as in the continuum (see the right-hand side of Fig. 7.1).

Since our formulation is translationally invariant, (7.63) is independent of $n$ and we can average over all lattice sites. Doing so and performing the Grassmann integration according to (5.54) we find

$$\Sigma^{\text{lat}}\left(a, m, |\Lambda|\right) = \frac{1}{a^4 |\Lambda|} \left\langle \text{tr}\left[\left(\mathbb{1} - \frac{a}{2}\, D\right) D_m^{-1}\right]\right\rangle_G , \tag{7.64}$$

where the trace runs over space–time, color, and Dirac indices, and

$$\langle X \rangle_G = \frac{1}{Z} \int \mathcal{D}[U]\, e^{-S_G[U]}\, \det[D_m]^{N_f}\, X \tag{7.65}$$

is the remaining gauge field integral. With the abbreviation (7.42) the propagator, i.e., the inverse of $D_m$ defined in (7.41), reads $D_m^{-1} = (\omega\, D + m\, \mathbb{1})^{-1}$, $\omega = 1 - am/2$, and after some algebra (7.64) turns into

$$\Sigma^{\text{lat}}\left(a, m, |\Lambda|\right) = \frac{1}{\omega\, a^4 |\Lambda|} \left\langle \text{tr}\left[(\omega\, D + m\, \mathbb{1})^{-1} - \frac{a}{2}\mathbb{1}\right]\right\rangle_G . \tag{7.66}$$

This expression is now ready for a numerical or analytical investigation of the chiral condensate.

At this point a few remarks on the necessary limits are in order. Spontaneous breaking of a symmetry requires an infinite system, therefore the limit $|\Lambda| \to \infty$ has to be taken. Subsequently the explicit breaking through the mass term has to be removed, i.e., the limit $m \to 0$ has to be performed.

Like in other cases of spontaneous symmetry breaking these limits cannot be interchanged. After these limits, (7.66) becomes a *bare condensate* at a fixed ultraviolet cutoff $1/a$. The definition of the *physical condensate* is regularization scheme dependent and its value must be converted to other schemes (like the continuum $\overline{\text{MS}}$ scheme) by renormalization with the scalar renormalization factor $Z_S$ which we discuss in Chap. 11. Subsequently one can analyze the scaling with $a$ and perform the continuum limit.

It is obvious that in a numerical approach the discussed sequence of limits is hard to establish. However, in [22] the chiral condensate has been studied in a 4D box of volume $V$ at finite mass $m$ using low-energy effective theory. An expression for the chiral condensate $\Sigma(m, V)$ as a function of $m$ and $V$ was obtained which contains the physical chiral condensate

$$\Sigma \equiv \lim_{m \to 0} \lim_{V \to \infty} \Sigma(m, V) \qquad (7.67)$$

as a parameter. Here both, $\Sigma$ and $\Sigma(m, V)$, are understood at finite cutoff. Since this is not necessarily a lattice cutoff we do not indicate it explicitly.

Once the functional form $\Sigma(m, V)$ is known, one can use lattice simulations to compute the condensate $\Sigma^{\text{lat}}(a, m, |\Lambda|)$ at finite volume $V = a^4 |\Lambda|$ and finite mass $m$ and fit the data to the function $\Sigma(m, V)$ to obtain the true bare condensate $\Sigma$ as a fit parameter. This procedure has first been carried out for the quenched case, which we discuss here. As always, the motivation for studying the quenched case is the hope that the quenched result for $\Sigma$ will be close to its value at $N_f = 3$.

Within partially quenched chiral perturbation theory one can derive the functional form for the chiral condensate [34, 35] for different *topological sectors* as a function of the volume $V$ and mass parameter $m$. We speak of the topological sector $\nu$ when in the path integral only gauge configurations with topological charge $Q_{\text{top}} = \nu$ are considered. On the lattice this is a concept which can, e.g., be implemented by sorting all configurations of a Monte Carlo ensemble with respect to $Q_{\text{top}}$ as determined from the index theorem (7.56).

The result from chiral perturbation theory mirrors a finding from random matrix theory, where one obtains universal properties of the spectra of certain symmetry classes of random matrices. The quenched result for the condensate in the topological sector $\nu$ is given by

$$\Sigma(m, V)_\nu = \Sigma z \left( I_{|\nu|}(z) K_{|\nu|}(z) + I_{|\nu|+1}(z) K_{|\nu|-1}(z) \right) + \frac{|\nu|}{mV}, \qquad (7.68)$$

where $z = mV\Sigma$ is a dimensionless scaling variable and $\Sigma$ is the bare condensate defined in (7.67). The lattice condensate $\Sigma^{\text{lat}}(a, m, |\Lambda|)$ is then computed for each sector $\nu$ separately and fitted to the functional form (7.68).

Let us discuss the strategy for a numerical determination of $\Sigma$ in more detail [36]. Expanding the Bessel functions in (7.68) one finds for $m \to 0$:

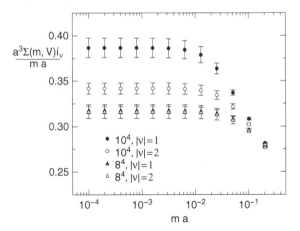

**Fig. 7.2.** Lattice results for the bare condensate for different volumes and different topological sectors [37]. (Reprinted from Hasenfratz et al., copyright (2002) with permission from Elsevier)

$$\frac{\Sigma(m, V)_\nu}{m} \quad \overset{m \to 0}{\longrightarrow} \quad \begin{cases} \Sigma^2 V \left(\frac{1}{2} - \ln(z/2) - \gamma\right) & \text{for} \quad \nu = 0 \,, \\[2mm] \Sigma^2 V \dfrac{1}{2\,|\nu|} + \dfrac{|\nu|}{m^2 V} & \text{for} \quad \nu \neq 0 \,. \end{cases} \tag{7.69}$$

The topologically trivial case ($\nu = 0$) contains a logarithmic divergence at small $m$ (= small $z$) and thus was found to be not very suitable for a numerical study at small $m$. For $\nu \neq 0$ an extra term from the zero modes appears which diverges even more strongly.[3] However, this contribution can be cleanly removed by hand in order to improve the signal. This is done by exploiting the chirality of the zero modes (7.54) and taking the trace in (7.66) only after the propagator has been multiplied with $(\mathbb{1} \pm \gamma_5)/2$. For each individual gauge configuration the sign is chosen opposite to the chirality of the zero modes. After this modification the trace can be computed using the stochastic estimator techniques discussed in Chap. 8.

Since we are computing the bare condensate at finite $V$ and $m$, cutoff effects can alter behavior (7.69). In leading order in $m$ this gives rise to an extra cutoff-dependent constant $c$. Thus for nontrivial topological sectors and subtracted zero mode contributions one finds the final form

$$\frac{\Sigma(m, V)'_\nu}{m} \quad \overset{m \to 0}{\longrightarrow} \quad \Sigma^2 \frac{V}{2\,|\nu|} + c \qquad (\nu \neq 0) \,. \tag{7.70}$$

The ratio $\Sigma^{\mathrm{lat}}(a, m, |\Lambda|)/m$ thus is expected to approach a constant at $m \to 0$, and comparison of different volumes and topological sectors allows one to extract the bare condensate $\Sigma$. The data in Fig. 7.2 were computed in [37] along these lines and nicely illustrate the behavior according to (7.70).

---

[3]For the case of dynamical fermions, zero modes are suppressed by the fermion determinant which vanishes at $m = 0$ and thus no such divergent term occurs.

The last step, which we will address in detail only in Chap. 11, is the renormalization with $Z_S$ to obtain the physical quenched condensate $\Sigma_{\text{phys}} = Z_S \Sigma$. Several independent calculations [36–39] determined the quenched chiral condensate and after renormalization obtained $\Sigma_{\text{phys}} \approx (270\,\text{MeV})^3$ (in the $\overline{\text{MS}}$ scheme at renormalization scale 2 GeV).

### 7.3.5 The Banks–Casher relation

So far we have discussed the numerical analysis of the chiral condensate for a finite system at finite quark mass $m$. However, one can at least formally perform the thermodynamic and chiral limits and parameterize the unknown nonperturbative information in terms of the eigenvalue density of the Dirac operator near the origin. This gives rise to an important relation, the so-called *Banks–Casher relation* [40], which we now derive.

The starting point of our derivation is (7.66). In a first step we express the trace as a sum over all eigenvalues $\lambda_i$ of $D$:

$$\text{tr}\left[(\omega\, D + m\,\mathbb{1})^{-1} - \frac{a}{2}\mathbb{1}\right] = \sum_{\lambda_i}\left(\frac{1}{\omega\,\lambda_i + m} - \frac{a}{2}\right)$$

$$= (n_+ + n_-)\left(\frac{1}{m} - \frac{a}{2}\right) + (n'_+ + n'_-)\left(\frac{1}{2\omega/a + m} - \frac{a}{2}\right)$$

$$+ \sum_{\lambda_i \neq 0, 2/a}\left(\frac{1}{\omega\,\lambda_i + m} - \frac{a}{2}\right)\,, \tag{7.71}$$

where in the second line we have split the sum over all eigenvalues in the contributions of the $n_+ + n_-$ eigenvalues 0, the $n'_+ + n'_-$ eigenvalues $2/a$, and a sum over the complex eigenvalues. Recalling definition (7.42) for $\omega$ one finds that the factor for the contribution of the eigenvalues at 0 reduces to $\omega/m$, whereas the contribution of the eigenvalues at $2/a$ vanishes. Expressing the eigenvalues $\lambda_i$ through the phases $\varphi_i$ introduced in (7.51) and combining the complex conjugate pairs of eigenvalues the last expression turns into

$$\frac{\omega}{m}(n_+ + n_-) + \frac{\omega}{2}\sum_{\varphi_i \neq 0, \pi}\frac{(1 + \cos(\varphi_i))\, m}{(2 - 2\cos(\varphi_i))\,(a\, m + \omega)\,\omega/a^2 + m^2}\,. \tag{7.72}$$

Thanks to the above discussed "absence of fine tuning" for a given gauge configuration only one of the numbers $n_+, n_-$ can be nonvanishing and we may replace $(n_+ + n_-) \to |Q_{\text{top}}|$, the number of zero modes. Inserting (7.72) for the trace in (7.66) one arrives at

$$\lim_{|\Lambda|\to\infty}\Sigma^{\text{lat}}(a, m, |\Lambda|) = \lim_{|\Lambda|\to\infty}\frac{1}{a^4|\Lambda|}\frac{\langle|Q_{\text{top}}|\rangle_G}{m} \tag{7.73}$$

$$+ \frac{1}{2\,a^4}\int_{-\pi}^{\pi}d\varphi\,\rho_A(\varphi)\frac{(1 + \cos(\varphi))\, m}{(2 - 2\cos(\varphi))\,(1 - a^2\, m^2/4)/a^2 + m^2}\,,$$

where we have defined the angular density $\rho_A(\varphi)$ for the angles $\varphi$ of the eigenvalues on the Ginsparg–Wilson circle as

$$\rho_A(\varphi) = \lim_{|\Lambda| \to \infty} \left\langle \frac{1}{|\Lambda|} \sum_{\varphi_i \neq 0, \pi} \delta(\varphi - \varphi_i) \right\rangle_G . \qquad (7.74)$$

In the definition of the density $\rho_A(\varphi)$ we have already performed the limit $|\Lambda| \to \infty$ (at fixed $a$) necessary for the spontaneous breaking of a continuous symmetry. In this limit the eigenvalues become dense on the Ginsparg–Wilson circle and $\rho_A(\varphi)$ as defined in (7.74) indeed becomes a density.

The first contribution (7.73) vanishes for $|\Lambda| \to \infty$, since $\langle |Q_{\text{top}}| \rangle_G$ does not grow faster than $\sqrt{V}$. This follows from the finiteness of $\chi_{\text{top}} = \langle Q_{\text{top}}^2 \rangle / V$. For the second term in (7.73) one can utilize that

$$\delta_m(X) = \frac{1}{\pi} \frac{m}{X^2 (1 + \mathcal{O}(m^2)) + m^2} \qquad (7.75)$$

is a $\delta$-sequence, i.e., $\delta_m(X) \to \delta(X)$ for $m \to 0$ (see e.g. [41]). Thus we obtain for the second part of (7.73), which remains finite for $m \to 0$:

$$\frac{\pi}{2 a^4} \int_{-\pi}^{\pi} d\varphi \ (1 + \cos(\varphi)) \ \delta\left(a^{-1} \sqrt{2 - 2 \cos(\varphi)}\right) \rho_A(\varphi)$$

$$= \frac{\pi}{2 a^3} \int_{-\pi}^{\pi} d\varphi \ (1 + \cos(\varphi)) \ \delta(\varphi) \rho_A(\varphi) = \frac{\pi}{a^3} \rho_A(0) \ , \ (7.76)$$

where we have used standard manipulations for the Dirac-delta with a function as argument. Equation (7.76) shows that after taking the infinite volume limit the chiral condensate is proportional to the angular density $\rho_A(\varphi)$ at $\varphi = 0$. We emphasize that in the definition of the density the exact zero eigenvalues are left out.

Usually one does not use the angular density $\rho_A$, but the density $\rho_\lambda$ of eigenvalues on the imaginary axis is $\rho_\lambda = \Delta n / \Delta y$, where $\Delta n$ is the number of eigenvalues per interval $\Delta y$ on the imaginary axis. For small angles $\varphi$ one has $\Delta y = \Delta \varphi / a$ (compare Fig. 7.1), implying that $\rho_\lambda = a \rho_A$. Thus we conclude

$$\Sigma^{\text{lat}}(a) \equiv \lim_{m \to 0} \lim_{|\Lambda| \to \infty} \Sigma^{\text{lat}}(a, m, |\Lambda|) = \frac{\pi}{a^3} \rho_A(0) = \frac{\pi}{a^4} \rho_\lambda(0) = \pi \rho(0) ,$$
$$(7.77)$$

where $\rho$ is the density of eigenvalues on the imaginary axis per unit volume.

While the numerical calculation presented in the last section shows the existence of a nonvanishing chiral condensate from first principles, it cannot provide insight about an underlying physical mechanism for chiral symmetry breaking. The Banks–Casher relation (7.77), however, has at least opened the door to understanding such a mechanism.

A very influential idea is that the chiral condensate is formed through a "fluid" of weakly interacting topological field configurations such as instantons (see [17, 18] for reviews). Often, as in the case of instantons, the topological configurations are localized and such structures have been nicknamed *topological lumps*. A gauge field configuration which carries topological charge gives rise to zero eigenvalues according to the index theorem (7.56). For a mixture of topological lumps with different charges one cannot assume that the zero eigenvalues survive unperturbed. Instead one expects that the eigenvalues move in the imaginary direction. A fluid of weakly interacting topological lumps gives rise to an accumulation of eigenvalues on the imaginary axis thus building up the density of eigenvalues $\rho(0)$ near the origin. This in turn gives rise to the nonvanishing chiral condensate via the Banks–Casher relation (7.77). This mechanism and the possible structure of the topological lumps have been studied extensively on the lattice and the reviews [42–46] may serve as a guide to further reading.

The density of eigenvalues can be studied in some detail within random matrix theory, which we addressed briefly in the preceding section. Fine details like distribution densities of the smallest eigenvalue, the next-to-smallest eigenvalue, and so on appear to be universal properties of the spectra of random matrices with the symmetry properties of the Dirac operator. There is a wealth of tools to analyze spectra in this way and thereby disentangle "universal" features from properties specific for the dynamics of the theory, such as the condensate value. More about these issues can be found in [47, 48].

## 7.4 The overlap operator

Up to now our presentation of chiral symmetry on the lattice did not refer to a particular solution of the Ginsparg–Wilson equation. In this section we close this gap and introduce the *overlap operator*. We analyze the locality properties of the overlap operator and discuss its numerical evaluation. A second solution of the Ginsparg–Wilson equation, the *fixed point Dirac operator*, will be presented in Chap. 9. Furthermore, in Sect. 10.2 we discuss domain wall fermions which are closely related to the overlap formulation.

### 7.4.1 Definition of the overlap operator

Originating from the initial papers on the overlap formulation [49–52], Neuberger presented the modern form of the overlap Dirac operator in [53] and showed in [11] that it is a solution of the Ginsparg–Wilson equation (7.29). Explicitly the overlap Dirac operator is given by

$$D_{\mathrm{ov}} = \frac{1}{a}\left(\mathbb{1} + \gamma_5\,\mathrm{sign}[H]\right)\,, \quad H = \gamma_5\,A\,, \tag{7.78}$$

where $A$ denotes some suitable $\gamma_5$-hermitian "kernel" Dirac operator, thus $H$ is hermitian with real eigenvalues. The operator sign function may be defined through the spectral representation of the operator (A.51). Alternatively one may write the overlap operator in the form

$$D_{\mathrm{ov}} = \frac{1}{a}\left(\mathbb{1} + \gamma_5 H \left(H^2\right)^{-1/2}\right) . \qquad (7.79)$$

The simplest choice is to use the Wilson Dirac operator for the kernel,

$$A = a\, D_W - \mathbb{1}(1+s) \,, \quad \text{such that} \quad D_{\mathrm{ov}} = \frac{1}{a}\left(\mathbb{1} + A \left(\gamma_5 A \gamma_5 A\right)^{-1/2}\right), \quad (7.80)$$

where $s$ is a real parameter with $|s| < 1$ which can be used to optimize locality (see below). $D_W$ is the massless Wilson Dirac operator (compare (5.51))

$$D_W(n|m)_{\substack{\alpha\,\beta \\ a\,b}} = \frac{4}{a}\delta_{\alpha\beta}\,\delta_{ab}\,\delta_{n,m} - \frac{1}{2a}\sum_{\mu=\pm1}^{\pm4}(\mathbb{1}-\gamma_\mu)_{\alpha\beta}\,U_\mu(n)_{ab}\,\delta_{n+\hat\mu,m} . \quad (7.81)$$

Since the Wilson Dirac operator is $\gamma_5$-hermitian (compare (5.76)) so is $A$, i.e., $\gamma_5 A \gamma_5 = A^\dagger$. Thus we have $\gamma_5 A \gamma_5 A = A^\dagger A$ and the square root in (7.80) is well defined through the spectral theorem.

In order to see that the overlap operator obeys the Ginsparg–Wilson equation we insert $D_{\mathrm{ov}}$ into (7.29). With the representation (7.78) we find

$$\begin{aligned}
a\, D_{\mathrm{ov}}\, D_{\mathrm{ov}}{}^\dagger &= \frac{1}{a}\left(\mathbb{1} + \gamma_5\,\mathrm{sign}[H]\right)\left(\mathbb{1} + \mathrm{sign}[H]\,\gamma_5\right) \\
&= \frac{1}{a}\left(\mathbb{1} + \gamma_5\,\mathrm{sign}[H] + \mathrm{sign}[H]\,\gamma_5 + \mathbb{1}\right) = D_{\mathrm{ov}} + D_{\mathrm{ov}}{}^\dagger . \quad (7.82)
\end{aligned}$$

We have taken into account hermiticity, $\mathrm{sign}\,H = \mathrm{sign}\,H^\dagger$, and the squared sign function as the unit operator: $(\mathrm{sign}\,H)^2 = \mathbb{1}$. Similar to the steps (2.40), (2.41), and (2.42) performed for the naive Dirac operator, one can expand the overlap operator (7.80) for small $a$ and show that it approaches the Dirac operator of the continuum (up to an irrelevant multiplicative constant: $\tilde{D}_{\mathrm{ov}} \approx i\gamma_\mu p_\mu/(1+s) + \mathcal{O}(p^2)$).

The Wilson operator $D_W$ used in the construction of the overlap operator can be replaced by some other lattice Dirac operator which is free of doublers [54]. This can improve the locality properties of the overlap operator (see next section) and it may speed up the convergence in the numerical evaluation of the overlap operator (Sect. 7.4.3) [38]. In this sense the overlap construction may also be viewed as a projection of a non-chiral lattice Dirac operator onto a solution of the Ginsparg–Wilson equation.

### 7.4.2 Locality properties of chiral Dirac operators

While the Wilson Dirac operator (7.81) and thus $A$ involve only nearest neighbor terms and therefore are sparse matrices, the inverse square root $(\gamma_5 A \gamma_5 A)^{-1/2}$ leads to a matrix that has nonvanishing entries for all pairs $n, m$ of lattice sites. Thus $D_{\mathrm{ov}}(n|m) \neq 0$ for all $n, m$ and the overlap operator is not an *ultralocal* operator. General arguments that this is true for all solutions of the Ginsparg–Wilson equation were given in [55].

However, locality is an essential concept for quantum field theories and its violation is punished by noncausal interactions. Thus one has to understand in which sense locality has to be manifest for a lattice field theory. A natural requirement is that the Dirac operator $D$ falls off exponentially independent of $\beta$, i.e., $D$ obeys the bound

$$\left| D(n|m)_{\substack{\alpha\beta \\ ac}} \right| \le C \exp\left(-\gamma \parallel n - m \parallel\right) , \tag{7.83}$$

where the constants $C$ and $\gamma$ are independent of the gauge field. If such a bound holds, then the interaction range $1/\gamma$ is a fixed distance in lattice units. However, when expressed in physical units, this distance decreases as we let $a \to 0$ in order to approach the continuum limit. Thus, the interaction range in physical units, $a/\gamma$, shrinks to zero and in the continuum limit we recover a local field theory. For the overlap operator locality in the sense of (7.83) was established in [56], both with analytical arguments and a numerical investigation. There it was also demonstrated that the parameter $s$ in (7.80) can be tuned to optimize locality.

### 7.4.3 Numerical evaluation of the overlap operator

Numerically the main problem is to compute the sign function of the operator $H$. Formally it is well defined through the spectral theorem:

$$\text{sign}[H] = \text{sign}\left[\sum_i \lambda_i |i\rangle\langle i|\right] = \sum_i \text{sign}(\lambda_i)|i\rangle\langle i| . \tag{7.84}$$

Only for exactly vanishing eigenvalues of $H$, a numerically highly improbable case, one introduces a "tie-breaker," e.g., a randomly chosen $\pm 1$ replacing sign $\lambda$. However, for the huge Dirac matrices exact complete diagonalization is prohibitively expensive and therefore methods based on (7.84) are not applied except for some few test cases.

In most computations one utilizes instead

$$\text{sign}[H] = H\,|H|^{-1} = H\left(H^2\right)^{-\frac{1}{2}} \tag{7.85}$$

and approximates $\left(H^2\right)^{-1/2}$ by either a polynomial in $H^2$ or a ratio of polynomials. The convergence of such an approximations depends on the actual matrix $H$, in particular on its eigenvalues, and therefore also on the current gauge field configuration. Smaller eigenvalues of $H$ will lead to worse convergence in general. As discussed in Sect. 6.2.5 the fluctuations of the real eigenvalues of the Wilson Dirac operator are nonnegligible. A real eigenvalue of $D_W$ in the vicinity of $(1 + s)/a$ will give rise to a small eigenvalue of $H$ leading to numerical problems. The situation can be improved by applying smearing techniques (compare Sect. 6.2.6) or by using for the overlap projection a different kernel operator which already approximates a solution of the Ginsparg–Wilson equation.

For the numerical evaluation of the sign function we discuss here the two most popular approximation methods and the so-called *small eigenmode reduction* technique. Comparative studies are, e.g., found in [57–59].

**Polynomial approximation:** The Chebyshev polynomials $T_n(x)$ are orthogonal polynomials with regard to the scalar product

$$(f, g) \equiv \int_{-1}^{1} dx \, \frac{f(x)g(x)}{\sqrt{1 - x^2}} , \tag{7.86}$$

normalized such that $(T_n, T_m) = \delta_{nm}$. A given function $r(x)$ has the series expansion

$$r(x) = \sum_{n=0}^{\infty} c_n \, T_n(x) \quad \text{with} \quad c_n = (r, T_n) . \tag{7.87}$$

The series converges pointwise for functions which are continuous in $[-1, 1]$ up to a finite number of discontinuities in that interval. For the truncated series

$$r(x) \approx \sum_{n=0}^{N-1} c_n \, T_n(x) , \tag{7.88}$$

the error is spread smoothly over the interval $-1 \leq x \leq 1$. The coefficients for the truncated series may be approximated by

$$c_n = \frac{\pi}{N} \sum_{k=1}^{N} r(x_k) \, T_n(x_k), \quad \text{where} \quad x_k = \cos\left(\left(k - \frac{1}{2}\right)\frac{\pi}{N}\right) . \tag{7.89}$$

Following [36, 60] we apply the method to computing the inverse square root of $H^2$. It can be shown that for $|s| \leq 1$, $\|H\| \leq 8$ (in lattice units). We denote the (in magnitude) smallest and the largest eigenvalues of $H$ with $\alpha$ and $\beta$. Consequently $H^2$ has eigenvalues $\lambda$ in an interval $[\alpha^2, \beta^2] \subset [0, 64]$. The general interval $\lambda \in [\alpha^2, \beta^2]$ can be mapped into the generic domain $x \in [-1, 1]$ by

$$x = \frac{2\lambda - (\beta^2 + \alpha^2)}{\beta^2 - \alpha^2} . \tag{7.90}$$

For the inverse square root function in question we then find

$$r(x) = \frac{1}{\sqrt{\lambda(x)}} = \left(\frac{1}{2}(\beta^2 + \alpha^2) + \frac{x}{2}(\beta^2 - \alpha^2)\right)^{-\frac{1}{2}} , \tag{7.91}$$

and with that function the coefficients are computed according to (7.89). The approximation of the sign function is then obtained by multiplication with $H$:

$$\text{sign}[H] = \frac{H}{\sqrt{H^2}} = H \sum_{n=0}^{N-1} c_n \, T_n(X) + \mathcal{O}\left(\exp\left(-2N\,|\alpha/\beta|\right)\right) ,$$

$$\text{with} \quad X \equiv \frac{2H^2 - (\beta^2 + \alpha^2)\mathbb{1}}{\beta^2 - \alpha^2} . \tag{7.92}$$

The error of the approximation series decreases $\propto \exp(-2\,N\,|\alpha/\beta|)$, i.e., the number of terms necessary for a requested accuracy grows proportional to $|\beta/\alpha|$. This ratio is just the *condition number* of $H$: $|\lambda_{\max}/\lambda_{\min}|$. To improve the condition number, the method is usually combined with a "removal" of small eigenmodes of $H$, which we discuss below. The series may have several hundred terms and therefore one should use error reducing techniques like the Clenshaw recursion relation for summing it [61].

The polynomial approximation is technically simple and quite robust and may be straightforwardly applied to the force calculations necessary when introducing dynamical fermions (cf. Chap. 8). Theoretical arguments, however, suggest that the polynomial approximation is less efficient than the rational approximation methods and indeed it is more sensitive to the condition number. In a comparison [58] for test examples the polynomial approximation typically needed between 1.5 and 4 times more matrix–vector multiplications with the operator $H$ than a rational approximation based on the Zolotarev method.

**Partial fractions á la Zolotarev:** In [57, 62] it was suggested to approximate $1/\sqrt{x^2}$ by a ratio of polynomials. Meanwhile it has become clear [58, 59, 63] that the best such approximation for the square root on an interval $x^2 \in [1, \beta^2/\alpha^2]$ is the Zolotarev approximation [64]. In particular,

$$\frac{1}{\sqrt{x^2}} \approx Q(x^2) = d \prod_{n=1}^{m} \frac{\left(x^2 + c_{2n}\right)}{\left(x^2 + c_{2n-1}\right)} \ , \quad c_n = \frac{\mathrm{sn}^2\left(n\,K(k')/(2m+1);\,k'\right)}{1 - \mathrm{sn}^2\left(n\,K(k')/(2m+1);\,k'\right)} \ ,$$
$$(7.93)$$

with $k' = \sqrt{1 - \alpha/\beta}$ and $K(k')$ is the so-called *complete elliptic integral* and $\mathrm{sn}(u;\,k)$ denotes a *Jacobi elliptic function* [65].

The normalization constant $d$ may be fixed by requiring $Q(1) = 1$. This results in an approximation to the sign function that is bounded by 1 from below (on the positive branch). Requesting instead that the approximation fluctuates symmetrically around 1 corresponds to setting

$$Q(1) + \left|\frac{\beta}{\alpha}\right| Q\left(\frac{\beta^2}{\alpha^2}\right) = 2 \ . \tag{7.94}$$

The rational function can be decomposed into partial fractions such that

$$Q(x^2) = d\,(x^2 + c_{2m}) \sum_{n=1}^{m} \frac{b_n}{x^2 + c_{2n-1}} \ , \tag{7.95}$$

where the parameters $b_n$ may be computed from (7.93) giving

$$b_n = \frac{\prod_{k=1}^{m-1}\left(c_{2k} - c_{2n-1}\right)}{\prod_{k=1, k\neq n}^{m}\left(c_{2k-1} - c_{2n-1}\right)} \ . \tag{7.96}$$

In order to apply this to the matrix $H$ we have to scale it by dividing by the absolute value of its smallest eigenvalue $|\alpha|$. The final formula for the sign function, obtained by multiplication with $H$, thus reads

$$\text{sign}[H] \approx d \, \frac{H}{|\alpha|} \left( \frac{H^2}{\alpha^2} + c_{2m} \right) \sum_{n=1}^{m} b_n \left( \frac{H^2}{\alpha^2} + c_{2n-1} \right)^{-1} . \tag{7.97}$$

Here $\alpha$ and $\beta$ denote the smallest and the largest eigenvalues of $H$ and consequently the matrix $(H/\alpha)^2$ has its spectrum in $[1, \beta^2/\alpha^2]$, as assumed for expansion (7.93).

Since in an application of (7.97) one always wants to compute the product of the overlap matrix with some vector, one may compute the terms of (7.97) simultaneously by a multi-shift conjugate gradient solver [66–68], as discussed already in Chap. 6 in the context of computing the quark propagators for several masses at once. The necessary overhead is small such that the computational effort is essentially governed by the smallest $|c_{2n-1}|$.

Compared to other rational approximations, for a given accuracy, the Zolotarev method drastically reduces the number of necessary terms in (7.97).

**Small eigenmode reduction:** In particular the polynomial approximation is quite sensitive to the condition number of $H$. For $|s| < 1$ the eigenvalues of $H$ are bounded from above, $|\lambda_{\max}| \leq 8$ (lattice units), and therefore the true source of trouble is the smallest eigenvalue, as discussed earlier. There is a method, however, to deal with cases with very small $|\lambda_{\min}|$. We know from the spectral theorem (see Appendix A.5) that

$$\begin{aligned} f[H] = \sum_i f(\lambda_i) \, v_i \, v_i^\dagger &= \sum_{\lambda_i < \lambda_c} f(\lambda_i) \, v_i \, v_i^\dagger + \sum_{\lambda_i \geq \lambda_c} f(\lambda_i) \, v_i \, v_i^\dagger \\ &\equiv \sum_{\lambda_i < \lambda_c} f(\lambda_i) \, v_i \, v_i^\dagger + f[H^{(\text{red})}] . \end{aligned} \tag{7.98}$$

Here $v_i$ denotes the eigenvector for eigenvalue $\lambda_i$. We may therefore split the problem into two parts. First we determine all eigenvalues below some value $\lambda_c$ as well as their eigenvectors. Then we define a reduced matrix by removing the small eigenvalue sector explicitly:

$$H^{(\text{red})} \equiv H - \sum_{\lambda_i < \lambda_c} \lambda_i \, v_i \, v_i^\dagger . \tag{7.99}$$

We compute $\text{sign} \, H^{(\text{red})}$ for this reduced operator, which has improved condition number, by some approximation scheme and invoke (7.98) to reconstruct

$$\text{sign}[H] = \text{sign}[H^{(\text{red})}] + \sum_{\lambda_i < \lambda_c} \text{sign}(\lambda_i) \, v_i v_i^\dagger . \tag{7.100}$$

The overhead due to the necessity to compute the low-eigenvalue sector is not as disastrous as it may seem. This has to be done only once for a given gauge configuration. The multiplication of the overlap Dirac operator (for that gauge configuration) with a vector, however, has to be computed quite often, as it is usually embedded in, e.g., a conjugate gradient solver used for calculating the quark propagator. The effort of computing the, say, lowest 20 eigenvalues is therefore in most situations small in comparison to the total cost.

# References

1. J. S. Bell and R. Jackiw: Nuovo Cimento **60A**, 47 (1969)
2. S. L. Adler: Phys. Rev. **177**, 2426 (1969)
3. C. Itzykson and J.-B. Zuber: *Quantum Field Theory* (McGraw-Hill, New York 1985)
4. M. E. Peskin and D. V. Schroeder: *An Introduction to Quantum Field Theory* (Addison-Wesley, Reading, Massachusetts 1995)
5. H. B. Nielsen and M. Ninomiya: Phys. Lett. B **105**, 219 (1981)
6. H. B. Nielsen and M. Ninomiya: Nucl. Phys. B **185**, 20 (1981)
7. H. B. Nielsen and M. Ninomiya: Nucl. Phys. B **193**, 173 (1981)
8. L. H. Karsten: Phys. Lett. B **192**, 315 (1981)
9. P. H. Ginsparg and K. G. Wilson: Phys. Rev. D **25**, 2649 (1982)
10. P. Hasenfratz: Nucl. Phys. B (Proc. Suppl.) **63A–C**, 53 (1998)
11. H. Neuberger: Phys. Lett. B **427**, 353 (1998)
12. M. Lüscher: Phys. Lett. B **428**, 342 (1998)
13. S. Chandrasekharan: Phys. Rev. D **60**, 074503 (1999)
14. M. Atiyah and I. M. Singer: Ann. Math. **93**, 139 (1971)
15. P. Hasenfratz, V. Laliena, and F. Niedermayer: Phys. Lett. B **427**, 125 (1998)
16. A. A. Belavin, A. M. Polyakov, A. S. Schwartz, and Y. S. Tyupkin: Phys. Lett. B **59**, 85 (1975)
17. T. Schäfer and E. Shuryak: Rev. Mod. Phys. **70**, 323 (1998)
18. D. Diakonov: Prog. Part. Nucl. Phys. **51**, 173 (2003)
19. R. Rajaraman: *Solitons and Instantons* (North Holland Publishers, Amsterdam 1982)
20. Y. Kikukawa and A. Yamada: Phys. Lett. B **448**, 265 (1999)
21. K. Fujikawa: Nucl. Phys. B **546**, 480 (1999)
22. H. Leutwyler and A. Smilga: Phys. Rev. D **46**, 5607 (1992)
23. C. Gattringer, R. Hoffmann, and S. Schaefer: Phys. Lett. B **535**, 358 (2002)
24. L. Del Debbio and C. Pica: JHEP **0402**, 003 (2004)
25. L. Del Debbio, L. Giusti, and C. Pica: Phys. Rev. Lett. **94**, 032003 (2005)
26. E. Witten: Nucl. Phys. B **156**, 269 (1979)
27. G. Veneziano: Nucl. Phys. B **159**, 213 (1979)
28. G. Veneziano: Phys. Lett. B **95**, 90 (1980)
29. L. Giusti, G. Rossi, M. Testa, and G. Veneziano: Nucl. Phys. B **628**, 234 (2002)
30. E. Seiler: Phys. Lett. B **525**, 355 (2002)
31. L. Giusti, G. Rossi, and M. Testa: Phys. Lett. B **587**, 157 (2004)
32. M. Lüscher: Phys. Lett. B **593**, 296 (2004)
33. M. Aguado and E. Seiler: Phys. Rev. D **72**, 094502 (2005)
34. J. Osborn, D. Toublan, and J. Verbaarschot: Nucl. Phys. B **540**, 317 (1999)
35. P. H. Damgaard, J. C. Osborn, D. Toublan, and J. J. Verbaarschot: Nucl. Phys. B **547**, 305 (1999)
36. P. Hernández, K. Jansen, and L. Lellouch: Phys. Lett. B **469**, 198 (1999)
37. P. Hasenfratz et al.: Nucl. Phys. B **643**, 280 (2002)
38. T. DeGrand: Phys. Rev. D **63**, 034503 (2000)
39. L. Giusti, C. Hoelbling, and C. Rebbi: Phys. Rev. D **64**, 114508 (2001)
40. T. Banks and A. Casher: Nucl. Phys. B **169**, 103 (1980)
41. F. Constantinescu: *Distributions and Their Applications in Physics* (Pergamon Press, Oxford 1980)

42. F. Bruckmann: PoS **LATTICE2007**, 006 (2007)
43. F. Bruckmann: in *Conceptual and Numerical Challenges in Femto- and Peta-Scale Physics*, edited by C. Gattringer et al., The European Physical Journal ST 152, p. 61 (Springer, Berlin, Heidelberg, New York 2007)
44. M. Engelhardt: Nucl. Phys. B (Proc. Suppl.) **140**, 92 (2005)
45. M. Garcia Pérez: Nucl. Phys. B (Proc. Suppl.) **94**, 27 (2001)
46. P. van Baal: Nucl. Phys. B (Proc. Suppl.) **63A–C**, 126 (1998)
47. P. H. Damgaard: Nucl. Phys. B (Proc. Suppl.) **106**, 29 (2002)
48. J. J. M. Verbaarschot and T. Wettig: Ann. Rev. Nucl. Part. Sci. **50**, 343 (2000)
49. R. Narayanan and H. Neuberger: Phys. Lett. B **302**, 62 (1993)
50. R. Narayanan and H. Neuberger: Phys. Rev. Lett. **71**, 3251 (1993)
51. R. Narayanan and H. Neuberger: Nucl. Phys. B **412**, 574 (1994)
52. R. Narayanan and H. Neuberger: Nucl. Phys. B **443**, 305 (1995)
53. H. Neuberger: Phys. Lett. B **417**, 141 (1998)
54. W. Bietenholz: Eur. Phys. J. C **6**, 537 (1999)
55. I. Horváth: Phys. Rev. Lett. **81**, 4063 (1998)
56. P. Hernández, K. Jansen, and M. Lüscher: Nucl. Phys. B **552**, 363 (1999)
57. R. G. Edwards, U. Heller, and R. Narayanan: Nucl. Phys. B **540**, 457 (1999)
58. J. van den Eshof et al.: Comput. Phys. Commun. **146**, 203 (2002)
59. A. D. Kennedy: Nucl. Phys. B (Proc. Suppl.) **128C**, 107 (2004)
60. P. Hernández, K. Jansen, and L. Lellouch: Nucl. Phys. B (Proc. Suppl.) **83**, 633 (2000)
61. W. H. Press, S. A. Teukolsky, W. T. Vetterling, and B. P. Flannery: *Numerical Recipes in C*, 2nd ed. (Cambridge University Press, Cambridge, New York 1999)
62. H. Neuberger: Phys. Rev. Lett. **81**, 4060 (1998)
63. T.-W. Chiu, T.-H. Hsieh, C.-H. Huang, and T.-R. Huang: Phys. Rev. D **66**, 114502 (2002)
64. E. I. Zolotarev: Zap. Imp. Akad. Nauk St. Petersburg **30**, 5 (1877)
65. I. S. Gradshteyn and I. M. Ryzhik: *Table of Integrals, Series, and Products*, 5th ed. (Academic Press, San Diego 1994)
66. A. Frommer et al.: Int. J. Mod. Phys. C **6**, 627 (1995)
67. B. Jegerlehner (unpublished) arXiv:hep-lat/9612014 (1996)
68. B. Jegerlehner: Nucl. Phys. B (Proc. Suppl.) **63**, 958 (1998)

# 8

# Dynamical fermions

QCD is the quantum field theory of gluons *and* quarks. In previous chapters we have found that for gluons the inclusion of the full dynamical behavior in Monte Carlo simulations is straightforward, but not so for quarks. In this chapter we discuss the still missing feature: dynamical quarks.

A calculation with dynamical fermions is a much more challenging enterprise than a quenched calculation, and new algorithmic ideas are needed for the inclusion of the fermion determinant. In this chapter we present some of these algorithms and techniques. Subsequently we address the coupling-mass phase diagram and discuss some results of dynamical QCD calculations.

## 8.1 The many faces of the fermion determinant

In the path integral, whenever the fermions occur bilinearly in the action, the Grassmann integral can be evaluated in closed form and leads to the fermion determinant as a factor. As an example we quote the expression for the two-point function of a flavor triplet meson from Sect. 6.1.5,

$$\langle O_T(n)\,\overline{O}_T(m)\rangle = -\frac{1}{Z}\int \mathcal{D}[U]\,\mathrm{e}^{-S_G[U]}\,\det[D_u]\det[D_d]$$
$$\times \mathrm{tr}\left[\Gamma D_u^{-1}(n|m)\,\Gamma D_d^{-1}(m|n)\right]\,,$$
$$Z = \int \mathcal{D}[U]\,\mathrm{e}^{-S_G[U]}\,\det[D_u]\det[D_d]\,. \qquad (8.1)$$

The meson interpolators are converted into products of quark propagators and each fermion flavor accounts for one fermion determinant factor. In our example we take into account the two lightest quarks $u$ and $d$. In the quenched approximation, discussed in Sect. 6.1.6, the determinants were put to unity by hand. This amounts to neglecting vacuum loops of quarks (see Fig. 6.2).

Here we now take into account the effect of the fermions. We stress that the fermion determinant is a functional of the gauge field and has to be

Gattringer, C., Lang, C.B.: *Dynamical Fermions.* Lect. Notes Phys. **788**, 185–211 (2010)
DOI 10.1007/978-3-642-01850-3_8          © Springer-Verlag Berlin Heidelberg 2010

computed anew for every gauge configuration. Even for lattices of moderate size the Dirac operator is a huge matrix with $N = 12|\Lambda|$ rows and columns, where 12 is the product of color and Dirac entries and $|\Lambda|$ the total number of lattice points. This number easily becomes larger than a million. The computational cost of calculating even only one such determinant in closed form, which formally has $N!$ contributing terms, is prohibitively high.

### 8.1.1 The fermion determinant as observable

Let us forget for a moment the technical problem of high computational cost. For small lattice size and a few configurations a direct determination is certainly possible. In this case one can try to treat the determinant as part of an observable and rewrite (8.1) as

$$\langle O\,[U, \psi, \overline{\psi}]\rangle = \frac{\langle\, \det[D_u]\,\det[D_d]\, O[U]\,\rangle_G}{\langle\, \det[D_u]\,\det[D_d]\,\rangle_G}\,. \tag{8.2}$$

The expectation value $\langle\ldots\rangle_G$ is computed with the path integral for pure gauge theory,

$$\langle A\rangle_G = \frac{1}{Z_G}\int \mathcal{D}[U]\,\mathrm{e}^{-S_G[U]}A[U]\,,\quad Z_G = \int \mathcal{D}[U]\,\mathrm{e}^{-S_G[U]}\,. \tag{8.3}$$

Although the idea of treating the determinant as an observable is appealing, it has a serious flaw: Depending on the gauge configuration the determinant may have widely different values, typically covering several orders of magnitude. In the sum over all configurations this leads to large fluctuations around the mean value and thus to an intrinsic instability. As a matter of fact, the distribution of the gauge configurations according to (8.3) is very different from that obtained when also the determinant is taken into account in their Monte Carlo generation.

For that reason, treating the determinant as an observable is numerically justified only for extremely large statistics. Only in lower dimensions (e.g., for the 2D Schwinger model) such an approach has led to acceptable results.

### 8.1.2 The fermion determinant as a weight factor

In order to obtain properly distributed gauge configurations, one tries to include the determinant as a probability weight factor when generating the Markov chain of gauge configurations. Thus the gauge fields $U$ are distributed according to the joint distribution (for two dynamical quark flavors)

$$\frac{1}{Z}\,\mathrm{e}^{-S_G[U]}\,\det[D_u]\,\det[D_d]\,. \tag{8.4}$$

Once the gauge configurations have been generated according to this distribution, the observables are computed on these gauge configurations as discussed in Chaps. 4 and 6.

There is, however, a potential problem: If we want to interpret the contribution of the fermion determinants as a factor in a probability weight it must be real and nonnegative. In most cases one can use the $\gamma_5$-hermiticity of $D$ (see (5.76)), i.e., $\gamma_5 D \gamma_5 = D^\dagger$, to show that the determinant is real:

$$\det[D]^* = \det[D^\dagger] = \det[\gamma_5 \, D \, \gamma_5] = \det[D] \,. \tag{8.5}$$

However, the determinant could still be real but negative. A possible physically justified (see (7.21)) solution is to assume that $u$ and $d$ quarks are mass degenerate and thus $D_u = D_d \equiv D$. More generally, for an even number of mass-degenerate quarks the fermion determinant is raised to an even power and the combined weight factor is nonnegative. No principal obstacle for an interpretation as probability weight is encountered then. For the case of two degenerate flavors we can use (8.5) and write

$$0 \leq \det[D] \, \det[D] = \det[D] \, \det[D^\dagger] = \det[D \, D^\dagger] \,. \tag{8.6}$$

In the last step we have rewritten the product of the two determinants for the two flavors into a single determinant of the hermitian matrix $D \, D^\dagger$, a form which will be useful later.

If in addition to $\gamma_5$-hermiticity the Dirac operator is also chiral (for vanishing quark mass), it obeys the Ginsparg–Wilson equation in the form (7.48). The spectrum for the massless case is then restricted to the Ginsparg–Wilson circle shown in Fig. 7.1. Since the eigenvalues come in complex conjugate pairs (cf. (7.45)), the determinant is real and nonnegative. If one introduces a mass term, the determinant is even strictly positive. Thus for a chiral $\gamma_5$-hermitian Dirac operator also odd powers of the fermion determinant give rise to a proper probability weight.

In special but important cases, e.g., when a chemical potential or a $\theta$-angle are introduced, $\gamma_5$-hermiticity does not hold and the determinant may be even complex. These cases need to be treated with different methods (see Chap. 12).

### 8.1.3 Pseudofermions

The central idea in introducing dynamical fermions in the Monte Carlo sampling of gauge fields is based on the analogy between fermionic and bosonic Gaussian integrals. Consider the vector $\phi = \phi_R + i \phi_I$ of $N$ complex variables and a matrix $A$ with eigenvalues $\lambda$ which all have a positive real part $\text{Re}\lambda > 0$. Then a Gaussian integral equivalent to (5.32) may be evaluated giving the generating functional

$$W[\chi, \chi^\dagger] \equiv \int_{\mathbb{R}^{2N}} \prod_{i=1}^{N} (d\phi_{R,i} \, d\phi_{I,i}) \, \exp\left(-\sum_{i,j=1}^{N} \phi_i^\dagger A_{ij} \phi_j + \sum_{i=1}^{N} \phi_i^\dagger \chi_i + \sum_{i=1}^{N} \chi_i^\dagger \phi_i\right)$$

$$= \frac{\pi^N}{|\det[A]|} \, \exp\left(\sum_{i,j=1}^{N} \chi_i^\dagger (A^{-1})_{ij} \chi_j\right) \,. \tag{8.7}$$

Up to an irrelevant factor $\pi^N$, the result agrees with the corresponding Grassmann integral, except that the determinant is now in the denominator. For simplicity we again introduce the abbreviation $\int \mathcal{D}[\phi_R]\mathcal{D}[\phi_I] \equiv \int \prod_i \mathrm{d}\phi_{R,i} \, \mathrm{d}\phi_{I,i}$.

This correspondence has led to the introduction of the so-called *pseudofermions*. These are bosons with the same number of degrees of freedom as the fermionic variables. One has used them in two different ways:

- With $A \equiv (D\,D^\dagger)^{-1}$ one may express the fermion determinant as Gaussian integral of a bosonic field. Used in that context we call the variables *pseudofermion fields* (see below).
- With $A \equiv (D\,D^\dagger)$ one may utilize the Gaussian integral in order to estimate matrix elements like $D^{-1}(n|m)$ or sums like $\mathrm{tr}[D^{-1}]$. This Monte Carlo sum introduces statistical noise and therefore we call these variables *noisy pseudofermions* used to construct *noisy pseudofermion estimators*. These techniques will be discussed in Sect. 8.4.

For a $D$ with nonvanishing eigenvalues we have $\det[D] = 1/\det[D^{-1}]$. Thus one may further utilize the boson-fermion analogy. In [1] the pseudofermion fields were introduced by observing that from (8.7) one obtains

$$\det[D\,D^\dagger] = \pi^{-N} \int \mathcal{D}[\phi_R]\mathcal{D}[\phi_I] \, \mathrm{e}^{-\phi^\dagger (D\,D^\dagger)^{-1}\phi} , \tag{8.8}$$

where we suppress color, Dirac, flavor, and lattice indices and use vector/matrix notation for $\phi$ and $D$. One uses this relation to replace the integral over the fermionic Grassmann variables for two mass-degenerate flavors by an integral over bosonic variables,

$$\int \mathcal{D}[\psi]\,\mathcal{D}[\overline{\psi}] \, \mathrm{e}^{-\overline{\psi}_u D\psi_u - \overline{\psi}_d D\psi_d} = \pi^{-N} \int \mathcal{D}[\phi_R]\mathcal{D}[\phi_I] \, \mathrm{e}^{-\phi^\dagger (D\,D^\dagger)^{-1}\phi} . \tag{8.9}$$

In this formulation the number of fermions of a given mass has to be even in order to guarantee positivity which is necessary for the convergence of the Gaussian integrals. The effective bosonic interaction contains the inverse matrix $(D\,D^\dagger)^{-1}$ and therefore is highly nonlocal.

### 8.1.4 Effective fermion action

A useful way of thinking about the fermion determinant is to interpret it as an additional contribution to the gauge action, the so-called *effective fermion action*. With (A.54) we write the determinant in the form of an exponential and obtain

$$\det[D] = \exp\left(\mathrm{tr}\left[\ln(D)\right]\right) \equiv \exp\left(-S_{\mathrm{F}}^{\mathrm{eff}}\right) ,$$
$$S_{\mathrm{F}}^{\mathrm{eff}}[U] = -\mathrm{tr}\left[\ln(D)\right] . \tag{8.10}$$

Here we have assumed that $\det[D]$ is real and positive, i.e., $D$ is either a chiral $\gamma_5$-hermitian matrix plus a mass term, or we work with two flavors and $D$ is replaced by $DD^\dagger$ according to (8.6).

In contrast to the local gauge action, the effective fermion action is a very nonlocal quantity and connects essentially all gauge variables of the system with each other. Here we use the terms nonlocal and local to distinguish whether coupling is to all variables of the system or only to field variables in the neighborhood. The latter is often called *ultralocal* to distinguish it from being local only in the continuum limit, for which it is sufficient to show that the coupling to distant field variables decreases exponentially with distance (compare the discussion of the locality properties for the Dirac operator in Sect. 7.4.2).

Generating gauge configurations distributed according to the combined Boltzmann weight factor

$$\exp(-S[U]) \quad \text{with} \quad S[U] = S_G[U] + S_F^{\text{eff}}[U] \tag{8.11}$$

is a possible way to include dynamical fermions. For generating sequences of gauge configurations according to the distribution (8.11) we need the change of the total action

$$\exp\left(-S[U'] + S[U]\right) = \exp\left(-S_G[U'] + S_G[U]\right) \exp\left(-S_F^{\text{eff}}[U'] + S_F^{\text{eff}}[U]\right) \tag{8.12}$$

for the Metropolis step, where we decide whether to accept the candidate configuration $U'$ or reject it. Computing the factor from the gauge action is simple (compare Sect. 4.1). The second term, coming from the effective fermion action, requires new techniques.

### 8.1.5 First steps toward updating with fermions

Before we describe modern fermion algorithms in more detail, let us briefly discuss some general ideas for the update. Different methods differ in the way the action change is determined and in the dependence of the computer time on the volume of the lattice and the fermion mass:

- Global changes of the gauge field, accepting with the determinant as weight factor: One estimates the ratio

$$\exp\left(-\Delta S_F^{\text{eff}}\right) \equiv \exp\left(-S_F^{\text{eff}}[U'] + S_F^{\text{eff}}[U]\right) = \det\left[\frac{D[U']}{D[U]}\right] \tag{8.13}$$

  by stochastic methods as discussed in Sect. 8.4.
- Small steps, estimating $\Delta S_F^{\text{eff}}$ by an approximation essentially linear in the change of the gauge variables. Due to this approximation errors may accumulate and one has to rely on corrective measures.

Both paths have been pursued.

Updating, e.g., a single gauge link variable in a Metropolis step requires the determination of the action change. For this one has to compute $D^{-1}$ acting on a vector, usually employing a conjugate gradient algorithm (see Sect. 8.2.3). Assuming that the necessary number of iterations depends only weakly on the volume, one such inversion needs computer time proportional to the lattice volume. The acceptance may be reasonable, but this computation has to be repeated for each change of a gauge variable, and therefore the full algorithmic cost grows with $V^2$.

On the other hand, considering a global update, where all link variables are altered (or a partial-global update, cf. [2]), the change of the action will be typically proportional to the volume and the acceptance rate drops exponentially with the volume. Thus one has to decrease the step size inversely proportional to the volume, which again leads to a computational effort growing with the volume squared.

However, there are algorithms with much better performance, based on the pseudofermion field concept. The idea is to choose changes of the gauge field not completely randomly but along directions determined by the action. This introduces a bias, but there are methods to effectively undo this bias in the updating such that the detailed balance is respected. The so-called *hybrid Monte Carlo* (HMC) algorithm is such a method and will be discussed next.

## 8.2 Hybrid Monte Carlo

The updating of the gauge field variables in a distribution $P_S(U) \propto \exp(-S)$ defined through the gauge field action and the effective fermionic action has two parts. First one has to find a reasonable candidate for a change of the gauge variables. This introduces an *a priori selection probability* factor $T_0(U'|U)$ (compare Sect. 4.1.3). In a second step one has to decide whether to accept or reject the new configuration, according to an *acceptance probability* $T_A(U'|U)$. Together the two steps provide the overall transition probability

$$T(U'|U) = T_A(U'|U)\, T_0(U'|U) . \tag{8.14}$$

The detailed balance condition (4.15) for $T(U'|U)$ may be obeyed with a Metropolis accept–reject probability following (4.16).

If one considers only the gauge field action both steps are simple since the gauge action is ultralocal (cf. Chap. 4). Thus computing the change of the gauge action is cheap and one makes small changes of variables, maybe even just for one link, and then decides whether to accept or reject the change.

Applying this to the full action including fermions is computationally expensive since the determinant leads to a nonlocal $S^{\mathrm{eff}}$ and thus computing the change of the action involves all link variables, even if only a single gauge link is altered. One therefore attempts to update many variables in one step.

Doing this in a naive manner typically leads to large changes of the action and to extremely small acceptance probability and thus is not efficient. Ideally, the acceptance probability should be large and only weakly volume dependent. At the same time the autocorrelation between subsequent configurations should be as small as possible.

There are various strategies to find a new configuration with a reasonable chance to be accepted. We now discuss the differential equation approach.

### 8.2.1 Molecular dynamics leapfrog evolution

Let us first look at the basic concept: A Hamiltonian process developing in computer time (Markov time) $\tau$. In order to focus on the central idea we discuss the method first for a real scalar field $Q$. For simplicity of presentation we suppress the lattice index for this field and employ vector/matrix notation. Using the gauge variables leads to technical complications which will be addressed later. Conjugate to the bosonic field variables $Q$ we introduce real momenta $P$ and write the formula for the vacuum expectation value of some observable $O$ as

$$
\begin{aligned}
\langle O \rangle_Q &= \frac{\int \mathcal{D}[Q] \, \exp(-S[Q]) \, O[Q]}{\int \mathcal{D}[Q] \, \exp(-S[Q])} \\
&= \frac{\int \mathcal{D}[Q] \, \mathcal{D}[P] \, \exp(-\frac{1}{2}P^2 - S[Q]) \, O[Q]}{\int \mathcal{D}[Q] \, \mathcal{D}[P] \, \exp(-\frac{1}{2}P^2 - S[Q])} = \langle O \rangle_{P,Q} \, .
\end{aligned}
\tag{8.15}
$$

To give an example for our use of vector/matrix notation, we remark that $P^2$ stands for

$$
P^2 = \sum_{n \in \Lambda} P(n)^2 \, .
\tag{8.16}
$$

The second form in (8.15) is equivalent since in the expectation value the Gaussian integrals for $P$ cancel. In the large volume limit this new form represents a microcanonical ensemble of a classical (nonrelativistic) system with Hamiltonian

$$
H[Q, P] \equiv \frac{1}{2}P^2 + S[Q] \, .
\tag{8.17}
$$

The classical equations of motion (in computer time) for this system are

$$
\begin{aligned}
\dot{P} &= -\frac{\partial H}{\partial Q} = -\frac{\partial S}{\partial Q} \, , \\
\dot{Q} &= \frac{\partial H}{\partial P} = P \, .
\end{aligned}
\tag{8.18}
$$

The notation used for the derivatives is a shorthand notation for the derivative with respect to the field value at a single lattice site in accordance with our vector/matrix convention.

The Eqs. (8.18) are called molecular dynamics equations since they determine the time evolution of a classical system of particles. The Hamiltonian is a constant of motion, and the path of the configurations $(P, Q)$ lies on a hypersurface of constant energy in phase space and thus would be always accepted if the evolution of (8.18) could be done exactly.

The Eqs. (8.18) can be evolved numerically, and according to (8.15) we may extract the requested expectation values $\langle O \rangle_Q$ from the $Q$–$P$ ensemble. We stress that one also has to show that the evolution gives rise to an update which is ergodic for the fields of interest, i.e., the fields $Q$ in this case.

Using the microcanonical evolution for pure gauge theories has been suggested in [3], for fermions in [4]. Such molecular dynamics evolution equations are also closely related to stochastic differential equations like the Langevin equation, where the classical evolution is disturbed by a noise term. The random noise gives rise to the quantum fluctuations of the quantum field theory [5, 6]. Changing the conjugate momenta randomly, according to the Gaussian distribution, and pursuing the microcanonical path in between these changes led to the so-called *hybrid algorithms* [7, 8].

The numerical implementation of (8.18) introduces a discrete step size $\varepsilon \equiv \Delta\tau$ and numerical errors are unavoidable. A simple linear evolution scheme introduces errors $\mathcal{O}(\varepsilon^2)$. Either one finds a way to extrapolate the results to vanishing step size or one introduces a corrective step. This second idea leads to the hybrid Monte Carlo algorithm.

We call a sequence of small steps following the approximate molecular dynamics evolution a *trajectory*. For the HMC method one first constructs a new configuration by evolution along such a trajectory and then decides in a Metropolis step whether to accept the new configuration.

It will turn out that for the detailed balance condition (4.15) we need to have two requirements on the molecular dynamics trajectory:

- *Area preservation* of the integration measure $\mathcal{D}[Q]\,\mathcal{D}[P]$.
- *Reversibility* of the trajectory, $T_{\mathrm{md}}(P', Q'|P, Q) = T_{\mathrm{md}}(-P, Q| - P', Q')$.

Reversibility can be obtained by using the so-called *leapfrog integration scheme* (assuming perfect arithmetic precision): Whereas $Q$ is evolved in $n$ steps of length $\varepsilon$, the conjugate momenta start with a half-step $\varepsilon/2$, then $(n - 1)$ full steps and finally again a half-step. We discuss the two required properties, area preservation and reversibility, for a single such step

$$
\begin{aligned}
Q(0) &\longrightarrow & Q(\varepsilon)\,, \\
P(0) \to P(\tfrac{\varepsilon}{2}) &\to & P(\varepsilon)\,,
\end{aligned}
\tag{8.19}
$$

and longer trajectories are built by combining several such steps. The simplest Euler integration scheme corresponds to the equations

$$Q(0) \qquad\qquad P(0)$$
$$\downarrow \qquad\qquad\quad \downarrow$$
$$\downarrow \qquad\quad P\left(\tfrac{\varepsilon}{2}\right) = P(0) - \left.\frac{\partial S}{\partial Q}\right|_{Q(0)}\frac{\varepsilon}{2} \qquad (8.20)$$
$$\qquad\qquad\qquad\quad \downarrow$$
$$Q(\varepsilon) = Q(0) + P\left(\tfrac{\varepsilon}{2}\right)\varepsilon \quad P(\varepsilon) = P\left(\tfrac{\varepsilon}{2}\right) - \left.\frac{\partial S}{\partial Q}\right|_{Q(\varepsilon)}\frac{\varepsilon}{2} \ .$$

Area preservation of the integration can be shown from the product of Jacobians along this sequence[1]:

$$\det\left[\frac{\partial(P_1,Q_1)}{\partial(P_0,Q_0)}\right] = \det\left[\frac{\partial(P_1,Q_1)}{\partial(P_{1/2},Q_1)}\frac{\partial(P_{1/2},Q_1)}{\partial(P_{1/2},Q_0)}\frac{\partial(P_{1/2},Q_0)}{\partial(P_0,Q_0)}\right] \ . \qquad (8.21)$$

The right-hand side factorizes into three determinants of triangular matrices, e.g.,

$$\det\left[\frac{\partial(P_{1/2},Q_0)}{\partial(P_0,Q_0)}\right] = \det\begin{bmatrix}1 & \cdots \\ 0 & 1\end{bmatrix} = 1 \ , \qquad (8.22)$$

and similarly for the other Jacobians. Thus the total Jacobian is 1 and the integration measure is invariant.

In order to inspect reversibility let us first combine the Eqs. (8.20) starting at time 0 from some initial value $(Q_0, P_0)$. We obtain

$$Q_1 = Q_0 + P_0\,\varepsilon - \frac{1}{2}\left.\frac{\partial S}{\partial Q}\right|_{Q_0}\varepsilon^2 \ ,$$
$$P_1 = P_0 - \frac{1}{2}\left(\left.\frac{\partial S}{\partial Q}\right|_{Q_0} + \left.\frac{\partial S}{\partial Q}\right|_{Q_1}\right)\varepsilon \ . \qquad (8.23)$$

Making a step backwards corresponds to starting at $(Q_1, P_1)$ and applying the equations with a negative step size $-\varepsilon$. We find

$$Q(\varepsilon - \varepsilon) = Q_0 = Q_1 - P_1\,\varepsilon - \frac{1}{2}\left.\frac{\partial S}{\partial Q}\right|_{Q_1}\varepsilon^2$$
$$= Q_1 - P_0\,\varepsilon + \frac{1}{2}\left(\left.\frac{\partial S}{\partial Q}\right|_{Q_0} + \left.\frac{\partial S}{\partial Q}\right|_{Q_1}\right)\varepsilon^2 - \frac{1}{2}\left.\frac{\partial S}{\partial Q}\right|_{Q_1}\varepsilon^2$$
$$= Q_1 - P_0\,\varepsilon + \frac{1}{2}\left.\frac{\partial S}{\partial Q}\right|_{Q_0}\varepsilon^2 = Q_0 \ , \qquad (8.24)$$
$$P(\varepsilon - \varepsilon) = P_0 = P_1 + \left(\left.\frac{\partial S}{\partial Q}\right|_{Q_1} + \frac{1}{2}\left.\frac{\partial S}{\partial Q}\right|_{Q_0}\right)\varepsilon = P_0 \ .$$

Thus we end up where we have started and reversibility is established. In the equation time always multiplies $P$ and thus this is equivalent to stating

---

[1] For notational convenience we use $Q(n\varepsilon) \equiv Q_n$ and $P(n\varepsilon) \equiv Q_n$.

that in the evolution $T_{\mathrm{md}}(P', Q'|P, Q) = T_{\mathrm{md}}(-P, Q| - P', Q')$, as wanted. Actually, the sign of $P$ may be reversed, since $P$ enters quadratically in the distributions, and thus $T_{\mathrm{md}}(P', Q'|P, Q) = T_{\mathrm{md}}(P, Q|P', Q')$.

One may combine $n$ leapfrog steps (8.23) to build a trajectory of length $n\varepsilon \approx 1$. This amounts to inserting a sequence of full steps for both $Q$ and $P$ in (8.20). The discretization error for the half steps is $\mathcal{O}(\varepsilon^2)$, for the normal steps $\mathcal{O}(\varepsilon^3)$. The number of such steps in typical calculations is $\mathcal{O}(100)$.

### 8.2.2 Completing with an accept–reject step

Once a new configuration proposal has been obtained as the endpoint of a leapfrog molecular dynamics trajectory, it is accepted only with a probability depending on the change of the total Boltzmann factor. This is necessary since the evolution of (8.18) is not exact due to the $\mathcal{O}(\varepsilon^2)$ errors. The full HMC step thus consists of the following parts:

- Given a configuration $Q$, generate a set of conjugate momenta $P$ from a Gaussian distribution $P_G(P) \propto \exp(-P^2/2)$.
- The molecular dynamics trajectory leads from a configuration $(P, Q)$ to a configuration $(P(\tau + n\varepsilon), Q(\tau + n\varepsilon)) \equiv (P', Q')$. As discussed above, this is a deterministic process obeying

$$T_{\mathrm{md}}(P', Q'|P, Q) = T_{\mathrm{md}}(-P, Q|-P', Q') . \qquad (8.25)$$

The new configuration is accepted with probability (see (4.16))

$$T_A(P', Q'|P, Q) = \min\left(1, \frac{\exp(-H[P', Q'])}{\exp(-H[P, Q])}\right) . \qquad (8.26)$$

We now prove that this is a valid Metropolis step in the Markov sequence of configurations, i.e., we show detailed balance for the algorithm. The total probability to move from $Q$ to $Q'$ results from integrating over all $P$ and $P'$,

$$T(Q'|Q) = \int \mathcal{D}[P]\,\mathcal{D}[P']\ T_A(P', Q'|P, Q)\, T_{\mathrm{md}}(P', Q'|P, Q)\, e^{-P^2/2} . \qquad (8.27)$$

We then transform

$$
\begin{aligned}
T_A(P', Q'|P, Q) &= \min\left(1, \frac{e^{-P'^2/2 - S[Q']}}{e^{-P^2/2 - S[Q]}}\right) \\
&= e^{-P'^2/2 - S[Q'] + P^2/2 + S[Q]}\,\min\left(\frac{e^{-P^2/2 - S[Q]}}{e^{-P'^2/2 - S[Q']}}, 1\right) \\
&= e^{-P'^2/2 - S[Q'] + P^2/2 + S[Q]}\,T_A(P, Q|P', Q') \\
&= e^{-P'^2/2 - S[Q'] + P^2/2 + S[Q]}\,T_A(-P, Q|-P', Q') .
\end{aligned}
\qquad (8.28)
$$

In the last step we have used that $T_A$ is obviously even in the momentum variables. Taking into account the reversibility (8.25) of the molecular dynamics evolution, the integral (8.27) becomes

$$T(Q'|Q) =$$
$$= \int \mathcal{D}[P]\mathcal{D}[P']\, T_A(-P,Q|-P',Q')\, T_{\mathrm{md}}(-P,Q|-P',Q')\, e^{-S[Q']+S[Q]-P'^2/2}$$
$$= \int \mathcal{D}[P]\mathcal{D}[P']\, T_A(P,Q|P',Q')\, T_{\mathrm{md}}(P,Q|P',Q')\, e^{-S[Q']+S[Q]-P'^2/2}\,.$$
$$(8.29)$$

The second step is possible, since the integration measure is invariant under a sign change of all momenta. Multiplication with $\exp(-S[Q])$ and comparing with (8.27) then gives

$$e^{-S[Q]}\, T(Q'|Q) = e^{-S[Q']}\, T(Q|Q')\,. \qquad (8.30)$$

This is the detailed balance equation (4.15) for the distribution probability $\exp(-S[Q])$ and the proof is complete.

If the molecular dynamics equations were implemented numerically exactly, then $H[P',Q'] = H[P,Q]$ and the *acceptance probability* (8.26) becomes equal to 1. The Monte Carlo step therefore corrects for numerical errors in the discretization. There is more to that, however. Sometimes it may be computationally advantageous to evolve the molecular dynamics trajectory with another, e.g., simpler action $\widehat{S}$ and to accept the change using the desired action $S$ in (8.26). This is perfectly legal. It will affect the acceptance rate, though, which may become very small if the actions differ in any significant manner.

### 8.2.3 Implementing HMC for gauge fields and fermions

Let us now apply the HMC-update to QCD with two dynamical, mass-degenerate flavors of quarks. Using pseudofermions as in (8.8) for rewriting the determinant, we have a system of pseudofermion fields $\phi$ and gauge fields $U$ distributed with the Boltzmann weight factor

$$\exp(-S[U]) \quad \text{with} \quad S[U] = S_G[U] - \phi^\dagger (D\,D^\dagger)^{-1}\phi\,. \qquad (8.31)$$

The pseudofermions $\phi$ have the same indices as the Dirac field $\psi$, i.e., color–, Dirac–, and space–time indices. The fields $\phi$ and $U$ are updated alternately according to the distribution (8.31).

The update of the fields $\phi$ is simple. In fact, $\phi$ may be easily created by generating a complex vector $\chi$ with Gaussian distribution $\exp(-\chi^\dagger\chi)$ and then determining $\phi = D\chi$. The Jacobian due to this transformation is not relevant since the gauge fields are not changed in this step.

Once $\phi$ has been constructed, the gauge fields $U$ are updated in a molecular dynamics trajectory treating $\phi$ as an external, constant field. This leads to a new candidate gauge configuration, which is then accepted or rejected in the final Metropolis step. This procedure of constructing a field $\phi$, evolving $U$, and closing with a Metropolis step, is iterated.

We have to take into account that the gauge fields $U$ are group elements whereas so far we have presented the molecular dynamics algorithm for variables in flat space. Before turning to SU(3) let us discuss the simpler abelian case of the gauge group U(1), where the link variables can be written as $U = \exp(iQ)$. We use as conjugate variables the real number $Q$ and a corresponding real momentum $P$. The Hamiltonian equations are identical to (8.18). Let us consider the discretizations of the second of those,

$$\dot{Q} = P \rightarrow Q(\varepsilon) = Q(0) + P(0)\,\varepsilon \tag{8.32}$$
$$\rightarrow -i \ln U(\varepsilon) = i \ln U(0) + P(0)\,\varepsilon \quad \rightarrow \quad U(\varepsilon) = e^{i\,P(0)\,\varepsilon}\,U(0)\,.$$

Integrating this differential equations might have led $Q$ out of the principal value interval of the group. For U(1) this is no problem, since the relation $U = \exp(iQ)$ automatically projects back.

Let us now implement the algorithm for the non-abelian gauge group SU(3). We can write each link variable as, cf. (A.2),

$$U = \exp\left(i\sum_{i=1}^{8} \omega^{(i)}\,T_i\right) \equiv \exp(iQ)\,. \tag{8.33}$$

As for the case of U(1) our $Q$ are identified with elements of the algebra, i.e., they are traceless hermitian matrices parametrized by the real variables $\omega^{(i)}$.

For each link variable $U_\mu(n)$ there are eight real momentum variables $P_\mu^{(i)}(n)$ conjugate to the parameters $\omega_\mu^{(i)}(n)$, where $n \in \Lambda$ is the space–time index. We may combine them to an element of the algebra of SU(3) by summing them with the generators $T_i$,

$$P_\mu(n) = \sum_{i=1}^{8} P_\mu^{(i)}(n)\,T_i\,. \tag{8.34}$$

Thus the $P_\mu(n)$ are also traceless and hermitian matrices. The Hamilton equations involve the sum (use (A.3))

$$\frac{1}{2}\sum_{n,\mu,i}\left(P_\mu^{(i)}(n)\right)^2 = \sum_{n,\mu} \mathrm{tr}\left[P_\mu(n)^2\right]\,. \tag{8.35}$$

We will formulate the algorithm in terms of the matrices $P_\mu(n)$ and $U_\mu(n)$.

For the molecular dynamics evolution we also need the derivative of the action with respect to the $Q$. Derivatives always live in the tangential space of a manifold. For unitary groups this is the algebra (see Appendix A.1) and derivatives thus are along some path in the algebra. The derivative of some function $f(U)$ of a group element in the algebra direction $T_i$ may be defined as,[2]

---

[2]This definition can be understood from differentiating $\ln(U) = i\sum_k \omega^{(k)}T_k$ with respect to $\omega^{(i)}$. We remark that this definition is not unique, but the outcome of (8.36) is, if $f(U)$ is a class function.

$$\nabla^{(i)} f(U) \equiv \frac{\partial f(U)}{\partial \omega^{(i)}} = \frac{\partial}{\partial \omega} f(e^{i\,\omega\,T_i}\,U)\Big|_{\omega=0} . \qquad (8.36)$$

In this way we find, e.g., $(A, B \in \mathrm{SU}(3))$

$$\nabla^{(i)} A\,U\,B = i\,A\,T_i\,U\,B ,$$
$$\nabla^{(i)} A\,U\,U\,B = i\,A\,T_i\,U\,U\,B + i\,A\,U\,T_i\,U\,B , \qquad (8.37)$$

and so on. The leapfrog discretized molecular dynamics equations (8.20) for altogether $n$ steps then become:

1. **Pseudofermions**
   *Generate the pseudofermion field $\phi = D\chi$, where $\chi$ is distributed according to $\exp(-\chi^\dagger \chi)$.*

2. **Conjugate fields**
   *Given a gauge configuration $U_0$ generate $P_0$ (eight real numbers for each $(n, \mu)$) according to the Gaussian distribution $\exp\left(-\operatorname{tr}\left[P^2\right]\right)$.*

3. **Initial step**
   $$P_{\frac{1}{2}} = P_0 - \frac{\varepsilon}{2}\,F[U, \phi]\Big|_{U_0} .$$

4. **Intermediate steps**
   *Full steps for $k = 1, \dots, n - 1$:*
   $$U_k = \exp\left(i\,\varepsilon\,P_{k-\frac{1}{2}}\right) U_{k-1} , \qquad P_{k+\frac{1}{2}} = P_{k-\frac{1}{2}} - \varepsilon\,F[U, \phi]\Big|_{U_k} .$$

5. **Final step**
   $$U_n = \exp\left(i\,\varepsilon\,P_{n-\frac{1}{2}}\right) U_{n-1} , \qquad P_n = P_{n-\frac{1}{2}} - \frac{\varepsilon}{2}\,F[U, \phi]\Big|_{U_n} .$$

6. **Monte Carlo step**
   *Accept–reject step at the end of the trajectory: accept the changes if a random number $r \in [0, 1)$ is smaller than*
   $$\exp\left(\operatorname{tr}\left[P^2\right] - \operatorname{tr}\left[P'^2\right] + S_G[U] - S_G[U'] \right. $$
   $$\left. + \phi^\dagger \left((D\,D^\dagger)^{-1} - (D'\,D'^\dagger)^{-1}\right) \phi\right) . \qquad (8.38)$$

In these equations

$$F[U, \phi] = \sum_{i=1}^{8} T_i\,\nabla^{(i)} \left(S_G[U] + \phi^\dagger (D\,D^\dagger)^{-1}\phi\right) \quad \in \mathrm{su}(3) \qquad (8.39)$$

denotes the driving force, an element of the algebra su(3).

For the pseudofermion contribution in the accept–reject step one re-uses the vector $(D'\,D'^\dagger)^{-1}\phi$ that has been computed for the force in step (5).

A technical problem is the exponentiation $\exp(i\,\varepsilon\,P)$ in step (4) of (8.38) which maps an element of the algebra to a group element. This may be done either by a series expansion or, faster, by utilizing the Cayley–Hamilton relation as done in [9].

The force term (8.39) involves derivatives with regard to the algebra elements. As an example let us determine the force due to the contribution from the Wilson gauge action (4.20) involving a particular link $U$. It is given by

$$-\frac{\beta}{6}\, \mathrm{tr}\left[U\,A + A^\dagger\, U^\dagger\right]\,, \tag{8.40}$$

where $A$ denotes the sum of adjacent staples (4.20). The derivative of the trace-term then is

$$\nabla^{(i)}\, \mathrm{tr}\left[U\,A + A^\dagger\, U^\dagger\right] = \mathrm{tr}\left[\mathrm{i}\,T_i\,U\,A - \mathrm{i}\,A^\dagger\,U^\dagger\,T_i\right] = \mathrm{tr}\left[\mathrm{i}\,T_i\left(U\,A - A^\dagger\,U^\dagger\right)\right]\,. \tag{8.41}$$

The resulting force is

$$-\frac{\beta}{6}\sum_{i=1}^{8} T_i\, \mathrm{tr}\left[\mathrm{i}\,T_i\left(U\,A - A^\dagger\,U^\dagger\right)\right] = -\frac{\beta}{12}\,\mathrm{i}\left(U\,A - A^\dagger\,U^\dagger\right)\,. \tag{8.42}$$

We have used that $\mathrm{i}\left(U\,A - A^\dagger\,U^\dagger\right)$ is traceless and hermitian and thus is a linear combination of the $T_j$. For such a sum one has the identity

$$\sum_j T_j\, \mathrm{tr}\left[T_j \sum_k c_k T_k\right] = \frac{1}{2}\sum_j c_j T_j\,. \tag{8.43}$$

The trace projects out the contribution of $T_j$ (introducing a factor of $1/2$), and the sum reconstructs the expression.

The contribution of the fermion action to the force requires the evaluation of (we use $\partial\,M^{-1}/\partial\,\omega = -M^{-1}(\partial\,M/\partial\,\omega)M^{-1}$ here)

$$\begin{aligned}
\nabla^{(i)}\left(\phi^\dagger\left(D\,D^\dagger\right)^{-1}\phi\right) &= -\phi^\dagger\left(D\,D^\dagger\right)^{-1}\left(\nabla^{(i)}D\,D^\dagger\right)\left(D\,D^\dagger\right)^{-1}\phi \\
&= -\left((D\,D^\dagger)^{-1}\phi\right)^\dagger\left(\frac{\partial D}{\partial\omega^{(i)}}\,D^\dagger + D\frac{\partial D^\dagger}{\partial\omega^{(i)}}\right)\left((D\,D^\dagger)^{-1}\phi\right)\,.
\end{aligned} \tag{8.44}$$

In each substep of the trajectory one therefore has to compute the derivative of $D$ and the inverse of $D\,D^\dagger$, which is the time-consuming part of the algorithm.

In computing the derivative of the Dirac operator one has to observe non-commutativity of the group elements. Some Dirac operators involve contributions, where a link variable occurs more than once. An obvious example is the overlap operator in a rational or series approximation (see Sect. 7.4.3). In such a case one has to be particularly careful in determining the derivative.

The Wilson Dirac operator (5.51) is simply linear in the gauge field variables and in this case computing the derivative is straightforward. We find for the contribution of a particular link $U_\mu(k)$

$$\frac{\partial D(n|m)}{\partial \omega_\mu^{(i)}(k)} = -\mathrm{i}\frac{\mathbb{1}-\gamma_\mu}{2a}\, T_i\, U_\mu(k)\, \delta_{n+\hat\mu,m}\delta_{n,k} + \mathrm{i}\frac{\mathbb{1}+\gamma_\mu}{2a}\, U_\mu(k)^\dagger\, T_i\, \delta_{n-\hat\mu,m}\delta_{m,k}\ .$$

(8.45)

Due to the Metropolis decision at the end of each trajectory the HMC algorithm is exact. Any numerical error done in the discretization is "repaired" by the probabilistic accept–reject step. However, the acceptance rate may become low. One therefore has to adjust the step size $\varepsilon$ to prevent too large aberrations. Typical numbers are 50–100 steps for a complete trajectory with an integrated time of $n\varepsilon \approx 1$. The estimated computational effort for the HMC algorithms grows only like $\propto V^{5/4}$ [10, 11].

## 8.3 Other algorithmic ideas

### 8.3.1 The R-algorithm

The HMC algorithm is exact. However, in the way presented here it is applicable only for positive definite fermion actions (like the overlap action) or for duplicated number of fermion species.[3] It evolved out of earlier updating methods, which also used microcanonical (or Langevin) type differential equations, without the correcting Monte Carlo step.

In all of the so-called *hybrid algorithms* [7] the system follows a molecular dynamics trajectory with refreshing the conjugate momenta regularly. This process is equivalent to an evolution according to a stochastic differential equation. It has been argued that systems out of equilibrium will be driven toward equilibrium and stay there during the updating process. The crucial problem is the necessary discretization, which introduces systematic errors in some order of the discretization time step.

One such method, the so-called *R-algorithm* [12], is advocated for Kogut–Susskind or staggered fermions (cf. Chap. 10). The central idea is to use the effective action discussed in (8.10). The molecular dynamics updating follows steps like those discussed earlier. The change of the effective action is estimated stochastically [13] with the help of noisy pseudofermions. The change of the effective fermion action due to a change of the gauge variable then is

$$\Delta S_{\mathrm{F}}^{\mathrm{eff}} = S_{\mathrm{F}}^{\mathrm{eff}}\,[U_\mu(n)'] - S_{\mathrm{F}}^{\mathrm{eff}}\,[U_\mu(n)]$$
$$= \sum_{n,\mu,i} \frac{\partial S_{\mathrm{F}}^{\mathrm{eff}}[U_\mu(n)]}{\partial \omega_\mu^{(i)}(n)}\, \Delta\omega_\mu^{(i)}(n) + \mathcal{O}\left(\Delta\omega_\mu^{(i)}(n)\right)^2\ .$$

(8.46)

For the derivative of the effective action we obtain

$$\frac{\partial S_{\mathrm{F}}^{\mathrm{eff}}}{\partial \omega_\mu^{(i)}(n)} = -\frac{\partial\,\mathrm{tr}[\ln(D[U])]}{\partial \omega_\mu^{(i)}(n)} = -\mathrm{tr}\left[D[U]^{-1}\frac{\partial D[U]}{\partial \omega_\mu^{(i)}(n)}\right]\ .$$

(8.47)

---

[3]Suggestions how to include single fermion species will be discussed in Sect. 8.3.3.

Note that the arguments are matrices and these are usually not commutative. However, the cyclicity property of the trace simplifies the expression. For actions where the link variables occur linearly, the derivative of $D$ is particularly simple.

As there is no correction MC step, the method will show a dependence on the chosen value of the discretization time step $\varepsilon$. By using clever modifications the errors can be made $\mathcal{O}(\varepsilon^2)$. This dependence has to be dealt with by repeating the computations for different values and of $\varepsilon$ extrapolating to 0.

At each time step one computes pseudofermions $\phi = M^\dagger \chi$ from Gaussian distributed random vectors $\chi$. More details can be found in [12].

### 8.3.2 Partial updates

For updating methods that are based on evolution equations (like the HMC and the R-algorithm or any such method working with Langevin-type differential equations) one has to calculate the force term. This becomes quite involved for Ginsparg–Wilson-type Dirac operators. There one has to use computationally costly algorithms already for evaluating $D$ itself.

It is therefore tempting to shortcut the update by updating larger chunks of gauge variables at once. However, as has been argued earlier, this may run into acceptance problems when working at large volumes: A naive implementation will lead to a numerical cost of $\mathcal{O}(V^2)$. One therefore has to invest some effort in providing better suggestions for the change of gauge variables, similar to the ideas originally leading to the hybrid algorithms.

It has been advocated [14, 15] to interleave gauge field updates on part of the lattice based on a local gauge action that approximates in some way the effect of the fermionic action. This may be done, e.g., by considering a set of closed gauge loops with coefficients properly tuned. This *partial-global* change of variables then undergoes an accept/reject step, where the ratio of determinants is estimated stochastically. Since the change of the effective action may be large, one has to be particularly careful in getting a reliable estimate of that ratio. One therefore invokes techniques like polynomial approximations, preconditioning, and multi-pseudofermions.

For real, positive determinants it is then possible to simulate single fermion species. This is true for Ginsparg–Wilson operators, which are normal matrices with eigenvalues that have positive real part. For such operators one may use a pseudofermion action of the form $(\phi, D^{-1}\phi)$ and Gaussian $\chi$ with $\phi = \sqrt{D}\chi$ to generate the correct distribution. For taking the square root one uses methods like that discussed for the construction of the overlap operator.

### 8.3.3 Polynomial and rational HMC

In the standard HMC algorithm the Dirac operator enters quadratically as $D\,D^\dagger$ to guarantee positivity of the real part of all eigenvalues necessary for the convergence of the pseudofermion Gaussian integral. Thus in that form only even numbers of fermions can be simulated.

If $\det[D]$ is positive, there is a way to circumvent this restriction. One approximates $D^{-1}$ by an operator $T\,T^\dagger$ for the HMC trajectory calculations and corrects for the approximation in the accept–reject step. The approximations suggested are either by polynomials [16–18] or by rational functions [19–22], in the spirit of the evaluations of the overlap operator in Chap. 7.

As an example we discuss the suggestions of [18]. The inverse of the Dirac operator is approximated by an even polynomial,

$$D^{-1} \approx P_{2n}(D) = \prod_{k=1}^{n} (D - z_{2k-1})(D - z_{2k-1}^*) \,, \tag{8.48}$$

with $z_k = 1 - \exp(i\,2\,\pi\,k/(2n+1))$. Assuming $\gamma_5$-hermiticity one has

$$\det[D - z_k^*] = \det[\gamma_5\,D^\dagger\,\gamma_5 - z_k^*] = \det[D - z_k]^* \,, \tag{8.49}$$

and we therefore may split $P_{2n}$ into a product

$$P_{2n}(D) = T_n(D)\,T_n^\dagger(D) \quad \text{with} \quad T_n(D) = \prod_{k=1}^{n} (D - z_{2k-1}) \,. \tag{8.50}$$

The correct determinant is

$$\det[D] = C\,\left(\det[T_n^\dagger\,T_n]\right)^{-1} \quad \text{with} \quad C = \det[D\,T_n^\dagger T_n] \,. \tag{8.51}$$

The HMC trajectory is then computed using a pseudofermion action

$$S = -\phi^\dagger\,T_n^\dagger(D)\,T_n(D)\,\phi \,, \tag{8.52}$$

thus providing for the second factor in (8.51), whereas the correction factor $C$ is dealt with in the Metropolis accept–reject step at the end of the trajectory.

Such methods allow one to work with odd numbers of quark flavors using the exact MC algorithm.

### 8.3.4 Multi-pseudofermions and UV-filtering

In the accept–reject Metropolis step, as it is necessary at the end of each HMC trajectory or in (partial) global algorithms, the stochastic determination with noisy pseudofermion estimation introduces additional noise. Although formally correct, the induced fluctuations of the fermionic force may destabilize the algorithm [2, 20, 23], enforcing smaller step sizes.

A possible way to reduce the noise of the stochastic estimator is to use more than one set of pseudofermion variables. Assuming positivity for $M$ one uses the identity

$$\det[M] = \left(\det[M^{1/n}]\right)^n . \tag{8.53}$$

The stochastic estimator then is

$$\det[M] \propto \prod_{k=1}^{n} \int \mathcal{D}[\phi_R]\mathcal{D}[\phi_I]e^{-\phi^\dagger M^{-1/n} \phi} , \tag{8.54}$$

where $M$ is either a positive definite Dirac operator or $M = D\,D^\dagger$. The matrix $M^{-1/n}$ can be approximated by a rational function or a polynomial in $M$.

A variant of this method [24] splits the determinant

$$\det[M] = \det[M'] \det[M'^{-1} M] \tag{8.55}$$

with separate pseudofermion fields for both factors. By choosing different parameters for the matrices $M$ and $M'$ one may improve the updating performance; the step size can be increased while the acceptance rate remains stable. In the applications one chooses different mass parameters for the two components; thus the method has also been called *mass preconditioning*.

Another method to improve the estimate of the ratio of determinants is to introduce an *UV-filter* [25]. The idea is to reduce the spread of the eigenvalues of the operator to be stochastically estimated. One defines a reduced matrix

$$D_{m,r} = D\,e^{f(D)} , \tag{8.56}$$

where $f(D)$ is chosen to be a suitable low-order polynomial in $D$ with real coefficients such that the eigenvalues of $D_{m,r}$ are concentrated around $z = 1$ in the complex plane, i.e., $e^{f(D)}$ roughly approximates $D^{-1}$. In the acceptance step one then computes (using $\gamma_5$-hermiticity)

$$\frac{\det[D^\dagger D]_{\mathrm{new}}}{\det[D^\dagger D]_{\mathrm{old}}} = \tag{8.57}$$

$$\exp\left(-2 \operatorname{tr}[f(D)_{\mathrm{new}}] + 2 \operatorname{tr}[f(D)_{\mathrm{old}}]\right) \frac{\det[D^\dagger_{m,r} D_{m,r}]_{\mathrm{new}}}{\det[D^\dagger_{m,r} D_{m,r}]_{\mathrm{old}}} .$$

The computation of the trace over polynomials of $D$ is possible for simple Dirac operators but may be an obstacle for extended ones.

### 8.3.5 Further developments

Both the multi-pseudofermion method and the polynomial approximation are in spirit related to the earlier suggestion of a multiboson algorithm [26].

Depending on the Dirac operator there are also various methods of preconditioning (see also Sect. 6.2.4). Mass preconditioning and variants thereof are one possible improvement. For operators involving only nearest neighbor couplings one may split the lattice into even and odd sites and in this way simplify the calculations. This method is called *even-odd preconditioning*.

Another hierarchically organized approach splits the lattice into non-overlapping domains and solves the system on the sub-domains. This so-called *Schwarz alternating procedure* [27–29] has been tested for both preconditioning and dynamical fermion algorithms. It is a domain decomposition method for solving elliptic differential equations (introduced already in the nineteenth century).

Although at least for HMC algorithms one expects the necessary computer time to grow with $V^{5/4}$ for a given set of couplings, the growth as a function of the lattice spacing or the quark mass is not as clear. Since we expect the quark mass to be proportional to the mass squared of the lightest pseudoscalar, it is useful to discuss the computational cost in such units. The expected functional form for the numerical cost is [30, 31]

$$\$\$\$ \propto \text{€€€} \propto \text{¥¥¥} \propto L^{z_L} a^{-z_a} M_\pi^{z_\pi} \ . \tag{8.58}$$

Various aspects enter such a consideration. As mentioned the volume scales with $z_L \simeq 5$. The condition number of the Dirac operator matrix decreases with the pion mass, thus the CGR solver needs more iterations. The step size in the HMC trajectory will have to be decreased and the autocorrelation time may increase. The various studies with, e.g., the Wilson fermion action have led to values of [30] $z_a \simeq 7$ and $z_\pi \simeq 6$, but also smaller values $\mathcal{O}(3-4)$ have been quoted [31].

The field of algorithms for dynamical fermions has been evolving over the years and the development has not yet ceased. Various ideas have been presented, which are not discussed here. Other ideas are certain to come and we advise to check the proceedings of the yearly lattice conferences.

## 8.4 Other techniques using pseudofermions

In Sect. 8.1.3 we have remarked that pseudofermion fields are used in noisy estimator techniques designed for evaluating fermionic observables. The starting point is the structural similarity between bosonic and fermionic generating functionals, (5.32) and (8.7).

Let us start discussing the idea of noisy estimators by using an example of high importance: the value of the fermion condensate $\langle \overline{\psi}\psi \rangle$. In Chap. 7 we have found that this is essentially given by $\text{tr}[D^{-1}]$. Straightforward calculation of the trace requires the determination of propagators for $12|\Lambda|$ point sources, clearly an impossible task. For the noisy estimator approach we consider a set of uncorrelated complex random numbers $\chi_i \equiv \chi_{R,i} + i\chi_{I,i}$ with unit variance,

$$\langle \chi_i \chi_j^* \rangle_\chi = \delta_{ij} \ . \tag{8.59}$$

For example, we may choose for the $\chi_i$ the Gaussian distribution

$$\langle \ldots \rangle_\chi = \pi^{-N} \int \mathcal{D}[\chi_R]\mathcal{D}[\chi_I] \ \exp\left(-\chi^\dagger \chi\right) \ldots \ . \tag{8.60}$$

We may approximate this integral by an average over $K$ random vectors $\chi^{(k)}$ with entries $\chi_i^{(k)}$ distributed according to the probability density (8.60),

$$\delta_{ij} = \langle \chi_i \chi_j^* \rangle_\chi \simeq \frac{1}{K} \sum_{k=1}^K \chi_i^{(k)} \chi_j^{(k)*} . \qquad (8.61)$$

Using this one can estimate the trace

$$\text{tr}[D^{-1}] = \sum_{i,j} (D^{-1})_{ji} \delta_{ij} \approx \frac{1}{K} \sum_{k=1}^K \chi^{(k)\dagger} D^{-1} \chi^{(k)} . \qquad (8.62)$$

Again the random variables agree in number with the fermionic degrees of freedom. The number $K$ of noise vectors considered depends on the desired accuracy of the estimate.

Gaussian noisy pseudofermions may also be utilized to evaluate the change of the fermionic effective action [32], which is required for the Monte Carlo update algorithm. For matrices $M$ with strictly positive real part of the eigenvalues we have

$$\det[M] = \frac{\int \mathcal{D}[\chi_R] \mathcal{D}[\chi_I] \exp\left(-\chi^\dagger \chi\right) \exp\left(-\chi^\dagger \left(M^{-1} - 1\right) \chi\right)}{\int \mathcal{D}[\chi_R] \mathcal{D}[\chi_I] \exp\left(-\chi^\dagger \chi\right)}$$

$$= \left\langle \exp\left(-\chi^\dagger \left(M^{-1} - 1\right) \chi\right)\right\rangle_\chi . \qquad (8.63)$$

This identity follows from (8.7) and may be used to determine directly the change of the effective 2-flavor action:

$$\exp(-\delta S_{F,\text{eff}}) = \det[(D' D'^\dagger)(D D^\dagger)^{-1}] = \det[D'^\dagger (D D^\dagger)^{-1} D']$$

$$= \left\langle \exp\left(-\chi^\dagger \left(D'^{-1} D D^\dagger (D'^\dagger)^{-1} - 1\right) \chi\right)\right\rangle_\chi \qquad (8.64)$$

$$= \left\langle \exp\left(-\chi^\dagger D'^{-1} \left(D D^\dagger - D' D'^\dagger\right) (D'^\dagger)^{-1} \chi\right)\right\rangle_\chi .$$

Again $\langle \ldots \rangle_\chi$ is approximated by a finite number of noise vectors.

It has been suggested in [33] to replace the sum over many $\phi$-ensembles by just one individual for the Monte Carlo acceptance decision. Determining that ratio for the updating step either way is correct in the average. However, this type of approach involves a Monte Carlo sum within a Monte Carlo sum and, as has been discussed in, e.g., [34], corresponds to an updating algorithm where the computer time grows with the square of the lattice volume at least, not considering other effects like critical slowing down.

Whenever one wants a precise calculation of quark propagators, one uses numerical inverters. However, it is expensive to compute the inverse matrix for many different sources. This is, e.g., necessary for sums over disconnected pieces (see Fig. 6.1). In such a case one might be willing to trade lower accuracy for a larger number of source points. Noisy pseudofermions provide a tool for approximating propagator elements. Exploiting (8.7) one may compute them as expectation values of $\phi_n \phi_k^*$ in a bosonic Gaussian distribution:

$$D_{nm}^{-1} = \left(D^\dagger D\right)_{nk}^{-1} D_{km}^\dagger = \frac{\int \mathcal{D}[\phi_R]\mathcal{D}[\phi_I]e^{-\phi^\dagger D^\dagger D \phi}\, \phi_n \phi_k^*}{\int \mathcal{D}[\phi_R]\mathcal{D}[\phi_I]e^{-\phi^\dagger D^\dagger D \phi}}\, D_{km}^\dagger . \tag{8.65}$$

As long as $D$ has nonvanishing eigenvalues this Gaussian integral is well-defined. It can be evaluated using a stochastic process for the $\phi$.

Formulas of this type have also been used in early suggestions to implement dynamical fermions [35] for computing the expressions necessary for the change of the effective fermionic action in (8.46).

## 8.5 The coupling-mass phase diagram

### 8.5.1 Continuum limit and phase transitions

In Sect. 3.5.4 we discussed the true continuum limit for pure gauge theory. We used the renormalization group analysis result (3.88) for the lattice spacing $a$ as a function of the inverse coupling $\beta$ to identify the limit $\beta \to \infty$ as the limit where $a \to 0$ and the continuum physics is recovered. Here we address the continuum limit a second time, but now reformulate it using statistical mechanics language.

To sketch the idea, we first consider the simple familiar 2D Ising model. Without external magnetic field, there is only a single coupling parameter. Depending on its value one finds two phases, a disordered "hot" phase and a phase with a nonvanishing magnetization, the "ordered" phase. A second-order phase transition separates the two phases. The phase transition occurs at a specific value of the coupling, the so-called *critical point* where the *correlation length* $\xi$ of the system diverges. The correlation length can be defined through the connected correlation function of two spin variables located at the origin and some site $n$. For large distances between the spins the correlation function decays exponentially

$$\langle s_n\, s_0 \rangle - \langle s_n \rangle\langle s_0 \rangle \propto \exp\left(-\frac{|n|}{\xi}\right) \quad \text{for} \quad n \to \infty . \tag{8.66}$$

The correlation length $\xi$ is a measure for the distance over which two spins are correlated. When the coupling is driven close to its critical value $\xi$ becomes large and the spins are correlated over long distances. The local structures extending only over sites close to each other become less and less important.

Why is the notion of a diverging correlation length relevant for lattice field theory? We have learned that the physical mass $M_H$ of some hadron can be computed from the two-point function of a hadron interpolator $O_H$,

$$\langle O_H(n_t)\, O_H(0)^\dagger \rangle \propto e^{-n_t a M_H} = e^{-n_t/\xi_H} . \tag{8.67}$$

Here we have defined the correlation length $\xi_H$ for the interpolator $O_H$ as

$$\xi_H = \frac{1}{aM_H} . \tag{8.68}$$

We stress that interpolators for different hadrons will in general give rise to different correlation lengths proportional to each other. Since $M_H$ is constant all correlation lengths $\xi_H$ of physically observable states should diverge in the continuum limit, $a \to 0$. As in the systems of statistical mechanics we must drive our couplings, the inverse gauge coupling $\beta$ and the quark mass $m$, toward critical values where the correlation length diverges. At these critical values we can construct a continuum limit. It is therefore of utmost importance to identify such critical points in the phase diagram.

### 8.5.2 The phase diagram for Wilson fermions

In QCD we expect that the correct continuum limit can be obtained at vanishing gauge coupling $g = 0$, corresponding to $\beta = \infty$, and a vanishing fermion mass parameter. However, also for other values of these couplings we have transitions separating phases of different physics.

For Wilson fermions it is usual to discuss the phase diagram in terms of the inverse gauge coupling $\beta = 6/g^2$ and the hopping parameter $\kappa$ introduced in (5.55). The hopping parameter is related to the bare quark mass in the continuum limit through $\kappa = 1/(2am+8)$. In the hopping parameter notation the Dirac operator is given by (5.57).

When discussing a phase diagram, it is always useful to first consider simple limiting cases. For $\kappa = 0$ the hopping term is turned off and the Dirac operator becomes a unit matrix. This limit corresponds to infinitely heavy quarks. This is the situation of the pure gauge theory (the quenched case) discussed already earlier. In that limit the string tension defined through the Wilson loop may be interpreted as order parameter. As we have discussed in Sect. 3.4.1, at $\beta = 0$ (infinite bare gauge coupling $g$) the theory is confining and Monte Carlo evidence is that this confinement phase extends over the whole range of values toward $\beta \to \infty$ ($g = 0$). Actually, since Monte Carlo calculations are limited, all we can really say is that for the pure gauge theory there is no numerical evidence for a phase transition to a deconfined phase for large enough lattices.

Another limiting case is that of free lattice fermions. This is obtained for $\beta = \infty$ ($g = 0$), where all plaquettes approach unity. On an infinite lattice this is equivalent to setting all link variables to $\mathbb{1}$. With the help of lattice Fourier transformation one can compute the quark propagators for the free case and find that the correlation length is given by the inverse of the quark mass. Thus the correlation length diverges for vanishing quark mass $m = 0$, which corresponds to a value of $\kappa = 1/8$.

For finite values of $\beta$ one finds that the smallest observed mass in the system (the mass of the pion) vanishes at some value $\kappa$ which we denote by

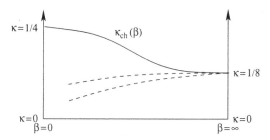

**Fig. 8.1.** The phase diagram for QCD with Wilson fermions. The coupling constants are $\beta = 6/g^2$ and the hopping parameter $\kappa$. The *dashed curves* are lines of constant mass ratios of, e.g., $M_\pi/M_\rho = $ const.

$\kappa_{ch}(\beta)$. Since we associate the pion with the Goldstone mode of spontaneous chiral symmetry breaking, this is the point where at finite $\beta$ the quark mass vanishes. In this indirect way we get our first information on the quark mass itself. This line of $\kappa = \kappa_{ch}(\beta)$ appears to continue from $\kappa_{ch}(\beta = \infty) = 1/8$ down to $\kappa_{ch}(\beta = 0)$. There are strong coupling expansion results for meson masses [36, 37] indicating that $\kappa_{ch}(\beta = 0) \approx 1/4$. The resulting phase diagram is sketched in Fig. 8.1.

At a given $\beta$ the quark masses thus assume values between infinity and 0 when $\kappa$ runs from 0 to $\kappa_{ch}(\beta)$. Driving $\kappa$ towards $\kappa_{ch}(\beta)$ the pion mass decreases toward 0, but other masses of the system, like the $\rho$ or the nucleon mass, approach nonzero values. At some value of $\kappa$ the mass ratio of, e.g., $M_\pi/M_\rho$ will assume its experimental value. At this value of $\kappa$ we then use dimensionless combinations $a\,M$ and the experimental values $M$ to determine the lattice spacing $a$ in order to set the scale.

The collection of such points for varying $\beta$ then defines a "physical" curve (dashed curves in Fig. 8.1). Along such a curve the lattice spacing decreases toward zero for increasing $\beta$ and the system approaches the continuum limit.

We emphasize that the dynamics of the sea quarks of course also influences the lattice spacing $a$, which thus is a function of both coupling parameters $a = a(\beta, \kappa)$. Curves of constant lattice spacing thus will not simply be lines of constant $\beta$. Also we point out that the lattice spacing is nonzero along the chiral curve $\kappa = \kappa_{ch}(\beta < \infty)$.[4] Approaching the continuum limit along the chiral curve is another alternative suitable for mass-independent renormalization schemes. In this situation the lattice spacing $a$ for given gauge coupling is defined in the chiral limit, i.e., on the chiral curve.

---

[4]The situation is similar to that of spontaneous symmetry breaking in the $N$-component $\phi^4$-theory. There the two couplings are the nearest neighbor-coupling $g$ (playing the role of our gauge coupling) and an external magnetic field $h$ (related to the fermion mass in QCD). In the broken phase, for $g > g_{crit}$ and $h = 0$ the order parameter $\langle \phi \rangle$ is nonzero and there are Goldstone modes as well as massive modes. The lattice spacing is zero only toward the critical point $g \to g_{crit}$. When the external field is switched on the Goldstone modes become massive, similar to the pions.

We have to add an important remark to the above discussion. As will be detailed in Chap. 12, by studying lattices with finite time extent (in physical units) one may simulate QCD at finite temperature. This introduces another parameter, the temperature, giving rise to a more complicated phase diagram. One finds a phase transition to a deconfined phase at some critical temperature, i.e., if the time (or space) extent becomes sufficiently small. Here we only stress that the above discussion of the phase structure is based on the assumption that the time extent of our lattice is large enough such that we always remain in the confined phase.

### 8.5.3 Ginsparg–Wilson fermions

The range of the bare mass parameter $a\,m$ for a Ginsparg–Wilson Dirac operator in the form of (7.41) is limited to $[0,2]$. The value $a\,m = 0$ defines the point of massless fermions and (spontaneously broken) chiral symmetry. At the maximum value $a\,m = 2$ the fermions decouple from the system; this corresponds to the limit of infinitely heavy quarks.

Since the curve of chiral symmetry is just a line at $m = 0$ and does not depend on the gauge coupling (unlike the discussed case of the Wilson action), the phase diagram looks simpler than that of Fig. 8.1. The variable $a\,m$ replaces $\kappa$ and ranges from $a\,m = 2$ for the pure gauge theory (quenched limit) to $m = 0$ for the chiral limit. Again one can find curves of constant mass ratios approaching the continuum limit at $\beta \to \infty$, $m = 0$.

Let us address another aspect of using a Dirac operator which for vanishing quark mass obeys the Ginsparg–Wilson equation. For such a Dirac operator the eigenvalue spectrum is restricted to a circle in the complex plane. The smallest distance of this circle to the origin is given by the bare quark mass and the smallest possible eigenvalue is given by this mass. For nonvanishing mass the inverse Dirac operator, i.e., the quark propagator can always be computed, and (in principle) the chiral line can be approached.

For non-chiral Dirac operators, e.g., the Wilson Dirac operator, one has no such restriction of the eigenvalues and they fluctuate depending on the gauge field. Thus even for nonvanishing quark mass zero eigenvalues of the Dirac operator may show up for exceptional configurations (compare Sect. 6.2.5). Consequently the quark propagator cannot be computed for such configurations. Thus the fluctuations of the Dirac eigenvalues often forbid a close approach to the chiral line $\kappa_{\mathrm{ch}}(\beta)$. The situation improves by using larger values of $\beta$, larger lattices, or more advanced Dirac operators with smaller fluctuations of their eigenvalues.

# 8.6 Full QCD calculations

In full QCD calculations the large mass quarks $c$, $b$, and $t$ contribute negligibly to the dynamics of the quark sea. It is therefore an excellent approximation to consider only two mass-degenerate light quarks $u$, $d$ and one heavier quark $s$ for the sea. Even that poses an algorithmic problem. For the non-exact R-type algorithms discussed in Sect. 8.3.1 the number of quark species of a kind may be easily fixed by a pre-factor; however the flavor mixing and the continuum limit are disputable. For the exact HMC-algorithm in its original form, on the other hand, positivity requires an even number of mass-degenerate fermions. There are, however, variants where odd numbers of fermions may be implemented with the help of polynomial approximations as mentioned in Sect. 8.3.

There are several international collaborations studying dynamical fermions for various types of Dirac operators. If we tried to mention them explictly this would imply omitting some of them even at the time of publication of this text. We therefore just remark that for the simple improved Wilson action spatial lattice sizes up to 50–100 are presently achieved for two light flavors and some even including the strange quark. Similar lattice sizes are being studied for twisted mass and staggered fermions (see Chap. 10). For the numerically more demanding domain wall formulation (Chap. 10) one typically works on lattices half that size (in lattice units). This roughly reflects the increased numerical cost, i.e., one to two orders of magnitude. First results for overlap fermions on smaller lattices, again a factor of two smaller than the domain wall sizes, are being produced, although with certain restrictions like fixed topological sectors due to algorithmic problems for the tunneling between

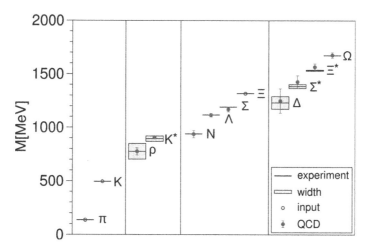

**Fig. 8.2.** Spectroscopy results from a fully dynamical simulation by the Budapest–Marseille–Wuppertal collaboration [38]. (From S. Dürr et al.: Science 322, 1224 (2008). Reprinted with permission from AAAS)

different sectors. In most of these collaborations a wide spectrum of physics questions is attacked by the participating scientists.

To give one example of a calculation with dynamical fermions we show in Fig. 8.2 the result of a spectroscopy calculation with two light dynamical fermions and a dynamical strange quark from the Budapest–Marseille–Wuppertal collaboration [38]. The simulation was done with the improved (compare Chap. 9) Wilson gauge- and Wilson fermion action on lattices with a spatial extent up to 4 fm. The gauge fields entering the Dirac operator were treated with six levels of stout smearing. Pion masses down to 190 MeV were reached and finite volume effects, as well as effects of resonances, were taken into account in the analysis. In Fig. 8.2 the masses for different mesons and baryons are shown, extrapolated to physical quark masses and vanishing lattice spacing.

Comparing these results with the quenched spectroscopy calculation of Fig. 6.5 one finds that the discrepancies between the lattice results and the experimental numbers have vanished. Figure 8.2 demonstrates that when taking into account the effects of light dynamical quarks, for a wide range of different quantum numbers a controlled ab initio lattice spectroscopy calculation leads to excellent agreement between numerical results and the experimental data.

# References

1. D. H. Weingarten and D. N. Petcher: Phys. Lett. B **99**, 333 (1981)
2. A. Alexandru and A. Hasenfratz: Phys. Rev. D **66**, 094502 (2002)
3. D. J. E. Callaway and A. Rahman: Phys. Rev. D **28**, 1506 (1983)
4. J. Polonyi and H. W. Wyld: Phys. Rev. Lett. **51**, 2257 (1983)
5. G. Parisi and Y.-S. Wu: Sci. Sin. **24**, 483 (1981)
6. J. R. Klauder: in *Recent Developments in High-Energy Physics; Acta Phys. Austriaca Suppl. 25*, edited by H. Mitter and C. B. Lang, p. 251 (Springer, Wien, New York 1983)
7. S. Duane and J. B. Kogut: Phys. Rev. Lett. **55**, 2774 (1985)
8. S. Duane and J. B. Kogut: Nucl. Phys. B **275**, 398 (1986)
9. C. Morningstar and M. Peardon: Phys. Rev. D **69**, 054501 (2004)
10. R. Gupta, G. W. Kilcup, and S. R. Sharpe: Phys. Rev. D **38**, 1278 (1988)
11. M. Creutz: Phys. Rev. D **38**, 1228 (1988)
12. S. Gottlieb et al.: Phys. Rev. D **35**, 2531 (1987)
13. G. G. Batrouni et al.: Phys. Rev. D **32**, 2736 (1985)
14. A. Hasenfratz and A. Alexandru: Phys. Rev. D **65**, 114506 (2002)
15. A. Hasenfratz, P. Hasenfratz, and F. Niedermayer: Phys. Rev. D **72**, 114508 (2005)
16. P. de Forcrand and T. Takaishi: Nucl. Phys. B (Proc. Suppl.) **53**, 968 (1997)
17. R. Frezzotti and K. Jansen: Phys. Lett. B **402**, 328 (1997)
18. T. Takaishi and P. de Forcrand: Int. J. Mod. Phys. C **13**, 343 (2002)
19. I. Horváth, A. D. Kennedy, and S. Sint: Nucl. Phys. B (Proc. Suppl.) **73**, 834 (1999)
20. M. A. Clark and A. D. Kennedy: Nucl. Phys. B (Proc. Suppl.) **129**, 850 (2004)

21. M. A. Clark, P. de Forcrand, and A. D. Kennedy: PoS **LAT2005**, 115 (2005)
22. M. A. Clark: PoS **LAT2006**, 004 (2006)
23. M. Hasenbusch: Phys. Rev. D **59**, 054505 (1999)
24. M. Hasenbusch: Phys. Lett. B **519**, 177 (2001)
25. P. de Forcrand: Nucl. Phys. B (Proc. Suppl.) **73**, 822 (1999)
26. M. Lüscher: Nucl. Phys. B **418**, 637 (1994)
27. M. Lüscher: JHEP **0305**, 052 (2003)
28. M. Lüscher: Comput. Phys. Commun. **165**, 199 (2005)
29. M. Lüscher: Comput. Phys. Commun. **156**, 209 (2004)
30. A. Ukawa: Nucl. Phys. B (Proc. Suppl.) **106**, 195 (2002)
31. M. Hasenbusch: Nucl. Phys. B (Proc. Suppl.) **129–130**, 27 (2004)
32. G. Bhanot, U. M. Heller, and I. O. Stamatescu: Phys. Lett. B **129**, 440 (1983)
33. M. Grady: Phys. Rev. D **32**, 1496 (1985)
34. M. Creutz: *Quantum Fields on the Computer* (World Scientific, Singapore 1992)
35. F. Fucito, E. Marinari, G. Parisi, and C. Rebbi: Nucl. Phys. B **180**, 369 (1981)
36. K. G. Wilson: Phys. Rev. D **10**, 2445 (1974)
37. K. G. Wilson: in *New Phenomena in Subnuclear Physics*, edited by A. Zichichi (Plenum, New York 1977)
38. S. Dürr et al.: Science **322**, 1224 (2008)

# 9

# Symanzik improvement and RG actions

When introducing the QCD action on the lattice we had to discretize the derivative terms that show up in the continuum action. It was pointed out then that any discretization, e.g., symmetric differences for the first derivative in the fermion action, gives rise to discretization effects. Typically the discretization effects are of $\mathcal{O}(a)$ for fermions and of $\mathcal{O}(a^2)$ for the gauge fields. These disappear only in the continuum limit when the lattice spacing $a$ is sent to zero. Performing the continuum limit is, however, a nontrivial task. As one decreases $a$, the number of lattice points has to increase, such that the physical volume remains constant (ideally one would first send the number of lattice points to infinity before sending $a$ to zero). Thus in a numerical simulation one always works with finite $a$ and the discretization errors have to be dealt with, e.g., by including them in the extrapolation to vanishing $a$.

An elegant way of approaching the problem is a systematical reduction of the discretization errors. We have already mentioned that the discretization we have chosen is not unique. Also other discretizations give rise to the same formal continuum limit. In particular one may combine different terms to obtain a lattice action with reduced discretization effects. Adding, e.g., an extra term to the Wilson fermion action and matching its coefficient appropriately, one can reduce the discretization error from $\mathcal{O}(a)$ to $\mathcal{O}(a^2)$. In a similar way it is possible (and necessary for a full improvement) to reduce also the discretization errors of the observables used. A systematic implementation of these ideas is the *Symanzik improvement program* which we discuss in Sect. 9.1.

A conceptionally different approach uses actions constructed from *renormalization group (RG) transformations*. Improvement is achieved by integrating out (blocking) short distance degrees of freedom and including their effect in the action for the blocked fields. This strategy is presented in Sects. 9.2–9.4. After discussing the blocking approach for free fermions where the lattice action can be computed in closed form (Sect. 9.2), we introduce the general setting for full QCD in Sect. 9.3. In Sect. 9.4 we finally address an important use of the RG approach, the identification of the lattice counterpart of

Gattringer, C., Lang, C.B.: *Symanzik Improvement and RG Actions*. Lect. Notes Phys. **788**, 213–242 (2010)
DOI 10.1007/978-3-642-01850-3_9 &copy; Springer-Verlag Berlin Heidelberg 2010

continuum symmetries, and as an application we derive the Ginsparg–Wilson equation.

## 9.1 The Symanzik improvement program

### 9.1.1 A toy example

Let us begin our discussion of improvement with a toy example which already contains most of the steps that will be taken when improving lattice QCD. We consider the (symmetric) discretization of the derivative $f'(x)$ for some function $f(x)$ of a single real variable $x$:

$$\frac{f(x+a) - f(x-a)}{2a} = f'(x) + a^2 C^{(2)}(x) + a^4 C^{(4)}(x) + \mathcal{O}(a^6) . \quad (9.1)$$

Note that due to the (anti-)symmetrical discretization on the left-hand side, only even powers of $a$ and odd derivatives of $f$ can appear on the right-hand side. We have introduced the abbreviations $C^{(2)}$ and $C^{(4)}$ for the correction terms. Since $a$ has dimension of length, in order to have the same dimensions on both sides the correction terms $C^{(k)}$ must have dimension $1+k$, i.e., $C^{(k)} \sim length^{-(1+k)}$ (assuming that $f$ is dimensionless).

Using the Taylor series expansion

$$f(x \pm a) = f(x) \pm a f'(x) + \frac{a^2}{2} f''(x) \pm \frac{a^3}{6} f'''(x) + \mathcal{O}(a^4) , \quad (9.2)$$

we can identify the leading correction term as

$$C^{(2)}(x) = \frac{1}{6} f'''(x) . \quad (9.3)$$

We stress at this point that this expression for the correction term is given in continuum language, i.e., as a higher derivative of $f$.

The strategy for improvement is to add to the left-hand side of (9.1) a discretized expression (i.e., an expression built from $f(x), f(x \pm a), f(x \pm 2a)$, ...) such that the correction terms on the right-hand side are canceled up to the requested order. For improvement of $\mathcal{O}(a^2)$ we thus use the ansatz

$$\frac{f(x+a) - f(x-a)}{2a} + c\, a^2 D^{(3)}[f](x) = f'(x) + \mathcal{O}(a^4) , \quad (9.4)$$

where $D^{(3)}[f]$ is a discretized expression obeying $D^{(3)}[f] \approx f''' + \mathcal{O}(a^2)$ and $c$ is some constant.

Employing again the power series (9.2), it is easy to see that

$$D^{(3)}[f](x) = \frac{f(x+2a) - 2f(x+a) + 2f(x-a) - f(x-2a)}{2a^3} , \quad c = -\frac{1}{6} \quad (9.5)$$

does the job and $\mathcal{O}(a^2)$ improvement is achieved. We remark, however, that the choice (9.5) is not unique, and, e.g., also terms including $f(x \pm 3a)$ could have been used.

Let us summarize the steps taken in our toy example, which already outline the approach for improving lattice QCD:

- We start from a simple discretized expression for the quantity of interest (the first derivative $f'$ in our example).
- Correction terms are identified using continuum language (higher derivatives in the toy example).
- The correction terms have certain symmetries (only odd derivatives in the example) and are ordered according to their dimension.
- In order to achieve improvement, discretized versions of the correction terms are added with suitable coefficients, such that corrections up to the desired order vanish.
- There is some arbitrariness in the choice of the discretized correction terms.

Exactly the same steps and features will appear in the improvement of lattice QCD. The main difference is the determination of the coefficients. In the toy example the coefficient $c$ followed from simple algebraic considerations. Due to the nonlinear nature of QCD and the necessary renormalization, the determination of the corresponding coefficients in QCD is much more involved and must be done perturbatively or through a nonperturbative matching procedure. The approach to improvement outlined here is known as *Symanzik improvement program* [1–4].

## 9.1.2 The framework for improving lattice QCD

When improving lattice QCD, we need to improve Euclidean correlation functions, e.g., two point functions of interpolators $O_1$ and $O_2$ located at two different space–time points $n, m \in \Lambda$,

$$\langle O_1(n)\, O_2(m) \rangle \,=\, \frac{1}{Z} \int \mathcal{D}[U] \mathcal{D}\left[\psi, \overline{\psi}\right] \, \mathrm{e}^{-S[U, \psi, \overline{\psi}]}\, O_1[U, \psi, \overline{\psi}; n] O_2[U, \psi, \overline{\psi}; m] \,,$$

(9.6)

which may be used to compute the energy spectrum and some matrix elements. To obtain improvement for all these observables in general we will have to improve both the lattice action $S$ and the interpolators $O_1, O_2$.

Let us begin the discussion with the improvement of the action. To be specific, we start on the lattice with Wilson's gauge action (2.49) and a fermion action with Wilson's Dirac operator (5.51). From the discussion of the previous chapters we expect discretization errors of $\mathcal{O}(a)$ for the fermion part and $\mathcal{O}(a^2)$ for the gauge part of the action.

Following the strategy illustrated in the toy example, we begin with identifying a continuum expression for the correction terms. These should be ordered according to their dimension and have the symmetries of the QCD action. In other words we write down an effective action which describes the behavior of Wilson's form of lattice QCD at finite $a$ and its approach to the continuum limit. Following [1, 2, 5–7] we write the effective action in the form

$$S_{\text{eff}} = \int \mathrm{d}^4 x \left( L^{(0)}(x) + a L^{(1)}(x) + a^2 L^{(2)}(x) + \ldots \right). \tag{9.7}$$

Here $L^{(0)}$ is the usual QCD Lagrangian as defined in (2.3) and (2.17). The terms $L^{(k)}, k \geq 1$ are the additional correction terms, which are built from products of quark and gluon fields such that they have dimensions $4 + k$, i.e., $L^{(k)} \sim length^{-(4+k)}$. Thus, compared to $L^{(0)}$, which is of dimension 4, these terms will contain additional derivatives or powers of the quark mass $m$. Requiring the symmetries of the lattice action (compare Sect. 5.4), one may show that the leading correction term $L^{(1)}(x)$ can be written as a linear combination of the following dimension-5 operators ($\sigma_{\mu\nu} \equiv [\gamma_\mu, \gamma_\nu]/2\mathrm{i}$):

$$\begin{aligned}
L_1^{(1)}(x) &= \overline{\psi}(x)\sigma_{\mu\nu} F_{\mu\nu}(x)\psi(x), \\
L_2^{(1)}(x) &= \overline{\psi}(x)\overrightarrow{D}_\mu(x)\overrightarrow{D}_\mu(x)\psi(x) + \overline{\psi}(x)\overleftarrow{D}_\mu(x)\overleftarrow{D}_\mu(x)\psi(x), \\
L_3^{(1)}(x) &= m \, \mathrm{tr}\left[F_{\mu\nu}(x)F_{\mu\nu}(x)\right], \\
L_4^{(1)}(x) &= m \left( \overline{\psi}(x)\gamma_\mu\overrightarrow{D}_\mu(x)\psi(x) - \overline{\psi}(x)\gamma_\mu\overleftarrow{D}_\mu(x)\psi(x) \right), \\
L_5^{(1)}(x) &= m^2 \, \overline{\psi}(x)\psi(x).
\end{aligned} \tag{9.8}$$

This list of operators may be further reduced by using the field equation $(\gamma_\mu D_\mu + m)\psi = 0$, which gives rise to the two relations

$$L_1^{(1)} - L_2^{(1)} + 2L_5^{(1)} = 0, \quad L_4^{(1)} + 2L_5^{(1)} = 0. \tag{9.9}$$

These relations may be used to eliminate the terms $L_2^{(1)}$ and $L_4^{(1)}$ from the set of operators that span the leading correction, and it can be shown [6, 7] that this is true beyond tree level (where the field equations hold). Thus it is sufficient to work with only the terms $L_1^{(1)}, L_3^{(1)}$, and $L_5^{(1)}$. Two of these, $L_3^{(1)}$ and $L_5^{(1)}$, are terms that up to some factor are already present in the original action, such that they can be accounted for by a redefinition of the bare parameters $m$ and $g$ (or $\beta = 6/g^2$).

Thus for $\mathcal{O}(a)$ improvement of the Wilson lattice action it is sufficient to add the Pauli term $L_1^{(1)}$ such that we obtain for the improved action

$$S_{\text{I}} = S_{\text{Wilson}} + c_{\text{sw}} \, a^5 \sum_{n \in \Lambda} \sum_{\mu < \nu} \overline{\psi}(n)\frac{1}{2}\sigma_{\mu\nu} \widehat{F}_{\mu\nu}(n)\psi(n). \tag{9.10}$$

The real coefficient $c_{\text{sw}}$ is often referred to as *Sheikholeslami–Wohlert coefficient* due to the authors of [5], where the improved action (9.10) was written down for the first time. As outlined in the toy example, $\widehat{F}_{\mu\nu}$ has to be a discretized version of the corresponding term in $L_1^{(1)}$, i.e., a lattice form of the field strength tensor. A convenient but not unique choice is

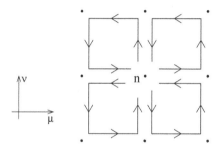

**Fig. 9.1.** Graphical representation of the sum $Q_{\mu\nu}(n)$ of plaquettes in the $\mu$–$\nu$ plane used for the discretization of the field strength in (9.11)

$$\widehat{F}_{\mu\nu}(n) \; = \; \frac{-i}{8a^2}\left(Q_{\mu\nu}(n) - Q_{\nu\mu}(n)\right) , \qquad (9.11)$$

where $Q_{\mu\nu}(n)$ is the sum of plaquettes $U_{\mu,\nu}(n)$ (compare (2.48)) in the $\mu$–$\nu$ plane as shown in Fig. 9.1,

$$Q_{\mu\nu}(n) \equiv U_{\mu,\nu}(n) + U_{\nu,-\mu}(n) + U_{-\mu,-\nu}(n) + U_{-\nu,\mu}(n) , \qquad (9.12)$$

which is a discretization of the continuum field strength tensor. Due to the shape of the terms which is reminiscent of a clover leaf, the last term in (9.10) is often referred to as *clover term* or *clover improvement*.

At this point we need to add a couple of comments: We saw that for $\mathcal{O}(a)$ improvement it is sufficient to add a single term to the fermion action. The only pure glue term $L_3^{(1)}$ was found to be proportional to the original gauge action and thus was absorbed in a redefinition of the bare gauge coupling. Relevant purely gluonic operators appear only at dimension 6, i.e., they contribute at $\mathcal{O}(a^2)$. This is in agreement with previous remarks that for bosonic fields the discretization errors are only of $\mathcal{O}(a^2)$. For the improvement of the gauge action we refer the reader to the original literature, e.g., [3, 8], where the so-called *Lüscher–Weisz gauge action* is presented.

The second comment concerns the discretization errors of chirally symmetric Dirac operators that obey the Ginsparg–Wilson equation (7.29). If one would add the order $\mathcal{O}(a)$-improving term, i.e., the clover term, the resulting Dirac operator would no longer obey the Ginsparg–Wilson equation. Thus we must conclude that a Ginsparg–Wilson Dirac operator is already $\mathcal{O}(a)$ improved. As a matter of fact the nonlinear right-hand side of the Ginsparg–Wilson equation (7.29) generates a lattice discretization of the Pauli term $L_1^{(1)}$ when the naive lattice Dirac operator (which necessarily is a building block of any lattice Dirac operator) is squared. This link between $\mathcal{O}(a)$ improvement and chiral symmetry hints already at a nonperturbative strategy [6] for determining the Sheikholeslami–Wohlert coefficient, which we will present below, following the discussion of the improvement of currents.

### 9.1.3 Improvement of interpolators

The improvement of the action, which we have discussed in the last section, is already sufficient for $\mathcal{O}(a)$ improvement of on-shell quantities such as hadron masses. However, for a full $\mathcal{O}(a)$ improvement of correlator (9.6), which is necessary for the improvement of the hadronic matrix elements computed from it, also the interpolators $O_i$ need to be improved.

As for the action the strategy is to find continuum expressions for the correction terms which are organized with respect to their dimension and are identified by requiring the symmetries of the operator $O_i$ one wants to improve. Since currents with arbitrary quantum numbers may be constructed, it is not possible to give an exhaustive list of terms needed for general $\mathcal{O}(a)$ improvement of $n$-point functions. Following [6, 7] we instead discuss two examples, the improvement of the isovector axial current $A_\mu^a$ and the pseudoscalar density $P^a$:

$$
\begin{aligned}
A_\mu^a(n) &= \frac{1}{2}\overline{\psi}(n)\gamma_\mu\gamma_5\tau^a\psi(n)\,, \\
P^a(n) &= \frac{1}{2}\overline{\psi}(n)\gamma_5\tau^a\psi(n)\,.
\end{aligned}
\tag{9.13}
$$

Here $\tau^a$ denotes one of the Pauli matrices $\tau^1, \tau^2, \tau^3$ acting in $N_f = 2$-flavor space. The dimension-4 operators needed for $\mathcal{O}(a)$ improvement of the current $A_\mu^a$ are given by

$$
\begin{aligned}
(A_1)_\mu^a(x) &= \frac{1}{2}\overline{\psi}(x)\gamma_5\sigma_{\mu\nu}\left(\overrightarrow{D}_\nu(x) - \overleftarrow{D}_\nu(x)\right)\tau^a\psi(x)\,, \\
(A_2)_\mu^a(x) &= \frac{1}{2}\partial_\mu\left(\overline{\psi}(x)\gamma_5\tau^a\psi(x)\right)\,, \\
(A_3)_\mu^a(x) &= \frac{m}{2}\overline{\psi}(x)\gamma_\mu\gamma_5\tau^a\psi(x)\,.
\end{aligned}
\tag{9.14}
$$

As for the improvement of the action we can apply the field equation, which can be used to show that $A_1$ can be written as a linear combination of $A_2$ and $A_3$. Furthermore $A_3$ has the form of the original current of (9.13) such that it can be absorbed in the overall multiplicative renormalization constant $Z_A$ for the axial vector current (compare Chap. 11). Thus we obtain for the $\mathcal{O}(a)$-improved axial current $A_I$ the expression

$$
(A_I)_\mu^a(n) = A_\mu^a(n) + c_A\, a\, \widehat{\partial}_\mu P^a(n)\,,
\tag{9.15}
$$

where the real coefficient $c_A$ still has to be determined. By $\widehat{\partial}_\mu$ we denote the symmetric difference operator $\widehat{\partial}_\mu f(n) = (f(n+\hat{\mu}) - f(n-\hat{\mu}))/2a$.

For the pseudoscalar density $P^a$ the only term that occurs is a normalization with an extra factor of the mass, which is again absorbed in the renormalization. Thus for $\mathcal{O}(a)$ improvement it is sufficient to work with $(P_I)^a = P^a$.

### 9.1.4 Determination of improvement coefficients

For the $\mathcal{O}(a)$ improvement of $n$-point functions of the axial currents we still need to determine the coefficients $c_{sw}$ and $c_A$. As already remarked, here QCD deviates from our toy example due to its nonlinearity and the necessary renormalization.

Using a traditional approach, one can compute the coefficients perturbatively [5, 10, 11]:

$$c_{sw} = 1 + 0.2659\,g^2 + \mathcal{O}(g^4) \ , \quad c_A = -0.00756\,g^2 + \mathcal{O}(g^4) \ . \tag{9.16}$$

However, for many important applications of lattice QCD one works in a regime where the coupling $g$ is not small and the perturbative result is not particularly useful. For an introductory discussion of another approach, *tadpole improvement* [12], see [13].

We have already remarked that for the nonperturbative determination of $c_{sw}$ and $c_A$ one can explore the chiral symmetry of QCD [6]. In Chap. 11 we will show that nontrivial relations between vacuum expectation values may be derived, the so-called *Ward identities*. Here we simply state their existence without proof. The particular Ward identity which we make use of here reads in the continuum

$$\big\langle \big(\partial_\mu A_\mu^a(x)\big)\, O \big\rangle = 2m\, \langle P^a(x)\, O \rangle \ . \tag{9.17}$$

In this equation $A_\mu^a$ and $P^a$ are the continuum counterparts of the corresponding lattice interpolators defined in (9.13). $O$ is an interpolator chosen such that $\langle P^a(x)\,O \rangle$ does not vanish, with the restriction that its support does not include the point $x$ where $A_\mu^a(x)$ and $P^a(x)$ are located. Equation (9.17) expresses the fact that the isovector axial current is conserved for a chiral theory where $m = 0$. Thus it is referred to as the *partially conserved axial current relation* (PCAC).

On the lattice we expect that after renormalization (see Chap. 11), the lattice interpolators approach the continuum relation (9.17), such that

$$\big\langle \big(\hat{\partial}_\mu A_\mu^{(r)\,a}(n)\big)\, O \big\rangle = 2m^{(r)} \big\langle P^{(r)\,a}(n)\, O \big\rangle + \mathcal{O}(a^z) \ . \tag{9.18}$$

The superscript $(r)$ in this equation expresses the fact that renormalized quantities are being used. The important observation is that the correction $\mathcal{O}(a^z)$ is different for the improved theory with properly chosen $c_{sw}$ and $c_A$, where one expects $z = 2$, while for the unimproved theory one expects $\mathcal{O}(a)$ corrections, i.e., $z = 1$. Thus we may use the corrections in the PCAC relation (9.18) to determine $c_{sw}$ and $c_A$.

A slight complication arises through the necessary renormalization. Using the improved axial current $(A_I)_\mu^a$ (for the pseudoscalar we work with the original definition, $(P_I)^a = P^a$), we write the renormalized interpolators as

$$A_\mu^{(r)\,a}(n) = Z_A \left(1 + b_A a m\right) \left(A_\mu^a(n) + c_A\, a\, \hat{\partial}_\mu P^a(n)\right) ,$$
$$P^{(r)\,a}(n) = Z_P \left(1 + b_P a m\right) P^a(n) \ . \tag{9.19}$$

Here $Z_A$ and $Z_P$ are the renormalization constants for the axial vector and the pseudoscalar, respectively (see Chap. 11). The real coefficients $b_A$ and $b_P$ parameterize their lattice corrections including correction terms such as $(A_3)^a_\mu$, which, as we have already remarked, were absorbed in the renormalization.

Based on (9.18) we define an unrenormalized quark mass $m_{\mathrm{AWI}}$, referred to as *PCAC mass* or *AWI (axial Ward identity) mass* through the ratio

$$m_{\mathrm{AWI}}(n) = \frac{\left\langle \left( \hat{\partial}_\mu A^a_\mu(n) + c_A\, a\, \hat{\triangle} P^a(n) \right) O^a \right\rangle}{2 \left\langle P^a(n) O^a \right\rangle} . \tag{9.20}$$

In this equation $O^a$ is an interpolator which we will define below and $\hat{\triangle}$ denotes the Laplacian operator on the lattice, $\hat{\triangle} f(n) = \sum_\mu (f(n + \hat{\mu}) - 2f(n) + f(n - \hat{\mu})) / a^2$. From the lattice PCAC relation (9.18) we obtain

$$m_{\mathrm{AWI}} = \frac{Z_P \left(1 + b_P a m\right)}{Z_A \left(1 + b_A a m\right)} m^{(r)} + \mathcal{O}(a^z), \tag{9.21}$$

where $z = 2$ is obtained when $c_{\mathrm{sw}}$ and $c_A$ have the correct nonperturbative values. For the determination of these values one evaluates $m_{\mathrm{AWI}}$ three times using three different $O^a$ in (9.20). This is done at fixed bare parameters such that one of the three values can be used to remove the ratio $(Z_P(1+b_P a m)) (Z_A(1+b_A a m))^{-1}$ and the other two determine $c_{\mathrm{sw}}$ and $c_A$.

The remaining problem is the choice of the operators $O^a$ that one inserts into (9.20). The operators $O^a$ must have the quantum numbers of $P^a$ and should be chosen such that they probe the PCAC relation (and thus the cutoff effect), but leave the renormalization factors in (9.21) unchanged. An operator with that properties [6] may be constructed using the so-called *Schrödinger functional*.

For the Schrödinger functional [14–18] one uses instead of periodic (or anti-periodic) boundary conditions in time direction, Dirichlet, i.e., fixed boundary conditions. For the spatial directions periodic boundary conditions are kept. Using a lattice with $N_T + 1$ lattice points in time direction, we need to specify the boundary values for the fields at $n_4 = 0$ and $n_4 = N_T$:

$$P_+\psi(\boldsymbol{n}, 0) = \rho(\boldsymbol{n}) \quad , \quad \overline{\psi}(\boldsymbol{n}, 0)P_- = \overline{\rho}(\boldsymbol{n}) ,$$
$$P_-\psi(\boldsymbol{n}, N_T) = \rho'(\boldsymbol{n}) \quad , \quad \overline{\psi}(\boldsymbol{n}, N_T)P_+ = \overline{\rho}'(\boldsymbol{n}) , \tag{9.22}$$

where $P_\pm$ are the projectors $P_\pm = (\mathbb{1} \pm \gamma_4)/2$. For the gauge fields we need to specify the values of the spatial gauge links at $n_4 = 0$ and $n_4 = N_T$:

$$U_j(\boldsymbol{n}, 0) = \mathrm{diag}\left( e^{i\phi_1/N}, e^{i\phi_2/N}, e^{i\phi_3/N} \right) ,$$
$$U_j(\boldsymbol{n}, N_T) = \mathrm{diag}\left( e^{i\phi'_1/N}, e^{i\phi'_2/N}, e^{i\phi'_3/N} \right) . \tag{9.23}$$

Here $N$ denotes the number of lattice points in spatial direction and the phases $\phi_i, \phi'_i$ are subject to the conditions $\phi_1 + \phi_2 + \phi_3 = 0$ and $\phi'_1 + \phi'_2 + \phi'_3 = 0$.

The dynamical variables, which are integrated over in the path integral constituting the Schrödinger functional in (9.25) below, are just the fermion and gauge degrees of freedom in the interior of our lattice. The boundary values, on the other hand, may be used as source terms. Making use of this feature we write down the interpolator $O^a$ which we insert in (9.21) as

$$O^a = -\frac{a^6}{2} \sum_{n,m} \frac{\partial}{\partial \rho(n)} \gamma_5 T^a \frac{\partial}{\partial \overline{\rho}(m)} . \tag{9.24}$$

The vacuum expectation value of, e.g., the correlator of $A_\mu^a$ and $O^a$ in the Schrödinger functional is then defined as

$$\langle A_\mu^a(n) O^a \rangle = \frac{1}{Z} \int \mathcal{D}[\psi, \overline{\psi}, U] \, A_\mu^a(n) \, O^a \, e^{-S[\psi, \overline{\psi}, U]} \bigg|_{\rho = \rho' = \overline{\rho} = \overline{\rho}' = 0} , \tag{9.25}$$

i.e., the boundary values for the fermions are set to 0 after the derivatives with respect to $\rho, \rho', \overline{\rho}$, and $\overline{\rho}'$ in the definition of $O^a$ have been performed. The boundary values for the gauge fields are kept at nontrivial values and now may be used to probe the system.

The operator $O^a$ corresponds to a quark source and sinks with zero spatial momentum placed at the timeslice $n_4 = 0$. Thus the vacuum expectation value (9.25) may be interpreted as a quark and an antiquark lines traveling into the interior of our lattice where they annihilate at the point $n$, i.e., the point where we placed the current $A_\mu^a(n)$.

For our application the key property of the Schrödinger functional is that in (9.21) the renormalization constants and the renormalized mass do not depend on the boundary conditions $\phi_i$ and $\phi_i'$. Thus these can be used to probe the exponent $z$ which parameterizes the discretization error. If $c_{sw}$ and $c_A$ are chosen correctly the exponent is $z = 2$, and the results for the PCAC mass $m_{AWI}$ as defined in (9.20) should be (essentially) independent of $\phi_i, \phi_i'$, and the coordinate $n_4$ where the current $A_\mu^a$ is inserted. For technical details of the strategy for determining $c_{sw}$ and $c_A$ from this condition we refer to the original literature [19–21].

The numerical result for the coefficient $c_{sw}$ for two flavors of dynamical Wilson fermions may be parameterized as [21]

$$c_{sw} = \frac{1 - 0.454g^2 - 0.175g^4 + 0.012g^6 + 0.045g^8}{1 - 0.720g^2} , \tag{9.26}$$

which is a reliable approximation for the range of couplings $g \in [0.0, 1.1]$.

# 9.2 Lattice actions for free fermions from RG transformations

In the last section we have achieved improvement by analyzing how different terms that may appear in a lattice discretization of the action depend on the lattice spacing $a$. We have seen that a linear combination of such terms with suitably chosen coefficients reduces the discretization errors. In the next three sections we now discuss a different approach, the so-called *renormalization group (RG) transformations*, in which the degrees of freedom above the lattice cutoff are integrated out and the effect of these UV modes is incorporated in suitable terms of the lattice action.

In this section we directly block free fermions from the continuum (see [22]). This warming-up exercise demonstrates some of the central concepts of RG transformations for an example that can be solved in closed form. The example is also of interest for the problem of chiral fermions on the lattice. We will see that the lattice action we obtain from the blocking procedure obeys the Ginsparg–Wilson equation, which we have already identified in Chap. 7 as the key to chiral symmetry on the lattice.

### 9.2.1 Integrating out the fields over hypercubes

The starting point is the continuum action for free massless fermions:

$$S_F\left[\phi,\overline{\phi}\right] = \int \mathrm{d}^4x\, \overline{\phi}(x)\,\gamma_\mu \partial_\mu\, \phi(x) = \int \mathrm{d}^4p\, \widehat{\overline{\phi}}(-p)\, \mathrm{i}\gamma_\mu p_\mu\, \widehat{\phi}(p)\,. \qquad (9.27)$$

For later use we have in the second step already transformed the action to momentum space using the symmetric convention

$$\widehat{f}(p) = \int_{-\infty}^{\infty} \frac{\mathrm{d}^4x}{(2\pi)^2}\, f(x)\, \mathrm{e}^{-\mathrm{i}p\cdot x}\,, \qquad f(x) = \int_{-\infty}^{\infty} \frac{\mathrm{d}^4p}{(2\pi)^2}\, \widehat{f}(p)\, \mathrm{e}^{\mathrm{i}p\cdot x} \qquad (9.28)$$

for the Fourier transformation in the continuum. We stress at this point that the Fourier transform in the continuum is denoted by $\widehat{f}$, while for the Fourier transform on the lattice $\widetilde{f}$ is used, as defined in Appendix A.3.

The continuum fields $\phi(x), \overline{\phi}(x)$ carry only Dirac indices, since no gauge field is coupled. In order to obtain blocked lattice fields $\phi_n^B, \overline{\phi}_n^B$ from the continuum fields $\phi(x), \overline{\phi}(x)$, we embed an infinite Euclidean lattice $\mathbb{Z}^4$ into our Euclidean space–time $\mathbb{R}^4$. The lattice constant $a$ is set to $a = 1$ for now, but will be displayed explicitly later when needed. We construct the blocked fields by integrating the continuum fields over hypercubes with sides of length 1, centered at the sites $n$ of our lattice, i.e., $n \in \mathbb{Z}^4$:

$$\phi_n^B = \left(\prod_{\mu=1}^{4} \int_{n_\mu-\frac{1}{2}}^{n_\mu+\frac{1}{2}} \mathrm{d}x_\mu\right) \phi(x) = \int_{-\infty}^{\infty} \frac{\mathrm{d}^4p}{(2\pi)^2}\, \widehat{\phi}(p)\, \pi(p)\, \mathrm{e}^{\mathrm{i}p\cdot n}\,. \qquad (9.29)$$

In this equation we have expressed the continuum field $\phi(x)$ through its Fourier transform $\widehat{\phi}(p)$ as defined in (9.28) and abbreviated the remaining space–time integral as

$$\pi(p) = \left(\prod_{\mu=1}^{4} \int_{-\frac{1}{2}}^{\frac{1}{2}} \mathrm{d}x_\mu\right) \mathrm{e}^{\mathrm{i}p\cdot x} = \prod_{\mu=1}^{4} \frac{2\sin(p_\mu/2)}{p_\mu}\,. \qquad (9.30)$$

The function $\pi(p)$ describes the blocking procedure in momentum space. The blocking of the conjugate field $\overline{\phi}$ proceeds in exactly the same way. Since the blocking procedure is trivial in Dirac space, the blocked fields simply inherit the Dirac indices of their continuum counterparts.

Equation (9.29) expresses the blocked field $\phi_n^B$ as a Fourier integral with the Fourier transform $\widehat{\phi}(p)$ of the continuum. For later use we rewrite (9.29) to obtain an alternative expression where the lattice Fourier transform $\widetilde{\phi}^B(p)$ of the blocked field is used:

$$\phi_n^B = \int_{-\pi}^{\pi} \frac{\mathrm{d}^4 p}{(2\pi)^2} \, \widetilde{\phi}^B(p) \, e^{ip\cdot n} \,, \quad \widetilde{\phi}^B(p) = \sum_{k\in\mathbb{Z}^4} \widehat{\phi}(p+2\pi k)\,\pi(p+2\pi k) \,. \quad (9.31)$$

Note that now the momentum integral runs only over the Brillouin zone, i.e., $p_\mu \in (-\pi, \pi]$, which is the momentum space of an infinite lattice, as can be seen from the formulas in Appendix A.3, when one sends the number of lattice points to infinity.

### 9.2.2 The blocked lattice Dirac operator

After defining the blocked fields $\phi_n^B, \overline{\phi}_n^B$ and evaluating their expression as Fourier integrals, we can now formally define the lattice theory obtained from the blocking procedure:

$$\exp\left(-\sum_{n,m} \overline{\psi}_n D_{nm} \psi_m\right) = \int \mathcal{D}\left[\phi, \overline{\phi}\right] \exp\left(-\sum_n (\overline{\psi}_n - \overline{\phi}_n^B)(\psi_n - \phi_n^B) - S_F\left[\phi, \overline{\phi}\right]\right).$$
$$(9.32)$$

In this equation $\psi_m, \overline{\psi}_n$ denote the fields of the target lattice theory, i.e., the fermions living on the infinite lattice $\mathbb{Z}^4$. Consequently the indices $n$ and $m$ run over all of $\mathbb{Z}^4$. The Dirac operator on the lattice is denoted by $D_{nm}$. The left-hand side thus is the Boltzmann factor $\exp(-S_{latt})$ of the lattice theory. Both the lattice fields $\psi_m, \overline{\psi}_n$ and the lattice Dirac operator $D_{nm}$ carry spinor indices which we suppress, i.e., we use vector/matrix notation in Dirac space.

The right-hand side of (9.32) defines how the lattice theory, or more precisely the lattice Dirac operator $D_{nm}$, is obtained from its continuum counterpart: The continuum fields $\phi, \overline{\phi}$ are integrated over in a path integral (we will comment on the continuum measure $\mathcal{D}\left[\phi, \overline{\phi}\right]$ later). In the Boltzmann factor we have the continuum action $S_F\left[\phi, \overline{\phi}\right]$ from (9.27) as usual and an additional quadratic piece which couples the blocked continuum fields $\phi^B, \overline{\phi}^B$ to the fields $\psi, \overline{\psi}$ of the target theory on the lattice.

Since the path integral on the right-hand side is Gaussian it can be solved in closed form: It is convenient [22] to rewrite the blocking term with an additional Gaussian integral over auxiliary Grassmann variables $\eta, \overline{\eta}$:

$$\exp\left(-\sum_n \left(\overline{\psi}_n - \overline{\phi}_n^B\right)\left(\psi_n - \phi_n^B\right)\right) = \tag{9.33}$$

$$\int \mathcal{D}[\eta,\overline{\eta}]\,\exp\left(\sum_n \left(\overline{\eta}_n\eta_n + \left(\overline{\psi}_n - \overline{\phi}_n^B\right)\eta_n + \overline{\eta}_n\left(\psi_n - \phi_n^B\right)\right)\right).$$

The auxiliary fields live on the lattice, carry a Dirac index, and the measure $\mathcal{D}[\eta,\overline{\eta}]$ is the product over the measures at all lattice points. The terms in (9.33) that mix the auxiliary fields and the blocked fields $\phi^B$ are written as

$$\sum_n \overline{\eta}_n \phi_n^B = \int_{-\infty}^{\infty}\frac{\mathrm{d}^4 p}{(2\pi)^2}\,\widehat{\overline{J}}(-p)\,\widehat{\phi}(p)\,,\quad \widehat{\overline{J}}(p) = \sum_n \overline{\eta}_n\,\pi(-p)\,\mathrm{e}^{-ip\cdot n}\,,$$

$$\sum_n \overline{\phi}_n^B \eta_n = \int_{-\infty}^{\infty}\frac{\mathrm{d}^4 p}{(2\pi)^2}\,\widehat{\overline{\phi}}(-p)\,\widehat{J}(p)\,,\quad \widehat{J}(p) = \sum_n \eta_n\,\pi(-p)\,\mathrm{e}^{-ip\cdot n}\,,\tag{9.34}$$

where we have used (9.29) to express the blocked fields through their continuum Fourier transforms.

Inserting (9.33) and (9.34) into the master formula (9.32), we obtain

$$\exp\left(-\sum_{n,m}\overline{\psi}_n D_{nm}\psi_m\right) = \int \mathcal{D}[\eta,\overline{\eta}]\,\exp\left(\sum_n \left(\overline{\eta}_n\eta_n + \overline{\psi}_n\eta_n + \overline{\eta}_n\psi_n\right)\right) I\left[J,\overline{J}\right]\,,\tag{9.35}$$

where we denote the remaining path integral over the continuum fields as $I\left[J,\overline{J}\right]$ given by

$$\int \mathcal{D}\left[\phi,\overline{\phi}\right]\exp\left(-\int_{-\infty}^{\infty}\frac{\mathrm{d}^4 p}{(2\pi)^2}\left((2\pi)^2\widehat{\overline{\phi}}(-p)\,\mathrm{i}\gamma_\mu p_\mu\,\widehat{\phi}(p) + \widehat{\overline{\phi}}(-p)\,\widehat{J}(p) + \widehat{\overline{J}}(-p)\,\widehat{\phi}(p)\right)\right)$$

$$= C\,\exp\left(\int_{-\infty}^{\infty}\frac{\mathrm{d}^4 p}{(2\pi)^4}\,\widehat{\overline{J}}(-p)\,\frac{-\mathrm{i}\gamma_\mu p_\mu}{p^2}\,\widehat{J}(p)\right)\,.\tag{9.36}$$

In the second step we have solved the path integral, as discussed in Chap. 5, by completing the square in the exponent and a shift of the integration variables in the path integral, $\widehat{\phi}'(p) = \widehat{\phi}(p) + \mathrm{i}\gamma_\mu p_\mu(2\pi)^{-2}p^{-2}\widehat{J}(p)$, and similarly for $\widehat{\overline{\phi}}$. The remaining term, which after the shift of the integration variables is independent of $J$ and $\overline{J}$, is denoted by the constant $C$. We stress at this point that we did not explicitly define the continuum measure $\mathcal{D}\left[\phi,\overline{\phi}\right]$ and only assumed, that it remains invariant under a shift of variables. As a matter of fact, for a proper definition of this measure, a regulator such as a finite lattice is necessary. If one removes this regulator the constant $C$ diverges which, however, is irrelevant for the evaluation of vacuum expectation values of operators, where such a constant cancels in the normalization of expectation values with the partition function $Z$. Thus we can ignore the constant $C$ and the path integral (9.36) is given by the term quadratic in $J$ which comes from the completion of the square.

The final step is to insert result (9.36) into the remaining integral (9.35) over $\eta$ and $\overline{\eta}$:

$$\exp\left(-\sum_{n,m}\overline{\psi}_n D_{nm}\psi_m\right)$$

$$= C\int \mathcal{D}[\eta,\overline{\eta}]\exp\left(\sum_{n,m}\overline{\eta}_n\,(\mathbb{1}+R)_{nm}\,\eta_m + \sum_n (\overline{\psi}_n\eta_n + \overline{\eta}_n\psi_n)\right)$$

$$= CC'\exp\left(-\sum_{n,m}\overline{\psi}_n\,(\mathbb{1}+R)_{nm}^{-1}\,\psi_m\right). \tag{9.37}$$

In the last step we have again solved the Gaussian path integral by completing the square in the exponent and the remaining term, which is independent of $\psi$ and $\overline{\psi}$, gives rise to another irrelevant constant $C'$. The matrix $R_{nm}$ was obtained by inserting the expressions for $\widehat{J}(p)$ and $\widehat{\overline{J}}(p)$ from (9.34) into the exponent on the right-hand side of (9.36). Explicitly the matrix $R$ reads

$$R_{nm} = \int_{-\infty}^{\infty}\frac{\mathrm{d}^4p}{(2\pi)^2}\,\pi(p)\,\mathrm{e}^{\mathrm{i}p\cdot n}\,\frac{-\mathrm{i}\gamma_\mu p_\mu}{p^2}\,\mathrm{e}^{-\mathrm{i}p\cdot m}\,\pi(-p)\,, \tag{9.38}$$

and by comparing the first and the last term in (9.37) we obtain the final result for the lattice Dirac operator (after dropping the constants $C$ and $C'$)

$$D_{nm} = (\mathbb{1}+R)_{nm}^{-1}\,. \tag{9.39}$$

The lattice Dirac operator $D$ is most conveniently expressed by switching to its Fourier transform:

$$\widetilde{D}(p|q) = \frac{1}{(2\pi)^4}\sum_{n,m}\mathrm{e}^{-\mathrm{i}p\cdot n}D_{nm}\mathrm{e}^{\mathrm{i}q\cdot m}\,. \tag{9.40}$$

For the necessary Fourier transform of $R$ one finds

$$\widetilde{R}(p|q) = \delta(p-q)\sum_{k\in\mathbb{Z}^4}\frac{\pi\,(q+2\pi k)^2}{(q+2\pi k)^2}(-\mathrm{i})\gamma_\mu(q+2\pi k)_\mu = -\mathrm{i}\,\delta(p-q)\gamma_\mu v_\mu(q)\,,$$

$$\text{with }v_\mu(q) = \sum_{k\in\mathbb{Z}^4}\frac{\pi\,(q+2\pi k)^2}{(q+2\pi k)^2}(q+2\pi k)_\mu\,. \tag{9.41}$$

Inserting this into (9.39) one obtains [22]

$$\widetilde{D}(p|q) = \delta(p-q)\,\widetilde{D}(q)\,,\ \ \widetilde{D}(q) = (a\,\mathbb{1}-\mathrm{i}\,\gamma_\mu v_\mu(q))^{-1} = \frac{a\,\mathbb{1}+\mathrm{i}\,\gamma_\mu v_\mu(q)}{a^2+v(q)^2}\,. \tag{9.42}$$

Here we have split off the Dirac-$\delta$ that ensures momentum conservation which is a consequence of translation invariance. We have also re-inserted the lattice spacing $a$ necessary for the analysis below.

### 9.2.3 Properties of the blocked action

The lattice propagator in momentum space, $\widetilde{D}(q)$, is the result of our blocking procedure for obtaining the effective lattice theory. It has a series of remarkable properties which we discuss now.

First we convince ourselves that $\widetilde{D}(q)$ has the correct naive continuum limit and that it is free of doublers. In other words we need to show that

$$\widetilde{D}(q) = \begin{cases} i\gamma_\mu q_\mu + \mathcal{O}(a) & \text{when all } q_\mu \approx 0 , \\ \mathcal{O}(1/a) & \text{when at least one } q_\mu \approx \pi/a . \end{cases} \tag{9.43}$$

When inspecting the final result (9.42) for the lattice Dirac operator, it is obvious that the properties (9.43) must come from the behavior of the functions $v_\mu(q)$ given in (9.41). As a matter of fact one can show (e.g., most quickly by analyzing the $v_\mu(q)$ numerically) that they behave as

$$v_\mu(q) \approx \begin{cases} q_\mu/q^2 & \text{when all } q_\mu \approx 0 , \\ 0 & \text{when at least one } q_\mu \approx \pi/a . \end{cases} \tag{9.44}$$

This behavior exactly implies the required properties of a lattice Dirac operator as stated in (9.43).

Having established that the blocked Dirac operator $\widetilde{D}$ gives rise to a proper lattice theory let us now investigate its properties further. When transforming back to real space it is found that the blocked Dirac operator is not ultralocal, in agreement with the general discussion in Sect. 7.4.2. It may, however, be demonstrated [22] that the blocked Dirac operator falls off exponentially as is necessary for a local quantum field theory.

Another important aspect is discretization errors. We have seen in (9.43) that in the limit of vanishing momentum $q$ we obtain the correct behavior $i\gamma_\mu q_\mu$. What, however, are the deviations from that ideal behavior as one increases the momentum? A physically interesting way of assessing this question is to analyze the dispersion relation of a free fermion described by $\widetilde{D}$. This may be done by computing the poles of $\widetilde{D}(q)^{-1}$ in the complex $q_4$ plane for different values of the spatial momentum $\boldsymbol{q}$. Plotting the energy $E(\boldsymbol{q})$, which is given by the modulus of the position $q_4$ of the pole, versus the spatial momentum, one finds that for the blocked Dirac operator the values fall exactly on top of the linear continuum dispersion relation, $E(\boldsymbol{q}) = |\boldsymbol{q}|$, essentially almost all the way up to the cutoff $\pi/a$. This indicates that the discretization errors are small.

Finally we comment on the chiral properties of the blocked lattice Dirac operator $\widetilde{D}(q)$. It is a simple exercise to show that it obeys the Ginsparg–Wilson equation in the form

$$\widetilde{D}(q)\,\gamma_5 + \gamma_5\,\widetilde{D}(q) = 2\,a\,\widetilde{D}(q)\,\gamma_5\,\widetilde{D}(q) . \tag{9.45}$$

We remark that here we encounter the Ginsparg–Wilson equation with a slightly different normalization than the original version (7.29), namely an extra factor of 2 on the right-hand side. This factor is, however, completely irrelevant, since the whole program for chiral symmetry on the lattice, which we presented in Chap. 7, can be implemented with the modified normalization as well. Thus we find that the blocked theory is also chiral.

To summarize this section, we have shown that blocking free fermions from the continuum gives rise to

- Proper lattice fermions without doublers and a local action.
- Fermions with a perfect dispersion relation almost up to the cutoff.
- A Dirac operator that obeys the Ginsparg–Wilson equation.

With these three properties it is obvious that the blocking procedure is a powerful method for constructing lattice discretizations of quantum field theories. In Sect. 9.4 we will show that the emerging chiral lattice symmetry is not a coincidence but is the consequence of a rather general mapping of continuum symmetries onto a corresponding lattice theory obtained by blocking. Before we come to this analysis, we need, however, to address the problem of generalizing the strategy, which we have here outlined for free fermions, to the case of full QCD where gluons and quark fields are interacting.

## 9.3 Real space renormalization group for QCD

In the last section we have demonstrated by an explicit calculation for free fermions that the RG approach of integrating out degrees of freedom above the cutoff $\sim 1/a$ is a powerful tool for constructing a lattice discretization of a quantum field theory. The resulting lattice action has beautiful properties, in particular discretization effects are absent almost up to the cutoff and the lattice Dirac operator obeys the Ginsparg–Wilson equation, which is a lattice manifestation of the continuum chiral symmetry.

For the case of full QCD, i.e., when gauge fields are coupled to the fermions, the situation is considerably more complicated. A serious obstacle prevents a direct blocking from the continuum like for free fermions considered in the last section: The continuum path integral on the right-hand side of (9.32) cannot be solved in closed form if non-Gaussian terms appear. Such terms are the interaction terms of the gauge field and the fermions, as well as the cubic and quartic terms of the gauge action.

The conceptual way out is to use a lattice version of QCD on a very fine lattice, i.e., a lattice close to the continuum limit where the lattice spacing $a$ is very small and discretization effects are tiny. This fine lattice replaces the continuum theory as a starting point for the blocking transformation. Instead of integrating hypercubes of the continuum we average the lattice fields on a block of sites of the fine lattice to construct the fields on a coarser lattice. The action on the coarse lattice is identified through an equation equivalent to (9.32). Iterating such a *real space renormalization group transformation* we may obtain the action of the target theory on a lattice with the coarseness

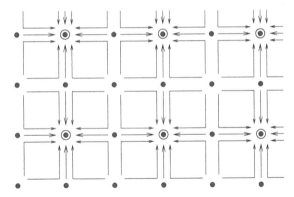

**Fig. 9.2.** Example of an overlapping discrete block spin transformation. We show for a plane of the lattice how the fermion fields on the fine lattice (*filled dots*) may be averaged to construct the blocked fermion fields living on the coarse lattice (*open circles* around the sites of the lattice with only even indices). Along the *arrows* gauge transporters are included to obtain gauge covariant spinors on the coarse lattice

we want to work at. Up to the cutoff this coarse action will have the same discretization errors as the theory on the fine lattice we started with. The physics of the short-distance UV modes, which are averaged over in the RG transformation, is encoded in terms of the action used on the coarse lattice. We remark that originally the real space renormalization group was developed as an approach for analyzing systems of statistical mechanics near criticality [23, 24]. As an introduction to its application in lattice QCD which goes beyond our brief sketch we recommend [25].

### 9.3.1 Blocking full QCD

Let us begin with describing how we can construct new fields living on a coarse lattice $\Lambda'$ with spacing $a' = 2a$ from fields on a fine lattice $\Lambda$ with lattice constant $a$. Since we now consider QCD, we have to block both the fermions and the gauge fields. Before we discuss the actual construction, we set the notational conventions: The space–time index of the fine lattice $\Lambda$ is denoted by $n$ and the space–time index of the coarse lattice $\Lambda'$ by $n'$. The fermion and the gauge fields on the fine lattice are $\psi_n, \overline{\psi}_n$ and $U_\mu(n)$, their blocked counterparts living on the coarse lattice are $\psi_{n'}^B, \overline{\psi}_{n'}^B$ and $U_\mu^B(n')$, and the genuine coarse fields are denoted by $\psi_{n'}', \overline{\psi}_{n'}'$ and $U_\mu'(n')$.

We start with discussing the blocking prescription of the fermions. Many different blocking schemes are possible. A rather simple one is illustrated in Fig. 9.2, where we show how the fields may be blocked in a plane of the lattice (the other planes are blocked equivalently). The fermion fields $\psi_n, \overline{\psi}_n$ live on the sites of the fine lattice (filled dots in the figure) and have to be averaged suitably for obtaining the blocked fermions $\psi_{n'}^B, \overline{\psi}_{n'}^B$, living on the coarse lattice which we choose to consist of the sites with only even indices of the fine lattice (marked by circles in the plot). The blocked fields $\psi_{n'}^B, \overline{\psi}_{n'}^B$ at a site $n'$ of the coarse lattice are obtained by averaging over all the fields $\psi_n, \overline{\psi}_n$ on all lattice sites $n$ which are connected to $n'$ by one or more oriented paths. Obviously the sites of the fine lattice are connected to several sites of the coarse lattice. Such a type of blocking prescription is referred to as *overlapping block spin transformation*.

It is mandatory to keep gauge covariance intact when combining the fermion fields $\psi_n, \overline{\psi}_n$ from different sites $n$ of the fine lattice. Thus we need to connect the fields at sites $n$ to the lattice site $n'$ where we put the blocked field through gauge transporters $U(n', n)$. Such transporters are obtained by multiplying the link variables $U_\mu(m)$ of the fine lattice along the oriented paths in the figure. If a link is run through in negative direction we apply our usual convention $U_{-\mu}(m) \equiv U_\mu(m - \hat{\mu})^\dagger$. In order to obtain a blocking prescription which respects the symmetries of the lattice, several sites $n$ are connected to $n'$ by a set of paths which are related by symmetries such as discrete rotations, reflections etc. We can summarize the blocking of the fermions as

$$\psi_{n'}^B = \sum_{n \in \mathcal{N}(n')} \left( \sum_{p \in \mathcal{P}(n', n)} s_p \prod_{(m, \mu) \in p} U_\mu(m) \right) \psi_n \equiv \sum_n \omega[U]_{n'n} \psi_n . \quad (9.46)$$

Here $\mathcal{N}(n')$ is a neighborhood of sites $n$ of the fine lattice around the target site $n'$ on the coarse lattice. By $\mathcal{P}(n', n)$ we denote the set of paths $p$ that we want to use for connecting the site $n$ of the fine lattice to $n'$, e.g., the choice depicted in Fig. 9.2. The product runs over all the links $(m, \mu)$ in the path $p$. Each path comes with a coefficient $s_p$. These coefficients must reflect the symmetries (e.g., paths related to each other by a rotation must have the same coefficient), and as a normalization condition, we require the coefficients of all paths connected to a target site $n'$ to sum up to 1. In the last step of (9.46) we have rewritten the blocking prescription by introducing a matrix $\omega[U]_{n'n}$ which depends on the gauge fields $U$ of the fine lattice and collects all the terms that connect $n'$ and $n$. The conjugate fields $\overline{\psi}(n)$ are blocked equivalently with the hermitian conjugate of $\omega[U]$. We point out that, as in the last section, the blocking procedure is trivial in Dirac space.

Concerning the blocking of the gauge link $U_\mu(n)$ one could simply multiply the two fine link variables that connect two sites of the coarse lattice in order to obtain the blocked variables. However, also here usually a more general gauge covariant averaging is used, with a simple version given by

$$U_\mu^B(n') = (1 - \alpha) U_\mu(n') U_\mu(n' + \hat{\mu}) \quad (9.47)$$
$$+ \frac{\alpha}{6} \sum_{\nu \neq \mu} \left( U_\nu(n') U_\mu(n' + \hat{\nu}) U_\mu(n' + \hat{\nu} + \hat{\mu}) U_\nu(n' + 2\hat{\mu})^\dagger \right.$$
$$\left. + U_\nu(n' - \hat{\nu})^\dagger U_\mu(n' - \hat{\nu}) U_\mu(n' - \hat{\nu} + \hat{\mu}) U_\nu(n' - \hat{\nu} + 2\hat{\mu}) \right) .$$

Here $\hat{\mu}$ and $\hat{\nu}$ are unit vectors on the fine lattice pointing in the $\mu$ and $\nu$ directions, and $\alpha$ is a real parameter of the blocking procedure. Essentially the blocking linearly combines paths of links on the fine lattice which connect neighboring sites on the coarse lattice. Along these paths the ordered products of the fine link variables are taken into account. Finding an optimal blocking prescription requires a certain amount of experimenting, and prescriptions more refined than the one in (9.47) are discussed in [26–29].

We remark that in the above definition the blocked field $U_\mu^B(n')$ is not an element of the gauge group, since the sum of two SU(3) matrices is not in SU(3). Below we will define how the blocked field $U_\mu^B(n')$ is connected to the link variables $U_\mu'(n')$ on the coarse lattice, which of course will then be elements of the gauge group.

Having defined how to obtain the blocked fields $\psi^B, \overline{\psi}^B$, and $U^B$, we can now formulate a blocking relation which generalizes (9.32) and defines the theory on the coarse lattice:

$$
\mathrm{e}^{-S_F'[\psi',\overline{\psi}',U'] - \beta' S_G'[U']} = \int \mathcal{D}[\psi,\overline{\psi}]\,\mathcal{D}[U]\, \mathrm{e}^{-S_F[\psi,\overline{\psi},U] - \beta S_G[U]}
$$
$$
\times\, \mathrm{e}^{-T_F[\psi',\overline{\psi}',\psi^B,\overline{\psi}^B,U] - \beta T_G[U',U^B]} . \qquad (9.48)
$$

In this equation $S_F[\psi,\overline{\psi},U]$ and $S_F'[\psi',\overline{\psi}',U']$ are the fermion actions for the fine and the coarse lattices, and $\beta S_G[U]$ and $\beta' S_G'[U']$ the corresponding gauge actions, where $\beta, \beta'$ denote the inverse gauge couplings for the two lattices. The right-hand side is now a well-defined lattice path integral over the degrees of freedom on the fine lattice. The blocking functions $T_F$ and $T_G$ connect the fermion and gauge fields $\psi', \overline{\psi}', U'$ of the target theory on the coarse lattice to the blocked quantities $\psi^B, \overline{\psi}^B, U^B$. For the fermions we use

$$
T_F[\psi',\overline{\psi}',\psi^B,\overline{\psi}^B,U] = \kappa_F \sum_{n'} (\overline{\psi}_{n'}' - b\,\overline{\psi}_{n'}^B)(\psi_{n'}' - b\,\psi_{n'}^B)
$$
$$
= \kappa_F \sum_{n'} \left( \overline{\psi}_{n'}' - b\sum_n \overline{\psi}_n \omega[U]_{nn'}^\dagger \right)
$$
$$
\times \left( \psi_{n'}' - b\sum_n \omega[U]_{n'n}\psi_n \right), \qquad (9.49)
$$

where $b$ is a real-valued normalization constant and $\kappa_F$ a real parameter that may be used for tuning properties of the action on the coarse lattice, in particular its locality. In the second line of (9.49) we have expressed the blocked fields with the transformation matrix $\omega[U]$ introduced in (9.46). For the gauge field we define

$$
T_G[U',U^B] = \sum_{n',\nu} \left( \frac{\kappa_G}{3}\mathrm{Re}\,\mathrm{tr}[U_\nu'(n')U_\nu^B(n')] - N[U_\nu^B(n')] \right),
$$
$$
\exp\left(N[U_\nu^B(n')]\right) \equiv \int \mathrm{d}W \exp\left(\beta\kappa_G\,\mathrm{Re}\,\mathrm{tr}[WU_\nu^B(n')^\dagger]/3\right). \qquad (9.50)
$$

The second equation defines the normalization $N\left[U_\nu^B(n')\right]$ through an integral over $W \in \mathrm{SU}(3)$, and $\kappa_G$ is again a real parameter of the transformation.

### 9.3.2 The RG flow of the couplings

In the last section we have defined the blocking prescriptions for fermions and gauge fields and thus have given a well-defined meaning to the renormalization group equation (9.48) which connects the actions for the theories on the coarse and the fine lattices. For an arbitrary coarse configuration $\psi', \overline{\psi}', U'$, the right-hand side of (9.48) defines the action on the coarse lattice in the exponent on the left-hand side.

We have already remarked that the key idea of the RG transformation is to include the physics of the modes on the fine lattice in suitable terms in the action used for the coarse lattice. Such terms can, e.g., look like the clover term (9.10) introduced in the first section of this chapter. All these terms of the action on the coarse lattice come with some coefficients $c_1', c_2' \ldots$, which for the example of the clover term correspond to the Sheikholeslami–Wohlert coefficient of the Symanzik improvement program. We assume that we can view the RG transformation (9.48) as a mapping in the space of the couplings which parameterize a lattice QCD action,

$$\beta\,,\,c_1\,,\,c_2 \ldots \;\xrightarrow{\;RG\;}\; \beta'\,,\,c_1'\,,\,c_2' \ldots \,, \tag{9.51}$$

where the $\beta, c_i$ are the couplings on the fine lattice (most of which vanish for the plain Wilson action which might be the action we start from). For labeling the couplings we do not distinguish between the fermion and gauge field parts of the action and only single out the inverse gauge couplings $\beta, \beta'$.

Starting from the theory with couplings $\beta, c_1, c_2 \ldots$, we block the fields and the couplings of the coarse lattice, $\beta', c_1', c_2', \ldots$, assume values such that the effects of the integrated-out fields of the fine lattice are taken into account. At the same time the lattice spacing increases by a scale factor of 2,

$$a \;\xrightarrow{\;RG\;}\; a' \;=\; 2a \,. \tag{9.52}$$

We remark that the scale factor of 2 is particular for the blocking transformation defined by Fig. 9.2 and (9.47), and other blocking prescriptions can have a different scale factor.

We now ask ourselves whether mapping (9.51) has a fixed point (FP), i.e., a set of couplings that is mapped onto itself under the RG transformation (9.51). Equation (9.52) suggests that two types of fixed points might exist, corresponding to $a = a' = \infty$ or to $a = a' = 0$. Obviously only the second possibility, which often is referred to as *critical fixed point*, is of interest to us since it corresponds to the continuum limit $a \to 0$. As we have discussed in Sect. 3.5.4, the continuum limit of lattice QCD, where $a \to 0$, is obtained in the limit $\beta \to \infty$. Thus we expect to find a suitable critical fixed point at

$\beta = \infty$ and parameter values $c_1^\star, c_2^\star, \ldots$ Under our RG transformation (9.51) these values are mapped onto themselves.

In a practical application we cannot start our RG transformation from $\beta = \infty$. However, a detailed analysis of the flow of couplings under repeated RG transformations (see, e.g., [25]) suggests that it is sufficient to start from a large but finite value of $\beta$ and a few steps of repeated RG transformations will drive the couplings $c_1, c_2, \ldots$ into the fixed point values $c_1^\star, c_2^\star, \ldots$

### 9.3.3 Saddle point analysis of the RG equation

So far we have specified the blocking prescription for full QCD, defined the RG transformation (9.48), and identified a possible fixed point which corresponds to the continuum physics we are interested in. What is left to do is to find a way of solving the RG transformation (9.48), such that we can construct the RG action on the lattice.

For this purpose we may use the fact that the critical fixed point we want to find corresponds to $\beta = \beta' = \infty$. In this limit the path integral on the right-hand side of (9.48) is dominated by its saddle point, i.e., a configuration (not necessarily unique) of gauge fields $U$ on the fine lattice which minimizes the exponent. On both sides of (9.48) the terms in the exponent which are multiplied with the inverse couplings $\beta, \beta'$ have to match as we send the inverse couplings to infinity. We read off the following equation for the gauge field part of the action:

$$S'_G [U'] = \min_U \left( S_G [U] + T_G^\infty [U', U^B] \right) . \tag{9.53}$$

The superscript $\infty$ attached to the blocking function $T_G$ indicates that it also has to be evaluated at $\beta = \infty$. Since we want to identify the so-called *fixed point (FP) action* $S_G^{FP}$ at the fixed point, which is defined by the coefficients $c_i^\star$ that are mapped onto themselves, we can set $S'_G = S_G = S_G^{FP}$ and obtain the equation that identifies the gauge part $S_G^{FP}$ of the fixed point action,

$$S_G^{FP} [U'] = \min_U \left( S_G^{FP} [U] + T_G^\infty [U', U^B] \right) \tag{9.54}$$

$$= \min_U \left( S_G^{FP} [U] + \sum_{n',\nu} \left( \frac{\kappa_G}{3} \mathrm{Re\ tr} \left[ U'_\nu(n') U_\nu^B(n') \right] - N^\infty \left[ U_\nu^B(n') \right] \right) \right) ,$$

where in the second step we have inserted the explicit expression of the blocking function $T_G$ as given in (9.50). The necessary limit $\beta \to \infty$ is easy to perform since $\beta$ enters only through the normalization constant $N^\infty$ which again can be evaluated through its saddle point giving

$$N^\infty \left[ U_\nu^B(n') \right] = \frac{\kappa_G}{3} \max_{W \in SU(3)} \mathrm{Re\ tr} \left[ W U_\nu^B(n')^\dagger \right] . \tag{9.55}$$

Matching the exponents with factors $\beta, \beta'$ in the saddle point analysis of (9.48) leads to Eq. (9.54) which determines the gauge part of the fixed point action. The requirement that also the remaining factors from the fermion contributions match gives rise to

$$e^{-S'_F[\psi',\overline{\psi}',U']} = \int \mathcal{D}[\psi,\overline{\psi}] \, e^{-S_F[\psi,\overline{\psi},U^{\min}] - T_F[\psi',\overline{\psi}',\psi^B,\overline{\psi}^B,U^{\min}]}, \quad (9.56)$$

where the right-hand side has to be evaluated for the configuration $U^{\min}$ which minimizes the right-hand side of the gauge field RG equation (9.54).

The integral on the right-hand side of (9.56) can be solved, since it is Gaussian. We write the coarse and fine fermion actions with the help of Dirac operators $D'$ and $D$ for the coarse and the fine lattices as

$$S'_F[\psi',\overline{\psi}',U'] = \sum_{n',m'} \overline{\psi}'_{n'} \, D'[U']_{n'm'} \, \psi'_{m'},$$

$$S_F[\psi,\overline{\psi},U^{\min}] = \sum_{n,m} \overline{\psi}_n \, D[U^{\min}]_{nm} \, \psi_m. \quad (9.57)$$

Using the fact that the blocking function $T_F[\psi',\overline{\psi}',\psi^B,\overline{\psi}^B,U^{\min}]$ is also bilinear in $\psi$ and $\overline{\psi}$ (see (9.49)) we solve the fermionic integral with the techniques of Sect. 5.1.4. The result of the integral on the right-hand side of (9.56) is the exponential of a bilinear form for the coarse fields $\psi'$, $\overline{\psi}'$. The kernel of this bilinear form has to match the kernel in the exponent of the bilinear form on the left-hand side of (9.56). We obtain

$$D'[U']_{n'm'} = \kappa_F \, \delta_{n'm'} \qquad (9.58)$$

$$-\kappa_F^2 b^2 \left( \omega[U^{\min}] \left( D[U^{\min}] + \kappa_F b^2 \omega[U^{\min}]^\dagger \omega[U^{\min}] \right)^{-1} \omega[U^{\min}] \right)_{n'm'},$$

which is the RG equation that relates the Dirac operator $D'$ on the coarse lattice to the Dirac operator $D$ of the fine lattice.

Equations (9.54) and (9.58) are the two RG equations that connect the action on the fine lattice to the action on the coarse lattice. It is interesting to note that the RG equation for the fixed point gauge action is independent of the problem for the fermions. The RG equation (9.58) for the fermions, however, needs as input a gauge configuration $U^{\min}$ on the fine lattice, which minimizes the right-hand side of (9.54). In the next section we discuss how solutions of (9.54) and (9.58) may be approximated.

### 9.3.4 Solving the RG equations

The two RG equations (9.54) and (9.58) are nonlinear equations which should hold for arbitrary coarse background gauge fields $U'$. In a first step one has to determine the fine configuration $U^{\min}$ that corresponds to a given $U'$ and minimizes the right-hand side of (9.54). This configuration is then used in the RG equation (9.58) for the fermions.

In order to determine the fixed point gauge action $S_G^{FP}$ and the Dirac operator $D'$, both these quantities have to be approximated and parameterized using a finite number of terms with couplings $c_i$. For the gauge action one may use a linear combination of traces over closed loops of link variables and powers thereof,

$$S_G^{FP}[U] = \sum_{l \in \mathcal{L}} \sum_m c_m(l) \left( \frac{1}{3} \operatorname{Re} \operatorname{tr} [\mathbb{1} - U_l] \right)^m . \qquad (9.59)$$

In this equation $\mathcal{L}$ is a collection of closed loops on the lattice and $U_l$ the ordered product of link variables along such a loop $l \in \mathcal{L}$. The couplings $c_m(l)$ are real parameters of the action. The set $\mathcal{L}$ of loops has to be chosen such that the lattice symmetries, e.g., discrete translations, rotations, are obeyed. Obviously, the parameterized action (9.59) is a generalization of the Wilson gauge action (2.49) and reduces to this action if we choose for $\mathcal{L}$ the set of all plaquettes and restrict ourselves only to powers $m = 1$ in (9.59).

The parameterization of the Dirac operator proceeds in a similar way [30, 31] and may also be viewed as a generalization of the Wilson Dirac operator (5.51). When we discussed the doubling problem in Sect. 5.2 we were forced to include terms that come with the unit matrix in Dirac space. In the ansatz for the parameterized Dirac operator now all possible 16 generators $\Gamma_k \in \{\mathbb{1}, \gamma_\mu, \sigma_{\mu\nu}, \gamma_\mu \gamma_5, \gamma_5\}$ of the Clifford algebra are used. Similar to the gauge action we allow for a whole collection of paths $p \in \mathcal{P}_{n,m}^{(k)}$ that connect two sites $n, m$ on the lattice and we obtain [30, 31]

$$D[U]_{nm} = \sum_{k=1}^{16} \Gamma_k \sum_{p \in \mathcal{P}_{n,m}^{(k)}} c_{n,m}^{(k,p)}[U] \, U_p , \qquad (9.60)$$

where $U_p$ denotes the product of link variables along the path $p$. Again the set of paths $\mathcal{P}_{n,m}^{(k)}$ has to respect the discrete lattice symmetries. In addition the set of paths $\mathcal{P}_{n,m}^{(k)}$ and the coefficients $c_{n,m}^{(k,p)}$ depend on the element of the Clifford algebra, since for the fermions the reflections and other symmetries such as charge conjugation (see Sect. 5.4) involve also matrices acting in Dirac space. The coefficients $c_{n,m}^{(k,p)}[U]$ may be gauge-invariant functions of the link variables $U_\mu$, e.g., they could depend on the local plaquette values.

For the gauge action $S_G^{FP}$ the couplings on both sides of (9.54) are the same and determine the fixed point action which then is used at arbitrary $\beta$. The actual determination is done via minimizing the difference between the left-hand and the right-hand sides of (9.54). This step actually involves two minimizations: The coefficients of the action and the fine configuration have to be varied. In a practical approach [26–29, 32] one first uses a simpler gauge action on the right-hand side of (9.54) which serves to determine a minimal gauge configuration $U^{\min}$ corresponding to a coarse configuration $U'$. In a second step this minimal configuration is then used to determine the coefficients $c_m(l)$ of the parameterized fixed point action (9.59). The resulting fixed point gauge action was tested nonperturbatively and was shown to have only very small cutoff effects [26–29].

For the fermionic RG equation (9.58) an approach, slightly different from the fixed point action for the gauge fields, was chosen [33–37]. Instead of finding a fixed point solution for the RG equation, a sequence of Dirac operators is determined using (9.58): One starts from a fine lattice, with very large $\beta$ and a Dirac operator $D$, for the fine lattice where the couplings are held fixed. The couplings of a new Dirac operator $D'$ for the coarse lattice are then determined from (9.58). Iterating this step the Dirac operator at the target $\beta$, which gives rise to the resolution one wants to work at, may be determined. In each step the coefficients are computed from (9.58) by minimizing a $\chi^2$ functional which may be chosen as

$$\chi^2 = \sum_i \sum_k \left\| \left( D' \left[ U'_{(i)} \right] - D^{\mathrm{rhs}} \left[ U^{\mathrm{min}}_{(i)} \right] \right) \psi^{\mathrm{rand}}_{(k)} \right\|^2 . \tag{9.61}$$

In this equation $D' \left[ U'_{(i)} \right]$ is the Dirac operator on the coarse lattice and $D^{\mathrm{rhs}} \left[ U^{\mathrm{min}}_{(i)} \right]$ denotes the right-hand side of the RG equation (9.58). The sum with index $i$ runs over an ensemble of coarse gauge configurations $U'_{(i)}$, $i = 1, 2, \ldots$, and the corresponding minimal configurations determined from the FP equation (9.54) are denoted as $U^{\mathrm{min}}_{(i)}$. The difference of the two Dirac operators then acts on a set of random vectors $\psi^{\mathrm{rand}}_{(k)}$ labeled by an index $k = 1, 2, \ldots$. The action of the difference of Dirac operators on the $\psi^{\mathrm{rand}}_{(k)}$ produces a set of vectors and by $\| \ldots \|$ we denote the norm of these vectors.

The resulting FP Dirac operator was tested in a series of papers [32–39] and it was established that the expected properties such as small cutoff effects can be captured by parameterizations (9.59) and (9.60). Probably even more important is the fact that the RG equations transport symmetries of the continuum onto the lattice. For chiral symmetry these properties were discussed in detail [35, 40, 41]. In particular the FP Dirac operator obeys the Ginsparg–Wilson equation and the properties which we discussed already in Chap. 7, such as the correct anomaly, index theorem, follow. We will revisit the fact that RG transformations transport symmetries onto the lattice in the next section.

We remark that the parameterization of the Dirac operator (9.60) has been used also in a different way [38, 42]. Inserting the parameterized Dirac operator (9.60) into the Ginsparg–Wilson equation (7.29), one can read off a system of coupled quadratic equations for the coefficients $c^{(k,p)}_{n,m}$. After truncation this system can be solved numerically and the resulting coefficients determine the so-called *chirally improved Dirac operator*. This operator was tested extensively [39, 43] and was found to have good chiral properties at a numerical cost considerably lower than necessary for the overlap operator.

## 9.4 Mapping continuum symmetries onto the lattice

In the last section we have seen that a practical implementation of the RG approach to full lattice QCD is a rather nontrivial enterprise. We conclude this chapter by presenting another very important application of the real space renormalization techniques: We show that the blocking transformation may be used to map symmetries of the continuum action onto a lattice version of the symmetry for the corresponding lattice action obtained by blocking. We discuss this connection between continuum and lattice symmetries here only for the case of fermions. This case is particularly important, since one of the symmetries of interest is chiral symmetry – a long-standing problem of the lattice formulation. If one maps the (vectorlike) chiral symmetry in the continuum onto the lattice, one finds that the emerging structure on the lattice is the Ginsparg–Wilson equation which we already discussed in detail in Chap. 7. However, the relation between continuum symmetries and their lattice counterparts is not restricted to chiral symmetries but may be worked out for general symmetries of a continuum theory.

### 9.4.1 The generating functional and its symmetries

Similar to the blocking of free lattice fermions discussed in Sect. 9.2, the starting point of our analysis is again the equation which defines the Dirac operator $D$ on the lattice through a blocking transformation from the fermionic continuum action $S$:

$$
e^{-\overline{\psi} D \psi} = \int \mathcal{D}\left[\phi, \overline{\phi}\right] e^{-\left(\overline{\psi} - \overline{\phi}^B\right) B \left(\psi - \phi^B\right) - S_F\left[\phi, \overline{\phi}\right]} . \tag{9.62}
$$

We denote the lattice fermions by $\psi, \overline{\psi}$ and use vector/matrix notation for all indices (space–time, color, Dirac). $\phi, \overline{\phi}$ are the fermion fields in the continuum which enter the path integral on the right-hand side. Through integrating over hypercubes which are centered at the points of the lattice, one constructs from them the blocked fields $\phi^B, \overline{\phi}^B$. These then live on the sites of the lattice and consequently they have the same indices as the lattice fields $\psi, \overline{\psi}$, in particular a discrete space–time index.

We stress that (9.62) is understood in a background gauge field which is not integrated over but also has to be blocked in a suitable way to maintain gauge covariance (see the last section). The continuum path integral in (9.62) is defined only formally, but could, e.g., be constructed by a lattice theory in the continuum limit, as done in the last section. Anyway, here we are only interested in symmetry properties of the fermion part of the action which is a bilinear form in the fermion fields. For this analysis a formal definition of the continuum path integral is sufficient.

Here we consider a more general blocking prescription than the one used in Sect. 9.2 and allow for a nontrivial blocking kernel $B$ which determines how the blocked fields $\phi^B, \overline{\phi}^B$ and the lattice fields $\psi, \overline{\psi}$ are mixed. At the moment we leave the blocking kernel $B$ unspecified, but will discuss its role in the end of this section.

The work by Ginsparg and Wilson [44] starts from (9.62) and analyzes its behavior under a chiral rotation of the lattice fields. Following [45] we use a slightly different approach and consider the generating functional on the lattice defined as

$$W\left[J,\overline{J}\right] = \int \mathcal{D}\left[\psi,\overline{\psi}\right] e^{\overline{\psi}J + \overline{J}\psi - \overline{\psi}D\psi} , \qquad (9.63)$$

where we have sources $J$ and $\overline{J}$ coupled to the lattice fermions. Inserting the exponential of the action from (9.62), we find an expression for the generating functional through a blocking prescription:

$$W\left[J,\overline{J}\right] = \int \mathcal{D}\left[\psi,\overline{\psi}\right] e^{\overline{\psi}J + \overline{J}\psi} \int \mathcal{D}\left[\phi,\overline{\phi}\right] e^{-\left(\overline{\psi} - \overline{\phi}^{B}\right) B \left(\psi - \phi^{B}\right) - S_F\left[\phi,\overline{\phi}\right]}$$

$$= \int \mathcal{D}\left[\phi,\overline{\phi}\right] e^{-\overline{\phi}^{B}B\phi^{B} - S_F\left[\phi,\overline{\phi}\right]} \int \mathcal{D}\left[\psi,\overline{\psi}\right] e^{-\overline{\psi}B\psi + \overline{\psi}\left(J + B\phi^{B}\right) + \left(\overline{J} + \overline{\phi}^{B}B\right)\psi}$$

$$= \det[B]\, e^{\overline{J}B^{-1}J} \int \mathcal{D}\left[\phi,\overline{\phi}\right] e^{\overline{J}\phi^{B} + \overline{\phi}^{B}J - S_F\left[\phi,\overline{\phi}\right]} . \qquad (9.64)$$

In the last step we have already formally solved the Gaussian integral over the lattice fields by completing the square in the exponent (compare the discussion in Sect. 9.2). The resulting expression (9.64) is the lattice generating functional in terms of a continuum path integral.

We now explore how a symmetry of the continuum fermion action $S_F$ affects the lattice generating functional $W\left[J,\overline{J}\right]$. In particular we consider a transformation of the continuum fields:

$$\phi \rightarrow \phi' = e^{i\varepsilon T}\phi , \quad \overline{\phi} \rightarrow \overline{\phi}' = \overline{\phi}\, e^{i\varepsilon \overline{T}} . \qquad (9.65)$$

The generators $T, \overline{T}$ of the transformation are matrices acting in Dirac space. More general transformations are possible as is discussed in [46]. Our choice implies that the blocked fields $\phi^B$ and $\overline{\phi}^B$ transform in the same way as the original fields, since the blocking from $\phi, \overline{\phi}$ to $\phi^B, \overline{\phi}^B$ is a purely scalar operation, in other words the blocked fields are essentially linear combinations of the continuum fields. We stress that the generators $T$ and $\overline{T}$ are independent transformations, i.e., $\phi$ and $\overline{\phi}$ need not transform in the same way.

We now assume that the transformation (9.65) is a symmetry of the action

$$S_F\left[\phi',\overline{\phi}'\right] = S_F\left[\phi,\overline{\phi}\right] . \qquad (9.66)$$

One can also evaluate the integral over the continuum fields in the last line of (9.64) using the transformed variables $\phi', \overline{\phi}'$. Doing so and exploring the invariance of the action, we obtain

$$\int \mathcal{D}\left[\phi', \overline{\phi}'\right]\, e^{\overline{J}\phi^{B\,\prime} + \overline{\phi}^{B\,\prime} J - S_F\left[\phi', \overline{\phi}'\right]} \tag{9.67}$$

$$= \int \mathcal{D}\left[e^{\mathrm{i}\varepsilon T}\phi, \overline{\phi}\, e^{\mathrm{i}\varepsilon \overline{T}}\right]\, e^{\overline{J}e^{\mathrm{i}\varepsilon T}\phi^{B} + \overline{\phi}^{B} e^{\mathrm{i}\varepsilon \overline{T}} J - S_F\left[\phi, \overline{\phi}\right]}$$

$$= \left(1 + \mathrm{i}\varepsilon \mathcal{A}_{\overline{T}T} + O(\varepsilon^2)\right)\int \mathcal{D}\left[\phi, \overline{\phi}\right]\, e^{\overline{J}e^{\mathrm{i}\varepsilon T}\phi^{B} + \overline{\phi}^{B} e^{\mathrm{i}\varepsilon \overline{T}} J - S_F\left[\phi, \overline{\phi}\right]}\, .$$

In the last step we have transformed the measure of the continuum path integral and taken into account that transformation (9.65) could be anomalous with the anomaly $\mathcal{A}_{\overline{T}T}$ showing up in the Jacobian of the transformation [47, 48]. Since later we will evaluate all expressions up to $O(\varepsilon)$, we have kept only the leading term of the Jacobian. For non-anomalous transformations $T, \overline{T}$ one has $\mathcal{A}_{\overline{T}T} = 0$.

Inserting result (9.67) back into the expression for the generating functional (9.64), we find that continuum symmetries (9.65) and (9.66) imply the following symmetry of the generating functional on the lattice:

$$W\left[J, \overline{J}\right] = e^{\overline{J}\left(B^{-1} - e^{\mathrm{i}\varepsilon T}B^{-1}e^{\mathrm{i}\varepsilon \overline{T}}\right)J}\left(1 + \mathrm{i}\varepsilon \mathcal{A}_{\overline{T}T} + O(\varepsilon^2)\right) W\left[e^{\mathrm{i}\varepsilon \overline{T}}J, \overline{J}e^{\mathrm{i}\varepsilon T}\right]\, . \tag{9.68}$$

Equation (9.68) summarizes how a continuum symmetry reflects itself in the lattice generating functional $W\left[J, \overline{J}\right]$ constructed through blocking.

### 9.4.2 Identification of the corresponding lattice symmetries

Having analyzed the manifestation of a continuum symmetry in our lattice generating functional, we now want to identify a corresponding symmetry on the lattice which also fulfills the symmetry condition (9.68). For that purpose we consider transformed lattice fields

$$\psi' = e^{\mathrm{i}\varepsilon M}\psi\, , \quad \overline{\psi}' = \overline{\psi}\, e^{\mathrm{i}\varepsilon \overline{M}}\, . \tag{9.69}$$

The two generators $M, \overline{M}$ for the lattice transformation are not yet known and we want to identify how they depend on the continuum generators $T, \overline{T}$ and the blocking kernel $B$. The fact that the transformation should be a symmetry of the lattice action implies

$$\overline{\psi}'\, D\,\psi' = \overline{\psi}\, e^{\mathrm{i}\varepsilon \overline{M}}\, D\, e^{\mathrm{i}\varepsilon M}\, \psi \overset{!}{=} \overline{\psi}\, D\,\psi\, . \tag{9.70}$$

For later use we remark that when expanded in $\varepsilon$ the invariance condition (9.70) at $O(\varepsilon)$ implies the commutation relation

$$\overline{M}\, D + D\, M = 0\, . \tag{9.71}$$

As we have done for the continuum expression in the last section, we now express the generating functional (9.63) in terms of the transformed fields $\psi', \overline{\psi}'$ and explore the implications of the symmetry (9.70):

$$W\left[J,\overline{J}\right] = \int \mathcal{D}\left[\psi',\overline{\psi}'\right]\, e^{-\overline{\psi}'D\psi'+\overline{\psi}'J+\overline{J}\psi'} , \tag{9.72}$$

$$= \det\left[e^{i\varepsilon\overline{M}}\right]\det\left[e^{i\varepsilon M}\right]\int \mathcal{D}\left[\psi,\overline{\psi}\right]\, e^{-\overline{\psi}D\psi+\overline{\psi}e^{i\varepsilon\overline{M}}J+\overline{J}e^{i\varepsilon M}\psi} .$$

The two Jacobi determinants on the right-hand side come from the transformation of the measure on the lattice. Using the relation $\det[A] = \exp(\operatorname{tr}\ln A)$ (see (A.54)) we can expand them as

$$\det\left[e^{i\varepsilon\overline{M}}\right]\det\left[e^{i\varepsilon M}\right] = 1 + i\varepsilon\,\operatorname{tr}\left[\overline{M}+M\right] + O(\varepsilon^2) . \tag{9.73}$$

Combining the last two equations we find the symmetry relation for the generating functional which is implied by lattice symmetry (9.70):

$$W\left[J,\overline{J}\right] = \left(1 + i\varepsilon\,\operatorname{tr}\left[\overline{M}+M\right] + O(\varepsilon^2)\right) W\left[e^{i\varepsilon\overline{M}}J, \overline{J}e^{i\varepsilon M}\right] . \tag{9.74}$$

Now we compare symmetry condition (9.74) from the lattice transformation to the corresponding continuum relation (9.68). By setting the two equal we obtain the equation

$$e^{\overline{J}\left(B^{-1}-e^{i\varepsilon\overline{T}}B^{-1}e^{i\varepsilon\overline{T}}\right)J}\left(1 + i\varepsilon\mathcal{A}_{\overline{T}T} + O(\varepsilon^2)\right) W\left[e^{i\varepsilon\overline{T}}J, \overline{J}e^{i\varepsilon T}\right] \tag{9.75}$$

$$= \left(1 + i\varepsilon\,\operatorname{tr}\left[\overline{M}+M\right] + O(\varepsilon^2)\right) W\left[e^{i\varepsilon\overline{M}}J, \overline{J}e^{i\varepsilon M}\right] ,$$

which we can use to identify the lattice transformation $M,\overline{M}$ that matches the continuum transformation $T,\overline{T}$. The last step is to insert the explicit form of the generating functional,

$$W\left[J,\overline{J}\right] = \det[D]\, e^{\overline{J}D^{-1}J} , \tag{9.76}$$

which is obtained by directly solving the Gaussian integral (9.63). When inserting (9.76), (9.75) becomes (we drop the factor $\det[D]$ on both sides)

$$e^{\overline{J}\left(B^{-1}-e^{i\varepsilon\overline{T}}B^{-1}e^{i\varepsilon\overline{T}}\right)J}\left(1 + i\varepsilon\mathcal{A}_{\overline{T}T} + O(\varepsilon^2)\right) e^{\overline{J}e^{i\varepsilon\overline{T}}D^{-1}e^{i\varepsilon T}J} \tag{9.77}$$

$$= \left(1 + i\varepsilon\,\operatorname{tr}\left[\overline{M}+M\right] + O(\varepsilon^2)\right) e^{\overline{J}e^{i\varepsilon\overline{M}}D^{-1}e^{i\varepsilon M}J} .$$

The last equation holds for arbitrary $\varepsilon$ and arbitrary sources $J,\overline{J}$. Thus the terms bilinear in $J$ and $\overline{J}$, as well as the $O(\varepsilon)$ terms independent of $J,\overline{J}$, have to match. For the latter term we conclude

$$\mathcal{A}_{\overline{T}T} = \operatorname{tr}\left[\overline{M}+M\right] , \tag{9.78}$$

and thus have identified the lattice counterpart of the continuum anomaly.

The terms bilinear in the sources $J,\overline{J}$ lead to a symmetry relation for the quark propagator,

$$B^{-1} - e^{i\varepsilon T} B^{-1} e^{i\varepsilon \overline{T}} + e^{i\varepsilon T} D^{-1} e^{i\varepsilon \overline{T}} = e^{i\varepsilon M} D^{-1} e^{i\varepsilon \overline{M}} . \tag{9.79}$$

When expanding in $\varepsilon$ one obtains at $O(\varepsilon)$

$$T(D^{-1} - B^{-1}) + (D^{-1} - B^{-1})\overline{T} = MD^{-1} + D^{-1}\overline{M} . \tag{9.80}$$

This equation is solved by

$$M = T(\mathbb{1} - B^{-1}D) , \qquad \overline{M} = (\mathbb{1} - DB^{-1})\overline{T} . \tag{9.81}$$

These are the generators of the lattice symmetry which we wanted to find. They depend on the continuum generators $T, \overline{T}$, the blocking kernel $B$, and the lattice Dirac operator $D$. Inserting $M, \overline{M}$ into the symmetry relation (9.71) one ends up with a nonlinear equation for the lattice Dirac operator $D$,

$$\overline{T}D + DT = D[B^{-1}\overline{T} + TB^{-1}]D , \tag{9.82}$$

which is a generalization of the Ginsparg–Wilson equation [45, 46].

When inserting the explicit form (9.81) into the anomaly equation (9.78) one obtains the final form for the anomaly

$$\mathcal{A}_{\overline{T}T} = \mathrm{tr}\left[T + \overline{T} - TB^{-1}D - DB^{-1}\overline{T}\right] . \tag{9.83}$$

It is easy to check that when blocking a vector-like continuum theory with a blocking kernel $B = 2 \cdot \mathbb{1}$ and considering a chiral rotation in the continuum, where $T = \overline{T} = \gamma_5$, Eq. (9.82) reduces to the usual Ginsparg–Wilson relation (7.29). The generators $M, \overline{M}$ of (9.81) are generators (7.2) of Lüscher's symmetry and the anomaly assumes the form $\mathcal{A} = -\mathrm{tr}[\gamma_5 D]$, which we have already discussed in Chap. 7.

We have shown that with the help of the blocking transformation we can identify the mathematical structures that implement chiral symmetry on the lattice, in particular the Ginsparg–Wilson equation, the chiral transformations on the lattice, and the form of the anomaly. All of these were developed individually in Chap. 7, but now follow in one go from a single calculation.

Recently the blocking transformations were explored further with the motivation of implementing properly Weyl fermions on the lattice as needed for the electroweak sector of the standard model [46, 49–51]. A central insight [50] concerns the role of the blocking matrix $B$:

1. The blocking matrix $B$ must break all symmetries that are anomalous in the target theory.
2. Other symmetries may be broken by $B$ if it is convenient.

For the case of QCD the choice of $B \propto \mathbb{1}$, which we have used in the above discussion, is sufficient since only the U(1) axial transformation is anomalous. For the case of the electroweak sector additional breaking is required due to the fermion number anomalies. For this case a suitable blocking matrix was proposed in a particular model [50] and the correct symmetry pattern was shown to emerge for the corresponding fixed point action. For the resulting Ginsparg–Wilson equations it is possible to find an explicit solution [51] using the overlap construction.

# References

1. K. Symanzik: Nucl. Phys. B **226**, 187 (1983)
2. K. Symanzik: Nucl. Phys. B **226**, 205 (1983)
3. M. Lüscher and P. Weisz: Commun. Math. Phys. **97**, 59 (1985)
4. M. Lüscher and P. Weisz: Erratum, Commun. Math. Phys. **98**, 433 (1985)
5. B. Sheikholeslami and R. Wohlert: Nucl. Phys. B **259**, 572 (1985)
6. M. Lüscher, S. Sint, R. Sommer, and P. Weisz: Nucl. Phys. B **478**, 365 (1996)
7. M. Lüscher: in *Probing the Standard Model of Particle Interactions, Proceedings Les Houches 1997,* edited by R. Gupta, A. Morel, E. de Rafael, and F. David, Vol. 1 (Elsevier, Amsterdam 1998)
8. G. Curci, P. Menotti, and G. Paffuti: Phys. Lett. B **130**, 205 (1983)
9. G. Curci, P. Menotti, and G. Paffuti: Erratum, Phys. Lett. B **135**, 516 (1984)
10. R. Wohlert (unpublished) DESY 87–069 (1987)
11. M. Lüscher and P. Weisz: Nucl. Phys. B **479**, 429 (1996)
12. G. P. Lepage and P. B. Mackenzie: Phys. Rev. D **48**, 2250 (1993)
13. T. DeGrand and C. DeTar: *Lattice Methods for Quantum Chromodynamics* (World Scientific, Singapore 2006)
14. K. Symanzik: Nucl. Phys. B **190**, 1 (1981)
15. M. Lüscher: Nucl. Phys. B **254**, 252 (1985)
16. M. Lüscher, R. Narayanan, P. Weisz, and U. Wolff: Nucl. Phys. B **384**, 168 (1992)
17. S. Sint: Nucl. Phys. B **421**, 135 (1994)
18. S. Sint: Nucl. Phys. B **451**, 416 (1995)
19. M. Lüscher et al.: Nucl. Phys. B **491**, 323 (1997)
20. R. G. Edwards, U. M. Heller, and T. R. Klassen: Nucl. Phys. (Proc. Suppl.) **63**, 847 (1998)
21. K. Jansen and R. Sommer: Nucl. Phys. B **530**, 185 (1998)
22. W. Bietenholz and U.-J. Wiese: Nucl. Phys. B **464**, 319 (1996)
23. K. G. Wilson and J. Kogut: Phys. Rep. **12C**, 76 (1974)
24. K. G. Wilson: Rev. Mod. Phys. **47**, 773 (1975)
25. P. Hasenfratz: in *Advanced School on Non-Perturbative Quantum Field Physics: Peniscola, Spain 2–6 June 1997,* edited by M. Asorey and A. Dobado, p. 137 (World Scientific, Singapore 1998)
26. P. Hasenfratz and F. Niedermayer: Nucl. Phys. B **414**, 785 (1994)
27. T. DeGrand, A. Hasenfratz, P. Hasenfratz, and F. Niedermayer: Nucl. Phys. B **454**, 587 (1995)
28. T. DeGrand, A. Hasenfratz, P. Hasenfratz, and F. Niedermayer: Nucl. Phys. B **454**, 615 (1995)
29. M. Blatter and F. Niedermayer: Nucl. Phys. B **482**, 286 (1996)
30. P. Hasenfratz et al.: Int. J. Mod. Phys. **C12**, 691 (2001)
31. C. Gattringer: Phys. Rev. D **63**, 114501 (2001)
32. C. B. Lang and T. K. Pany: Nucl. Phys. B **513**, 645 (1998)
33. P. Hasenfratz et al.: Nucl. Phys. B (Proc. Suppl.) **94**, 627 (2001)
34. P. Hasenfratz et al.: Nucl. Phys. B (Proc. Suppl.) **106**, 799 (2002)
35. P. Hasenfratz et al.: Nucl. Phys. B **643**, 280 (2002)
36. S. Hauswirth: Ph.D. thesis, Universität Bern, Switzerland (2002)
37. T. Jörg: Ph.D. thesis, Universität Bern, Switzerland (2002)
38. C. Gattringer: Nucl. Phys. B (Proc. Suppl.) **119**, 122 (2002)

39. C. Gattringer et al. [BGR (Bern-Graz-Regensburg) collaboration]: Nucl. Phys. B **677**, 3 (2004)
40. P. Hasenfratz, V. Laliena, and F. Niedermayer: Phys. Lett. B **427**, 125 (1998)
41. P. Hasenfratz: Nucl. Phys. B **525**, 401 (1998)
42. C. Gattringer, I. Hip, and C. B. Lang: Nucl. Phys. B **597**, 451 (2001)
43. C. Gattringer et al.: Phys. Rev. D **79**, 054501 (2009)
44. P. H. Ginsparg and K. G. Wilson: Phys. Rev. D **25**, 2649 (1982)
45. C. Gattringer and M. Pak: PoS **LATTICE2007**, 081 (2007)
46. P. Hasenfratz, F. Niedermayer, and R. von Allmen: JHEP **0610**, 010 (2006)
47. K. Fujikawa: Phys. Rev. Lett. **42**, 1195 (1979)
48. K. Fujikawa: Phys. Rev. D **21**, 2848 (1980)
49. P. Hasenfratz and M. Bissegger: Phys. Lett. B **613**, 57 (2005)
50. P. Hasenfratz and R. von Allmen: JHEP **0802**, 079 (2008)
51. C. Gattringer and M. Pak: Nucl. Phys. B **801**, 353 (2008)

# 10

# More about lattice fermions

The implementation of chiral symmetry on the lattice discussed in Chap. 7 was an unsolved puzzle for many years. The problem has continuously inspired the lattice community and led to a variety of formulations for fermions on the lattice. So far we have focused on two kinds of fermions in our presentation, Wilson-type and overlap fermions. Wilson fermions provide a simple robust and easy-to-use discretization, while overlap fermions implement chiral symmetry in the most transparent way.

Here we introduce some of the other ideas for discretizing fermions on the lattice; most of them rooted in the problem of chiral symmetry. The exceptions are heavy quark effective formulations designed for simulations with heavy quarks like charm and bottom. We stress that we can only cover the basic ideas for the various lattice fermions and refer to the original literature for more detailed accounts and the current status of numerical simulations.

## 10.1 Staggered fermions

*Staggered fermions*, often referred to as *Kogut–Susskind fermions* [1], are a formulation where the 16-fold degeneracy of the naive discretization is reduced to only four quarks, while at the same time a remnant chiral symmetry is maintained. This is achieved by a transformation which mixes spinor and space–time indices, distributing the quark degrees of freedom on the hypercubes of the lattice.

### 10.1.1 The staggered transformation

In order to construct staggered fermions we go back to the naive discretization already presented in (2.29). The free naive fermion action reads

$$
S_F[\psi, \overline{\psi}] = a^4 \sum_{n \in \Lambda} \overline{\psi}(n) \left( \sum_{\mu=1}^{4} \gamma_\mu \frac{\psi(n+\hat{\mu}) - \psi(n-\hat{\mu})}{2a} + m \, \psi(n) \right) . \quad (10.1)
$$

Gattringer, C., Lang, C.B.: *More About Lattice Fermions*. Lect. Notes Phys. **788**, 243–266 (2010)
DOI 10.1007/978-3-642-01850-3_10      © Springer-Verlag Berlin Heidelberg 2010

As we have found in Sect. 5.2, due to the doublers, this action gives rise to 16 quark flavors. In the Wilson Dirac operator 15 of these are removed by adding the Wilson term at the cost of explicitly breaking chiral symmetry.

However, action (10.1) has a symmetry which can be used to reduce the number of quark flavors without having to break chiral symmetry. To make explicit this symmetry one performs a space–time-dependent variable transformation of the fermion fields $\psi(n), \overline{\psi}(n)$. This so-called *staggered transformation* eliminates the matrices $\gamma_\mu$ in (10.1). To see this we define new field variables $\psi(n)'$ and $\overline{\psi}(n)'$ by setting

$$\psi(n) = \gamma_1^{n_1}\, \gamma_2^{n_2}\, \gamma_3^{n_3}\, \gamma_4^{n_4}\, \psi(n)' \;, \quad \overline{\psi}(n) = \overline{\psi}(n)'\, \gamma_4^{n_4}\, \gamma_3^{n_3}\, \gamma_2^{n_2}\, \gamma_1^{n_1} \;. \tag{10.2}$$

The transformation matrices are simply products of the $\gamma_\mu$ raised to a power given by the corresponding component $n_\mu$ of the site index $n = (n_1, n_2, n_3, n_4)$. The staggered transformation thus mixes space–time and Dirac indices.

Since the $\gamma$-matrices obey $\gamma_\mu^2 = \mathbb{1}$, it is obvious that the mass term is invariant, i.e., $\overline{\psi}(n)\psi(n) = \overline{\psi}(n)'\psi(n)'$. In the kinetic term of (10.1) the $\psi$-field is shifted by one lattice constant with respect to the $\overline{\psi}$-field. Thus transformations (10.2) for the fields on the two sites differ by one power of $\gamma_\mu$. This extra $\gamma_\mu$ cancels the one which is explicit in the kinetic term of (10.1). Taking into account the necessary reordering of the $\gamma$-matrices one finds for the terms of the derivative in, e.g., the 3-direction

$$\overline{\psi}(n)\,\gamma_3\,\psi(n \pm \hat{3}) = (-1)^{n_1+n_2}\,\overline{\psi}(n)'\mathbb{1}\psi(n \pm \hat{3})' \tag{10.3}$$

and similar for the terms in the other directions. The action (10.1) turns into

$$S_F\left[\psi', \overline{\psi}'\right] = a^4 \sum_{n \in \Lambda} \overline{\psi}(n)'\mathbb{1} \left(\sum_{\mu=1}^{4} \eta_\mu(x)\frac{\psi(n+\hat{\mu})' - \psi(n-\hat{\mu})'}{2\,a} + m\,\psi(n)'\right),$$
$$\tag{10.4}$$

where we have introduced the staggered sign functions

$$\eta_1(n) = 1 \;,\; \eta_2(n) = (-1)^{n_1} \;,\; \eta_3(n) = (-1)^{n_1+n_2} \;,\; \eta_4(n) = (-1)^{n_1+n_2+n_3} \;,$$
$$\tag{10.5}$$

which take over the role of the matrices $\gamma_\mu$. Obviously action (10.4) is diagonal in Dirac space and has the same form for all four Dirac components.

The staggered fermion action is obtained by keeping only one of the four identical components. Coupling the gauge fields we end up with

$$S_F[\chi, \overline{\chi}] = a^4\sum_{n \in \Lambda}\overline{\chi}(n)\left(\sum_{\mu=1}^{4}\eta_\mu(x)\frac{U_\mu(n)\chi(n+\hat{\mu})-U_\mu^\dagger(n-\hat{\mu})\chi(n-\hat{\mu})}{2\,a} + m\chi(n)\right),$$
$$\tag{10.6}$$

where $\chi(n)$ and $\overline{\chi}(n)$ are Grassmann-valued fields with only color indices but without Dirac structure. Having discarded 3 of the 4 identical copies, we can expect that from the 16 quark degrees of freedom of the naive action only 4 have survived. This hypothesis will be analyzed in the next section.

We have claimed that the staggered transformation leaves a subset of the chiral symmetry intact. In order to see this, we first must identify $\gamma_5$ in the new spinor basis obtained by the transformation (10.2). The new form of $\gamma_5$ can be read off when transforming a pseudoscalar bilinear according to (10.2):

$$\overline{\psi}(n)\,\gamma_5\,\psi(n) = \eta_5(n)\,\overline{\psi}(n)'\,\mathbb{1}\,\psi(n)' \,, \tag{10.7}$$

where we have defined

$$\eta_5(n) = (-1)^{n_1+n_2+n_3+n_4} \,. \tag{10.8}$$

This site-dependent sign plays the role of $\gamma_5$ in the staggered world. It is obvious that for vanishing mass $m$ the staggered action (10.6) is invariant under the continuous ($\alpha \in \mathbb{R}$) transformation

$$\chi(n) \longrightarrow e^{i\alpha\eta_5(n)}\,\chi(n) \,, \qquad \overline{\chi}(n) \longrightarrow \overline{\chi}(n)\,e^{i\alpha\eta_5(n)} \,. \tag{10.9}$$

This is the global chiral symmetry for the staggered theory, where the quark degrees of freedom are distributed on the hypercube.

### 10.1.2 Tastes of staggered fermions

The staggered transformation mixes lattice and Dirac indices. This raises conceptual as well as technical questions. An example for the latter is the problem of constructing hadron interpolators with definite spin and parity. For details of this construction we refer the reader to the literature (e.g., [2]), although some of the ideas can be found in the subsequent discussion.

In order to understand better the conceptual status of staggered fermions, we here address the question of how many quarks the staggered action (10.6) describes and which symmetries relate them. For simplicity this study is done in the free case ($U_\mu(n) \equiv \mathbb{1}$).

For the analysis of staggered fermions we follow the strategy developed in [3, 4]. The idea is to group together the 16 sites of a hypercube and to interpret the corresponding degrees of freedom as 4 species of quarks, each of them with the familiar 4-spinor structure. Let us assume that the $N_\mu$ are even, i.e., we have an even number of sites in all directions. We consider the non-intersecting hypercubes with origins separated by multiples of two lattice spacings. We write our site labels $n_\mu = 0, 1, \ldots N_\mu - 1$ in terms of labels $h_\mu$ for these hypercubes and labels $s_\mu$ for the corners of the hypercube as

$$n_\mu = 2\,h_\mu + s_\mu \quad \text{with} \quad h_\mu = 0, 1, \ldots N_\mu/2 - 1 \,, \quad s_\mu = 0, 1 \,. \tag{10.10}$$

With this labeling of sites, the staggered sign function $\eta_\mu(n)$ defined in (10.5) becomes independent of $h$ and is a function of only the vector $s$:

$$\eta_\mu(n) = \eta_\mu(2h + s) = \eta_\mu(s) \,. \tag{10.11}$$

We now want to combine all degrees of freedom, $\chi(2h+s)$, $\overline{\chi}(2h+s)$ ($s_\mu = 0, 1$), sharing a common hypercube label $h$ into a new set of fields $q(h)_{ab}$, $\overline{q}(h)_{ab}$ with indices $a, b = 1, 2, 3, 4$. In order to sneak in the familiar $\gamma$-matrices we define $s$-dependent $4 \times 4$ matrices $\Gamma^{(s)}$ as products of the $\gamma_\mu$ (compare (10.2)),

$$\Gamma^{(s)} = \gamma_1^{s_1} \, \gamma_2^{s_2} \, \gamma_3^{s_3} \, \gamma_4^{s_4} \, . \tag{10.12}$$

It is easy to check that they obey the orthogonality and completeness relations familiar from Fierz transformation,

$$\frac{1}{4} \operatorname{tr} \left[ \Gamma^{(s)\dagger} \Gamma^{(s')} \right] = \delta_{s,s'} \, , \quad \frac{1}{4} \sum_s \Gamma_{ba}^{(s)*} \Gamma_{b'a'}^{(s)} = \delta_{a,a'} \delta_{b,b'} \, . \tag{10.13}$$

We define new quark fields $q(h)$ and $\overline{q}(h)$ as linear combinations of the fields $\chi(2h + s)$, $\overline{\chi}(2h + s)$:

$$q(h)_{ab} \equiv \frac{1}{8} \sum_s \Gamma_{ab}^{(s)} \chi(2h + s) \, , \quad \overline{q}(h)_{ab} \equiv \frac{1}{8} \sum_s \overline{\chi}(2h + s) \, \Gamma_{ba}^{(s)*} \, . \tag{10.14}$$

Using (10.13), the linear transformations can be inverted to obtain

$$\chi(2h + s) = 2 \operatorname{tr} \left[ \Gamma^{(s)\dagger} q(h) \right] \, , \quad \overline{\chi}(2h + s) = 2 \operatorname{tr} \left[ \overline{q}(h) \, \Gamma^{(s)} \right] \, . \tag{10.15}$$

With the help of these equations we can express the mass term of (10.6) in terms of the $q$ and $\overline{q}$ fields:

$$a^4 \sum_n \overline{\chi}(n) \, \chi(n) = a^4 \sum_h \sum_s \overline{\chi}(2h + s) \, \chi(2h + s) \tag{10.16}$$

$$= 4 \, a^4 \sum_h \sum_s \overline{q}(h)_{ba} \, \Gamma_{ab}^{(s)} \Gamma_{b'a'}^{(s)\dagger} \, q(h)_{a'b'}$$

$$= (2a)^4 \sum_h \operatorname{tr} \left[ \overline{q}(h) q(h) \right] \, ,$$

where in the second step the completeness relation of (10.13) was used. The kinetic term of (10.6) is somewhat more tricky since the shifted fields $\chi(2h + s \pm \hat{\mu})$ mix contributions from different hypercubes. We first must assign the proper hypercube to the individual fields $\chi(2h + s \pm \hat{\mu})$ and only after that we can represent them in terms of the $q(h)_{ab}$. For the forward shifted field one finds

$$\chi(2h + s + \hat{\mu}) = \begin{cases} \chi(2h + s + \hat{\mu}) = 2 \operatorname{tr} \left[ \Gamma^{(s+\hat{\mu})\dagger} q(h) \right] & \text{for } s_\mu = 0 \, , \\ \chi(2(h+\hat{\mu}) + s - \hat{\mu}) = 2 \operatorname{tr} \left[ \Gamma^{(s-\hat{\mu})\dagger} q(h+\hat{\mu}) \right] & \text{for } s_\mu = 1 \, . \end{cases} \tag{10.17}$$

From definition (10.12) of the $\Gamma^{(s)}$ follows $\Gamma^{(s\pm\hat{\mu})} = \eta_\mu(s)\gamma_\mu \Gamma^{(s)}$, which brings (10.17) into the form

$$\chi(2h + s + \hat{\mu}) = 2\eta_\mu(s)\,\mathrm{tr}\left[\Gamma^{(s)\dagger}\gamma_\mu\left(q(h)\delta_{s_\mu,0} + q(h + \hat{\mu})\delta_{s_\mu,1}\right)\right]. \quad (10.18)$$

Performing the equivalent steps for the field shifted in negative $\mu$-direction, $\chi(2h + s - \hat{\mu})$, we can write the $\mu$-contribution to the kinetic term of (10.6) as

$$4a^4 \sum_h \frac{1}{2a} \sum_s \mathrm{tr}\left[\overline{q}(h)\Gamma^{(s)}\right] \quad (10.19)$$

$$\times\, \mathrm{tr}\left[\Gamma^{(s)\dagger}\gamma_\mu\left(q(h)\delta_{s_\mu,0} + q(h + \hat{\mu})\delta_{s_\mu,1} - q(h - \hat{\mu})\delta_{s_\mu,0} - q(h)\delta_{s_\mu,1}\right)\right].$$

Unfortunately we cannot yet sum over $s$ and employ the completeness relation of (10.13), since in the second factor $\Gamma^{(s)\dagger}$ is combined with $s$-dependent terms. This problem is overcome by realizing that in the summation over hypercubes, defined in (10.10), we could have started also at $\hat{\mu}$ instead of the origin. If one writes the sums over $h$ and $s$ in (10.16) with this shifted convention, some of the $q$ and $\overline{q}$ are shifted by $\pm\hat{\mu}$ and the hypercube indices $s_\mu = 0$ and $s_\mu = 1$ are interchanged. The latter change provides the $s_\mu$ contributions missing in (10.19). Thus we can average the two ways of writing (10.19) and obtain after some algebra

$$S_F[q,\overline{q}] = (2a)^4 \sum_h \left( m\,\mathrm{tr}\left[\overline{q}(h)q(h)\right] \right. \quad (10.20)$$

$$\left. + \sum_\mu \mathrm{tr}\left[\overline{q}(h)\gamma_\mu\nabla_\mu q(h)\right] - a \sum_\mu \mathrm{tr}\left[\overline{q}(h)\gamma_5\triangle_\mu q(h)\gamma_\mu\gamma_5\right] \right),$$

where we have defined derivative operators on the blocked lattice (note that here the sites are separated by $b \equiv 2a$),

$$\nabla_\mu f(h) = \frac{f(h+\hat{\mu}) - f(h-\hat{\mu})}{2b}\,,\quad \triangle_\mu f(h) = \frac{f(h+\hat{\mu}) - 2f(h) + f(h-\hat{\mu})}{b^2}.$$
$$(10.21)$$

Guided by the form of the first two terms of (10.20), we identify Dirac indices $\alpha$ and quark species labels $t$ for our fermion fields by setting

$$\psi^{(t)}(h)_\alpha \equiv q(h)_{\alpha t}\,,\quad \overline{\psi}^{(t)}(h)_\alpha \equiv \overline{q}(h)_{t\alpha}\,. \quad (10.22)$$

Expressed in terms of these new fields the free staggered action reads

$$S_F[\psi,\overline{\psi}] = b^4 \sum_h \left( \sum_{t=1}^{4} \left( m\,\overline{\psi}^{(t)}(h)\psi^{(t)}(h) + \sum_{\mu=1}^{4} \overline{\psi}^{(t)}(h)\,\gamma_\mu\nabla_\mu\psi^{(t)}(h) \right) \right.$$

$$\left. - \frac{b}{2} \sum_{t,t'=1}^{4} \sum_{\mu=1}^{4} \overline{\psi}^{(t)}(h)\,\gamma_5\,(\tau_5\tau_\mu)_{tt'}\,\triangle_\mu\psi^{(t')}(h) \right). \quad (10.23)$$

Here we have introduced the matrices $\tau_\mu = \gamma_\mu^T, \mu = 1, 2, \ldots 5$, and $b$ is the lattice spacing on the lattice of hypercubes. In order to distinguish the $N_t = 4$

different species of quarks labeled by $t = 1, 2, 3, 4$ from usual flavor, they are referred to as *tastes of staggered fermions*. The first two terms in (10.23) are diagonal in taste space and represent the mass and kinetic terms for the four tastes of fermions expected to be described by (10.6). The third term is reminiscent of a Wilson term, but mixes the different tastes. This taste symmetry-breaking term reduces the symmetry of the kinetic term which is invariant under independent vector and axial rotations for each of the four tastes. The taste-breaking term is only invariant under the remaining symmetry $U(1) \times U(1)$ given by the rotations

$$\psi' = e^{i\alpha}\psi \ , \quad \overline{\psi}' = \overline{\psi} e^{-i\alpha} \ ,$$
$$\psi' = e^{i\beta \Gamma_5}\psi \ , \quad \overline{\psi}' = \overline{\psi} e^{i\beta \Gamma_5} \ , \tag{10.24}$$

where we have defined the taste-mixing generator $\Gamma_5 = \gamma_5 \otimes \tau_5$. This symmetry may be identified with a subgroup of the axial taste symmetry group $\mathrm{SU}(N_t)_A$.

### 10.1.3 Developments and open questions

The symmetry of the four tastes, described by the staggered action, is not the full symmetry of four independent (massless) flavors of quarks, but instead reduced to (10.24). Although the taste-breaking term vanishes in a naive continuum limit, its presence in numerical simulations which are necessarily done at finite lattice spacing raises several questions.

On a more fundamental level one has to understand whether effects of taste breaking survive the continuum limit as it is performed in an actual calculation. As we have discussed in Sect. 3.5.4, the "true continuum limit" is obtained by driving the couplings to a critical point where the correlation length diverges. The open question thus is whether the taste-breaking terms change the behavior of the theory at the critical point, or in more fancy words, whether the taste-breaking terms change the universality class of the theory.

Ignoring these more fundamental issues one can try to minimize the effect of the taste-breaking terms. In order to come up with a strategy to do so, it is important to remember that the different tastes are combinations of the staggered field variables $\chi$ and $\overline{\chi}$ on a hypercube. Thus the different tastes see different link variables $U_\mu(n)$. Consequently strongly fluctuating gauge links will give rise to large taste symmetry-breaking effects. A possible strategy to ameliorate these effects is to use gauge actions that suppress such strong fluctuations, see, e.g., [5-7]. An alternative approach, already addressed in Sect. 6.2.6, is to apply blocking or smearing to the gauge fields which locally averages out large fluctuations [8, 9].

The effect of such blocking or smearing steps can, e.g., be assessed by analyzing the spectrum of the Dirac operator for staggered fermions (given here for the $\chi$-basis (10.6)):

$$D^{\mathrm{st}}(n|m) = m\,\delta_{n,m} + \sum_{\mu=1}^{4} \eta_\mu(n) \frac{U_\mu(n)\delta_{n+\hat{\mu},m} - U_\mu(n-\hat{\mu})^\dagger \delta_{n-\hat{\mu},m}}{2a} \ . \tag{10.25}$$

It is easy to see that for the massless case this Dirac operator is anti-hermitian and obeys a staggered $\gamma_5$-hermiticity:

$$D^{\mathrm{st}\,\dagger}(n|m) = -D^{\mathrm{st}}(n|m) = \eta_5(n)D^{\mathrm{st}}(n|m)\eta_5(m) \,. \qquad (10.26)$$

These two properties guarantee that the eigenvalues come in complex conjugate pairs and all have the same real part given by the quark mass. Thus fluctuations of the real part of the eigenvalues are impossible and no exceptional configurations (compare the discussion in Sect. 6.2.5) can occur. Consequently the determinant (product of all eigenvalues) is always real and nonnegative and, for nonzero quark mass, strictly positive.

Furthermore, if taste symmetry breaking was absent, then each eigenvalue would be 4-fold degenerate. In a systematic comparison [10, 11] it was indeed found that a sufficient amount of blocking or smearing drives the eigenvalues toward 4-fold degeneracy.

Staggered fermions are widely used for dynamical simulations. The reason is that due to the reduced number of degrees of freedom (no Dirac structure) staggered fermions are numerically cheaper to simulate and at the same time are chirally symmetric. However, a problem is that the action describes four tastes of quarks, while in a realistic QCD simulation one would like to have two light mass-degenerate $u$ and $d$ quarks and one heavier strange quark. In order to suitably reduce the number of degrees of freedom it has been proposed to simulate QCD with an effective action given by

$$\exp(-S_{\mathrm{eff}}) = \exp(-S_G)\, \det[D^{\mathrm{st}}(m_{ud})]^{\frac{1}{2}}\, \det[D^{\mathrm{st}}(m_s)]^{\frac{1}{4}} \,, \qquad (10.27)$$

where $m_{ud}$ is the average of $u$ and $d$ quark masses and $m_s$ the strange quark mass. From a mathematical point of view, taking the square or quartic roots of the determinant is unproblematic, since as we have shown it is real and positive. However, this procedure is quite nontrivial from a conceptual perspective and we must ask ourselves if the universality class remains the same in this approach. Probably even more important is the question whether the effective action can be expressed in the form of a local lattice field theory. For a snapshot of the ongoing debate about these issues see [12–16] and references therein. Although the conceptual problems are not all resolved, simulations with staggered fermions have found good agreement with experimental results. Examples are found in [17–20].

## 10.2 Domain wall fermions

The overlap formulation of lattice QCD, discussed in Sect. 7.4, is related to another version of chiral lattice QCD, so-called *domain wall fermions*. Domain wall fermions make use of a 5D lattice and construct chiral Dirac fermions on a 4D interface of the 5D lattice. The action of the 5D theory is simple, i.e., rather similar to the usual Wilson formulation. This implies that already established numerical methods can be applied with only minor adaptions.

### 10.2.1 Formulation of lattice QCD with domain wall fermions

The basic concepts for domain wall fermions were outlined in the seminal paper [21] by Kaplan. The ideas were developed further in [22–24] giving rise to the formulation of domain wall fermions mainly used now.

The idea is to work with a 5D lattice $\Lambda_5 = \Lambda \times Z_{N_5}$, where $\Lambda$ is our 4D lattice and $N_5$ denotes the number of lattice points in the auxiliary 5-direction. On this lattice we use fermion fields which are described by the 4-component spinors

$$\Psi(n,s)_{\substack{\alpha \\ c}} \, , \; \overline{\Psi}(n,s)_{\substack{\alpha \\ c}} \; \text{with } n \in \Lambda \, , \, s = 0, \ldots, N_5-1 \, , \, \alpha = 1, \ldots, 4 \, , \, c = 1,2,3 \, . \tag{10.28}$$

Obviously the Dirac (index $\alpha$) and color (index $c$) structures are unchanged and we have simply added an additional direction which we label with $s$. The fermion action for the 5D domain wall theory is given by (setting $a \equiv 1$)

$$S_F^{\mathrm{dw}}[\Psi, \overline{\Psi}, U] = \sum_{n,m \in \Lambda} \sum_{s,r=0}^{N_5-1} \overline{\Psi}(n,s) \, D^{\mathrm{dw}}(n,s|m,r) \, \Psi(m,r) \, , \tag{10.29}$$

where we use vector/matrix notation for the Dirac and color indices. The 5D domain wall Dirac operator $D^{\mathrm{dw}}$ is given by

$$D^{\mathrm{dw}}(n,s|m,r) = \delta_{s,r} \, D(n|m) + \delta_{n,m} \, D_5^{\mathrm{dw}}(s|r) \, . \tag{10.30}$$

The first term is diagonal in the auxiliary direction and is built with the usual 4D Wilson Dirac operator (compare (5.51) where we write explicitly also color and Dirac indices)

$$D(n|m) = (4 - M_5) \, \delta_{n,m} - \frac{1}{2} \sum_{\mu=\pm 1}^{\pm 4} (\mathbb{1} - \gamma_\mu) \, U_\mu(n) \, \delta_{n+\hat{\mu},m} \, . \tag{10.31}$$

We have introduced a new mass parameter $M_5$, which is not to be confused with the quark mass parameter $m$ of the 4D theory. The avoidance of doublers and the positivity of the transfer matrix restrict this parameter to $0 < M_5 < 1$. The link variables $U_\mu(n), \mu = 1,2,3,4$ are elements of the gauge group as usual. They do not depend on the coordinate in the fifth dimension, i.e., identical copies are used for the different 4D slices.

The second contribution to (10.30) is diagonal in $\Lambda$ and the operator $D_5^{\mathrm{dw}}(s|r)$, acting in the 5-direction, is given by

$$D_5^{\mathrm{dw}}(s|r) = \delta_{s,r} - (1 - \delta_{s,N_5-1}) \, P_- \, \delta_{s+1,r} - (1 - \delta_{s,0}) \, P_+ \, \delta_{s-1,r}$$
$$+ m \, (P_- \, \delta_{s,N_5-1}\delta_{0,r} + P_+ \, \delta_{s,0}\delta_{N_5-1,r}) \, . \tag{10.32}$$

Here $P_\pm = (\mathbb{1} \pm \gamma_5)/2$ are the usual chiral projectors acting on the Dirac indices. The parameter $m$ will turn out to be the mass parameter of the 4D target theory. For the fermions at mass $m = 0$ the boundary conditions in the 5-direction are fixed, since hopping from $s = N_5 - 1$ in positive 5-direction is blocked by the factor $(1 - \delta_{s, N_5 - 1})$, and the hopping from $s = 0$ in negative direction is eliminated by $(1 - \delta_{s,0})$. Only the terms containing the mass $m$ connect the slices at $s = 0$ and $s = N_5 - 1$. On the 4D lattice $\Lambda$ the conventional boundary conditions are used: periodic in the spatial directions 1,2,3 and anti-periodic in time (4-direction). The link variables $U_\mu(n)$ are periodic in all four directions of $\Lambda$ and, as mentioned, one uses identical copies for all values of the 5-coordinate $s$. We stress that in the 5-direction there is no component $U_5$.

Having presented the action for the 5D fermion fields $\Psi, \overline{\Psi}$, we can now construct the 4D physical fields $\psi, \overline{\psi}$ which live on the 4D boundary of $\Lambda_5$. These are defined as

$$\psi(n) = P_- \, \Psi(n,0) + P_+ \, \Psi(n, N_5 - 1) \, , \; \overline{\psi}(n) = \overline{\Psi}(n, N_5 - 1) \, P_- \; + \overline{\Psi}(n,0) \, P_+ \, , \tag{10.33}$$

where $n \in \Lambda$. The physical fields $\psi$ and $\overline{\psi}$ thus are built from the degrees of freedom on the first ($s = 0$) and last ($s = N_5 - 1$) 4D slice. The spinors $\psi, \overline{\psi}$ inherit their Dirac and color indices from $\Psi, \overline{\Psi}$.

The physical fields $\psi$ and $\overline{\psi}$ can now be used to construct the physical observables of interest. For example the 4D scalar density $\overline{\psi}(n)\psi(n)$ assumes the form

$$\overline{\psi}(n)\psi(n) = \overline{\Psi}(n, N_5 - 1) \, P_- \, \Psi(n,0) + \overline{\Psi}(n,0) \, P_+ \, \Psi(n, N_5 - 1) \, , \tag{10.34}$$

when (10.33) is used to express the physical density $\overline{\psi}(n)\psi(n)$ in terms of the 5D fields. We remark that (10.34) corresponds exactly to the term in (10.32) which is proportional to $m$. Thus $m$ can be identified as the mass parameter of the 4D theory.

The transition to the 4D world can be implemented in a transparent way by using generating functionals (compare (5.32)). For our purpose we define the generating functional as

$$W\left[J, \overline{J}\right] = \left\langle \exp\left( \sum_{n \in \Lambda} \left( \overline{J}(n)\psi(n) + \overline{\psi}(n)J(n) \right) \right) \right\rangle \tag{10.35}$$

$$= \left\langle \exp\left( \sum_{n \in \Lambda} \left( \overline{J}(n) \left( P_- \, \Psi(n,0) + P_+ \, \Psi(n, N_5 - 1) \right) \right. \right.\right.$$

$$\left.\left.\left. + \left( \overline{\Psi}(n, N_5 - 1) \, P_- \; + \overline{\Psi}(n,0) \, P_+ \right) J(n) \right) \right) \right\rangle_5 .$$

Here $J(n)$ and $\overline{J}(n)$ are Grassmann-valued source fields which have the same indices (Dirac and color) as the physical fields $\psi(n), \overline{\psi}(n)$ they couple to. In this equation $\langle \dots \rangle$ denotes the expectation value of the 4D target theory and $\langle \dots \rangle_5$ denotes the vacuum expectation value in the 5D world, which we discuss in detail in the next section.

Once the generating functional $W\left[J, \overline{J}\right]$ is defined, one can work with it like in the regular 4D formulation. In particular $W\left[J, \overline{J}\right]$ can be differentiated with respect to individual components of $J(n)$ or $\overline{J}(n)$ which bring the corresponding components of the physical fields $\psi, \overline{\psi}$ down from the exponent. After the physical observables are built from these, one sets the sources to $J = \overline{J} = 0$ (compare Sect. 5.1.6). As a result the vacuum expectation values of the 4D target theory are formulated as the expectation values in the 5D formulation. The latter is then used to evaluate the expressions.

### 10.2.2 The 5D theory and its equivalence to 4D chiral fermions

So far we have introduced the fields $\Psi$ and $\overline{\Psi}$ which live on our 5D lattice and their action and discussed how to construct from them the fields of the 4D theory. We still need to present the detailed definition of the 5D vacuum expectation values $\langle \dots \rangle_5$ and finally we should convince ourselves that for $m = 0$ the 5D theory gives rise to 4D chiral fermions.

In order to make connection to a 4D overlap-type theory, in addition to the 5D fermion fields $\Psi, \overline{\Psi}$ the 5D path integral also contains the fields $\Phi, \overline{\Phi}$ which have the same indices as the fermion fields (10.28) but are bosonic variables (not Grassmann numbers). Thus they are often referred to as *pseudofermion fields* (see Sect. 8.1.3) or *Pauli–Villars fields*. From a physical point of view their role is the removal of heavy degrees of freedom which appear for large $N_5$.

The definition of the 5D path integral for the vacuum expectation value $\langle \dots \rangle_5$, needed for the generating functional $W[J, \overline{J}]$ given in (10.35), reads

$$\langle O \rangle_5 = \frac{1}{Z} \int \mathcal{D}[\Psi, \overline{\Psi}, \Phi, \overline{\Phi}, U] \, e^{-S_F^{\mathrm{dw}}[\Psi,\overline{\Psi},U] - S_B^{\mathrm{pf}}[\Phi,\overline{\Phi},U] - S_G[U]} \, O[\Psi, \overline{\Psi}, U] \,. \tag{10.36}$$

$S_F^{\mathrm{dw}}[\Psi, \overline{\Psi}, U]$ is the domain wall action and $S_G[U]$ is some lattice version of the 4D gauge action, e.g., the Wilson action (2.49). The path integral is over the 5D fields $\Psi$ and $\overline{\Psi}$ and the 4D gauge fields $U$. We have also introduced the integral over the bosonic pseudofermions $\Phi$ and $\overline{\Phi}$. The action for the pseudofermions is given by

$$S_F^{\mathrm{dw}}[\Phi, \overline{\Phi}, U] = \sum_{n,m \in \Lambda} \sum_{s,r=0}^{N_5-1} \overline{\Phi}(n, s) \, D^{\mathrm{pf}}(n, s|m, r) \, \Phi(m, r) \,,$$

$$\text{with} \quad D^{\mathrm{pf}}(n, s|m, r) = \delta_{s,r} \, D(n|m) + \delta_{n,m} \, D_5^{\mathrm{pf}}(s|r) \,. \tag{10.37}$$

The first part is identical to the term in the Dirac operator (10.30), i.e., given by the Wilson operator (10.31). Concerning the difference operator $D_5^{\mathrm{pf}}$ for the 5-direction different choices are possible. A variant convenient for numerical simulations is [25]

$$D_5^{\mathrm{pf}}(s|r) = D_5^{\mathrm{dw}}(s|r)\Big|_{m=1} \,. \tag{10.38}$$

A direct connection of the 5D theory (10.36) to a variant of the overlap formulation was given in [26]. Using an elegant chain of transformations Neuberger showed (for a slightly different choice of $D^{\mathrm{pf}}$) that (see also [27])

$$\det\left[D^{\mathrm{dw}}\right] = \det\left[D^{\mathrm{ov}}_{N_5}\right] \det\left[D^{\mathrm{pf}}\right] . \qquad (10.39)$$

Applying this result to our path integral (10.36) we find

$$\int \mathcal{D}[\Psi,\overline{\Psi},\Phi,\overline{\Phi}]\, e^{-S^{\mathrm{dw}}_F[\Psi,\overline{\Psi},U] - S^{\mathrm{pf}}_B[\Phi,\overline{\Phi},U]} = \frac{\det[D^{\mathrm{dw}}]}{\det[D^{\mathrm{pf}}]} = \det[D^{\mathrm{ov}}_{N_5}] . \quad (10.40)$$

Here we have used that for the path integral over the bosonic pseudofermions the determinant appears in the denominator (compare Sect. 5.1.4). Obviously the pseudofermion fields in (10.36) are needed to cancel the factor $\det[D^{\mathrm{pf}}]$ in (10.39). This step can also be viewed as a transformation of the integration variables with $\det[D^{\mathrm{pf}}]$ being the corresponding Jacobian.

The Dirac operator $D^{\mathrm{ov}}_{N_5}$ is often referred to as the *truncated overlap* operator and is given by (we drop an irrelevant overall factor of $1/2$ compared to the notation in [26])

$$D^{\mathrm{ov}}_{N_5} = \mathbb{1} + \gamma_5 \tanh\left(\frac{N_5}{2}\,\widetilde{H}\right) \overset{N_5 \to \infty}{\longrightarrow} \mathbb{1} + \gamma_5 \operatorname{sign}[\widetilde{H}] . \qquad (10.41)$$

The operator $\widetilde{H}$ is a (nonlocal) variant of the operator $H$ which we used in Sect. 7.4 to construct the overlap operator (see [26] for the exact definition). Equation (10.41) shows that in the limit $N_5 \to \infty$, where the tanh approaches the sign function, the operator assumes the form of the overlap operator (7.78), and a quick calculation similar to (7.82) shows that $D^{\mathrm{ov}}_\infty$ obeys the Ginsparg–Wilson equation (7.29).

The truncated overlap operator describes the 4D theory and in the limit $N_5 \to \infty$ gives rise to massless fermions. For finite $N_5$ the chirality-violating effects are exponentially suppressed with an exponent $\propto N_5$. In practical simulations typical values used are $N_5 = 10-30$. The parameter $M_5$ can be tuned to minimize the chirality-violating effects.

Working with domain wall fermions has the big advantage that numerical algorithms for 4D Wilson fermions can be easily adapted for the 5D domain wall formulation. As the Wilson formulation, the domain wall operator also contains only nearest neighbor terms and no evaluation of matrix-valued functions as for the overlap operator (see Sect. 7.4.3) is required. For an example of dynamical simulations with domain wall fermions, see [28].

## 10.3 Twisted mass fermions

*Twisted mass QCD* (tmQCD) is a formulation which in its simplest form is for QCD with two mass-degenerate quark flavors of Wilson fermions (QCD with isospin). The isospin degree of freedom is used to introduce an additional mass term with a nontrivial isospin structure. This *twisted mass term* provides a useful infrared regulator and furthermore can be utilized to obtain $\mathcal{O}(a)$ improvement of the lattice formulation. Twisted mass QCD was first outlined in [29–31]. In this section we give a brief introduction. For more detailed recent reviews, which cover also the present status of the numerical analysis of tmQCD, we recommend [32, 33].

### 10.3.1 The basic formulation of twisted mass QCD

We begin our discussion with presenting the defining equations. As already stated, tmQCD is a theory for two mass-degenerate quarks.[1] Thus from now on we denote by $\chi$ and $\overline{\chi}$ quark fields which in addition to Dirac and color indices carry also a flavor index which may assume $N_f = 2$ values. For notational convenience we will suppress the indices and use matrix/vector notation instead. The fermion action for lattice tmQCD with Wilson fermions then reads

$$S_F^{\text{tw}}[\chi, \overline{\chi}, U] = a^4 \sum_{k,n \in \Lambda} \overline{\chi}(k) \left( D(k|n)\, \mathbb{1}_2 + m\, \mathbb{1}_2\, \delta_{k,n} + i\mu\gamma_5\tau^3\, \delta_{k,n} \right) \chi(n) \,.$$
$$(10.42)$$

We display unit matrices $\mathbb{1}_2$ only for flavor space and omit them for color and Dirac. $D(k|n)$ denotes the massless Wilson Dirac operator for a single flavor (compare (5.51)):

$$D(k|n) = \frac{4}{a}\delta_{k,n} - \frac{1}{2a}\sum_{\mu=\pm 1}^{\pm 4} (\mathbb{1} - \gamma_\mu)\, U_\mu(k)\, \delta_{k+\hat{\mu},n} \,. \qquad (10.43)$$

Action (10.42) differs from the usual action for two mass-degenerate flavors with mass $m$ by adding the term $i\mu\gamma_5\tau^3$ to the Dirac operator. The real parameter $\mu$ is called the *twisted mass*. We stress, however, that the conventional mass term is trivial in color, Dirac, and flavor space, while the twisted mass term is trivial only in color space, has a $\gamma_5$ in Dirac space, and the third Pauli matrix $\tau^3 = \text{diag}(1, -1)$ acts in flavor space.

One of the original [29] motivations for introducing the twisted mass term[2] was its use as an infrared regulator which removes exceptional configurations (compare Sect. 6.2.5). Using just the standard mass these are absent only for sufficiently large values of $m$. The fact that the twisted mass term is a save remedy against exceptional configurations follows from the following chain of identities:

$$\det[D\mathbb{1}_2 + m\mathbb{1}_2 + i\mu\gamma_5\tau^3] = \det[D + m + i\mu\gamma_5]\,\det[D + m - i\mu\gamma_5] \qquad (10.44)$$
$$= \det[D + m + i\mu\gamma_5]\,\det[\gamma_5(D + m - i\mu\gamma_5)\gamma_5]$$
$$= \det[D + m + i\mu\gamma_5]\,\det[D^\dagger + m - i\mu\gamma_5]$$
$$= \det[(D + m + i\mu\gamma_5)\,(D^\dagger + m - i\mu\gamma_5)]$$
$$= \det[(D + m)(D + m)^\dagger + \mu^2] > 0 \quad \text{for } \mu \neq 0 \,.$$

---

[1] For the generalization of tmQCD to the case of quarks with different masses we refer the reader to [34, 35].

[2] Another motivation being the simplification of renormalization properties (see the following discussion and Chap. 11).

The twisted mass Dirac operator is diagonal in flavor space and thus the determinant for the 2-flavor operator was written as a product of two determinants for a single flavor. Subsequently the $\gamma_5$-hermiticity (5.76) of the Wilson Dirac operator was employed. The inequality in the last line holds because the eigenvalues of the product $(D + m)(D + m)^\dagger$ are real and nonnegative. Thus a nonvanishing value of the twisted mass $\mu$ ensures that the determinant of the twisted mass Dirac operator is strictly positive and consequently zero eigenvalues, which cause the exceptional configurations, are excluded. In more detail one can show that the spectrum of the twisted mass Dirac operator in the complex plane is expelled from a strip of width $2\mu$ along the real axis [36]. Equation (10.44) establishes that the determinant is real and positive for arbitrary gauge configurations and thus standard Monte Carlo techniques are applicable.

The above discussion shows that the twisted mass term may be used as an alternative infrared regulator which can be combined with the usual mass term. Since it is possible to work with both, nonvanishing $m$ and $\mu$, it is convenient to introduce the so-called *polar mass* $M$ and the *twist angle* $\alpha$:

$$M = \sqrt{m^2 + \mu^2}, \qquad \alpha = \arctan(\mu/m). \qquad (10.45)$$

With these definitions the mass terms of tmQCD can be written as

$$m\mathbb{1}_2 + i\mu\gamma_5\tau^3 = Me^{i\alpha\gamma_5\tau^3} \quad \text{with} \quad m = M\cos(\alpha), \ \mu = M\sin(\alpha). \quad (10.46)$$

The case of $\alpha = \pi/2$, i.e., $m = 0$, $\mu > 0$, is often referred to as *maximal* or *full twist*. Below it will be shown that choosing full twist is special since it implies $\mathcal{O}(a)$ improvement. The value $\alpha = 0$ corresponds to zero twist.

It is instructive to perform a simple transformation of the fermion fields to new variables $\psi, \overline{\psi}$ defined as

$$\psi = R(\alpha)\,\chi, \quad \overline{\psi} = \overline{\chi}\,R(\alpha), \quad R(\alpha) = e^{i\alpha\gamma_5\tau^3/2}. \qquad (10.47)$$

A few lines of algebra show that the fermion action (10.42) turns into

$$S_F[\psi, \overline{\psi}, U] = a^4 \sum_{k,n\in\Lambda} \overline{\psi}(k) \left( D^{\mathrm{tw}}(k|n) + M\,\mathbb{1}_2\,\delta_{k,n} \right) \psi(n). \qquad (10.48)$$

Obviously the twisted mass term has disappeared and is replaced by a conventional mass term with a mass parameter given by the polar mass $M$ defined in (10.45). The twisted Dirac operator $D^{\mathrm{tw}}$ is now a genuine 2-flavor operator and reads

$$D^{\mathrm{tw}}(k|n) = \frac{4}{a}\,e^{-i\alpha\gamma_5\tau^3}\,\delta_{k,n} - \frac{1}{2a}\sum_{\mu=\pm1}^{\pm4}\left(e^{-i\alpha\gamma_5\tau^3} - \gamma_\mu\right)U_\mu(k)\,\delta_{k+\hat{\mu},n}. \qquad (10.49)$$

We observe that the naive parts of the lattice Dirac operator (the terms with $\gamma_\mu$) are not affected by the twist and only the Wilson term is rotated. Thus in the basis $\psi, \overline{\psi}$, which is referred to as the *physical basis*, the mass term and the kinetic terms assume their conventional form and only the term needed to remove the doublers is introduced in a twisted form. In the naive continuum limit this latter term is of $\mathcal{O}(a)$ and thus vanishes as $a \to 0$. Our old basis $\chi, \overline{\chi}$, where the twist affects the mass term, is known as the *twisted basis*.

We finally remark that the symmetries of QCD assume a different form in the twisted basis. To give an example, it is easy to see that the modified parity transformations (the gauge field still transforms as stated in (5.72))

$$\chi(\boldsymbol{n}, n_4) \overset{\mathcal{P}}{\to} \gamma_4 \, \tau^{1,2} \, \chi(-\boldsymbol{n}, n_4) \,, \quad \overline{\chi}(\boldsymbol{n}, n_4) \overset{\mathcal{P}}{\to} \overline{\chi}(-\boldsymbol{n}, n_4) \, \gamma_4 \, \tau^{1,2} \quad (10.50)$$

leave the action (10.42) invariant. Note that we have two different choices and can use either of the two Pauli matrices $\tau^1$ or $\tau^2$. A collection of the twisted forms of other symmetries can, e.g., be found in the appendix of [33].

### 10.3.2 The relation between twisted and conventional QCD

Having introduced a new type of infrared regulator, one of course has to establish that in some limit the twisted formulation describes conventional QCD. At the end of the last section we have seen that when changing to the physical basis $\chi, \overline{\chi}$, the twist affects only the Wilson term which in the naive continuum limit vanishes anyway. This is already a good sign, but the equivalence of conventional and tmQCD has to be established also for the proper continuum limit (compare Sect. 3.5.4).

We begin with studying the relation between conventional and twisted mass QCD by analyzing the situation in the continuum. The conventional action for two mass-degenerate fermions in the continuum is given by

$$S_F[\psi, \overline{\psi}, A] = \int d^4x \, \overline{\psi}(x) \, (\gamma_\mu D_\mu(x) \, \mathbb{1}_2 \, + \, M \, \mathbb{1}_2 \,) \, \psi(x) \,, \qquad (10.51)$$

where $D_\mu(x) = \partial_\mu + i A_\mu(x)$. Vacuum expectation values of some operator $O$ are computed with the path integral which is formally defined as ($S_G[A]$ denotes the action for the gauge fields)

$$\langle O \rangle = \frac{1}{Z} \int \mathcal{D}[\psi, \overline{\psi}, A] \, e^{-S_F[\psi, \overline{\psi}, A] - S_G[A]} \, O[\psi, \overline{\psi}, A] \,. \qquad (10.52)$$

Relating this conventional form of the vacuum expectation value to the corresponding expression in tmQCD is now merely an exercise in changing integration variables. The transformation we need is the one given in (10.47). Under this transformation the conventional continuum action (10.51) changes to the continuum tmQCD action given by

$$S_F^{\text{tw}}[\chi, \overline{\chi}, A] = \int d^4x \, \overline{\chi}(x) \, (\gamma_\mu D_\mu(x) \, \mathbb{1}_2 \, + \, m \, \mathbb{1}_2 \, + \, i\mu\gamma_5\tau^3) \, \chi(x) \,, \quad (10.53)$$

where the mass parameters $m$ and $\mu$ are related to the mass parameter $M$ of the conventional action (10.51) through (10.46). Since transformation (10.47) is non-anomalous, the integration measure remains invariant and, somewhat formally, we can write $\mathcal{D}\left[\psi, \overline{\psi}\right] = \mathcal{D}\left[\chi, \overline{\chi}\right]$. Thus we define vacuum expectation values in the twisted basis as

$$\langle O^{\mathrm{tw}}\rangle_{\mathrm{tw}} = \frac{1}{Z_{\mathrm{tw}}} \int \mathcal{D}[\chi, \overline{\chi}, A] \, \mathrm{e}^{-S_F[\chi, \overline{\chi}, A] - S_G[A]} \, O^{\mathrm{tw}}[\chi, \overline{\chi}, A] \,. \tag{10.54}$$

We are left with the task of relating an operator $O[\psi, \overline{\psi}, A]$ in the physical basis to its counterpart $O^{\mathrm{tw}}[\chi, \overline{\chi}, A]$ in the chiral basis. This is a simple exercise as we illustrate in an example, where we choose $O$ to be the nonsinglet axial–pseudoscalar correlator,

$$O\left[\psi, \overline{\psi}\right] = A^1_\mu(x) \, P^1(y) \quad \text{where} \quad A^a_\mu = \frac{1}{2} \overline{\psi} \gamma_\mu \gamma_5 \tau^a \psi \,, \; P^a = \frac{1}{2} \overline{\psi} \gamma_5 \tau^a \psi \,. \tag{10.55}$$

A brief calculation shows that transformation (10.47) gives rise to

$$A^1_\mu \;\to\; \cos(\alpha) \, A^{1\,\mathrm{tw}}_\mu - \sin(\alpha) \, V^{2\,\mathrm{tw}}_\mu \,, \quad P^1 \;\to\; P^{1\,\mathrm{tw}} \quad \text{with}$$

$$A^{a\,\mathrm{tw}}_\mu = \frac{1}{2} \overline{\chi} \gamma_\mu \gamma_5 \tau^a \chi \,, \; V^{a\,\mathrm{tw}}_\mu = \frac{1}{2} \overline{\chi} \gamma_\mu \tau^a \chi \,, \; P^{a\,\mathrm{tw}} = \frac{1}{2} \overline{\chi} \gamma_5 \tau^a \chi \,. \tag{10.56}$$

Inserting these, we find for our correlator the following relation between the vacuum expectation value in the conventional and the twisted form:

$$\langle O \rangle = \left\langle A^1_\mu(x) \, P^1(y) \right\rangle \tag{10.57}$$

$$= \cos(\alpha) \left\langle A^{1\,\mathrm{tw}}_\mu(x) P^{1\,\mathrm{tw}}(y) \right\rangle_{\mathrm{tw}} - \sin(\alpha) \left\langle V^{1\,\mathrm{tw}}_\mu(x) P^{1\,\mathrm{tw}}(y) \right\rangle_{\mathrm{tw}} = \langle O^{\mathrm{tw}}\rangle_{\mathrm{tw}} \,.$$

In an equivalent way arbitrary $n$-point functions of conventional QCD can be mapped onto linear combinations of $n$-point functions of tmQCD.

The above discussion is based on the (formal) expressions in the continuum. To carry the arguments for the relation between conventional and tmQCD over to the lattice one has to consider the renormalized theory in the continuum limit using a mass-independent renormalization scheme [30]. The renormalized mass and twisted mass parameters are

$$m^{(r)} = Z_m \left(m - m_c\right), \quad \mu^{(r)} = Z_\mu \, \mu \,, \tag{10.58}$$

and the twist angle in the renormalized theory must be defined as

$$\alpha = \arctan(\mu^{(r)}/m^{(r)}) \,. \tag{10.59}$$

Maximal twist, i.e., $\alpha = \pi/2$, corresponds to $m^{(r)} = 0$. This case is particularly simple, since setting the bare mass parameter to its critical value, $m = m_c$, which can, e.g., be identified by a vanishing PCAC mass (see Sect. 11.1.2), already leads to $m^{(r)} = 0$.

In the renormalized theory we can use the arguments given for the continuum discussion above and conclude that standard QCD correlation functions can be expressed as linear combinations of correlators in tmQCD. This relation remains valid for finite lattice spacing up to discretization errors [30].

### 10.3.3 $\mathcal{O}(a)$ improvement at maximal twist

Although an important initial motivation for the introduction of tmQCD was the cure for exceptional configurations, today the property of $\mathcal{O}(a)$ improvement is considered more important. We now briefly address this feature of tmQCD at maximal twist [37].

With the introduction of the twisted mass term we now have two parameters $m, \mu$, which define the physics we want to describe. The relative size of the two mass parameters is characterized by the twist angle $\alpha$. We have already observed that among the possible values of $\alpha$, the case of $\alpha = \pi/2$, i.e., maximal twist is singled out. For maximal twist the discretization effects of $\mathcal{O}(a)$ vanish and the leading corrections appear only at $\mathcal{O}(a^2)$. This property is referred to as $\mathcal{O}(a)$ *improvement* of tmQCD and will be discussed now.

The special role of $\alpha = \pi/2$ can already be seen in the free case. In momentum space the twisted mass Dirac operator reads (compare (5.48) for the case of Wilson fermions with conventional mass term)

$$
\frac{\mathrm{i}}{a} \sum_{\mu=1}^{4} \gamma_\mu \sin(p_\mu a) + \frac{1}{a} \sum_{\mu=1}^{4} (1 - \cos(p_\mu a)) + M \cos(\alpha) + \mathrm{i} M \sin(\alpha)\gamma_5\tau^3 ,
$$

(10.60)

where we have dropped all unit matrices. It is straightforward to check that the inverse of this matrix, i.e., the propagator in momentum space, is given by

$$
\frac{-\frac{\mathrm{i}}{a} \sum_\mu \gamma_\mu \sin(p_\mu a) + \frac{1}{a} \sum_\mu (1-\cos(p_\mu a)) + M \cos(\alpha) - \mathrm{i} M \sin(\alpha)\gamma_5\tau^3}{\frac{1}{a^2} \sum_\mu \sin(p_\mu a)^2 + \left(\frac{1}{a} \sum_\mu (1-\cos(p_\mu a)) + M \cos(\alpha)\right)^2 + M^2 \sin(\alpha)^2} .
$$

(10.61)

The energy of the fermion described by this propagator is given by the position of the pole. Expanding the denominator in $a$ gives

$$
p^2 (1 + a M \cos(\alpha)) + M^2 + \mathcal{O}(a^2) ,
$$

(10.62)

and for the two poles we find ($M^2/P^2$ is held fixed)

$$
\mathrm{i} p_4 = \pm \sqrt{p^2 + M^2} \mp a \cos(\alpha) \frac{M^3}{2\sqrt{p^2 + M^2}} + \mathcal{O}(a^2) .
$$

(10.63)

It is evident that the $\mathcal{O}(a)$ correction to the dispersion relation in the continuum, $\mathrm{i} p_4 = \pm\sqrt{p^2 + M^2}$, comes with a factor of $\cos(\alpha)$. This allows to turn off the $\mathcal{O}(a)$ term by choosing maximal twist, i.e., by setting $\alpha = \pi/2$. From an algebraic point of view this result can be understood by noting that at $\alpha = \pi/2$ the Wilson term and the mass term, which now consist of only the twisted part, are orthogonal in isospin space. Thus a mixed term, which is of $\mathcal{O}(a)$, cannot emerge.

Having illustrated that the spectrum of the free theory is $\mathcal{O}(a)$ improved at maximal twist is certainly interesting but of course one would like to establish $\mathcal{O}(a)$ improvement for full tmQCD. This problem can be addressed by performing the Symanzik improvement program (compare Sect. 9.1) for tmQCD. In order to have maximal twist in the renormalized theory we need a vanishing renormalized quark mass parameter $m^{(r)}$. Thus we consider the case where the bare standard mass parameter $m$ is tuned to its critical value, $m = m_c$, which may be defined by the requirement of a vanishing PCAC quark mass (compare Sects. 9.1.4 and 11.1.2). Setting $m = m_c$ implies for the renormalized quark mass parameter, $m^{(r)} = 0$, as needed for maximal twist. In this case the relevant contribution to the effective continuum action is (we show only the leading term – for a complete list see [31])

$$S_F^{\mathrm{eff}} = S_0 + a\, S_1 + \dots + \mathcal{O}(a^2)\,, \tag{10.64}$$

with

$$S_0 = \int \mathrm{d}^4 x\, \overline{\chi}\left(\gamma_\mu D_\mu + i\mu\gamma_5\tau^3\right)\chi\,, \quad S_1 = c_{\mathrm{sw}} \int \mathrm{d}^4 x\, \overline{\chi}\,\sigma_{\mu\nu}F_{\mu\nu}\chi\,. \tag{10.65}$$

The correction $S_1$ to the continuum action $S_0$ is treated as an insertion in expectation values, such that we obtain for the vacuum expectation value of some operator $O$:

$$\langle O \rangle = \langle O \rangle_0 + a\,\langle \Delta O \rangle_0 - a\,\langle O\, S_1 \rangle_0 + \mathcal{O}(a^2)\,. \tag{10.66}$$

The expectation value $\langle \dots \rangle_0$ is with respect to the action $S_0$ and by $\Delta O$ we denote the counterterms for the operator $O$. The operators $O$ may be classified according to their symmetry under the discrete chirality transformation

$$\chi \longrightarrow i\gamma_5\tau^1\chi\,, \quad \overline{\chi} \longrightarrow i\overline{\chi}\gamma_5\tau^1\,, \tag{10.67}$$

which is one of the symmetries of massless QCD with two flavors (see Sect. 7.1.2). Operators can be decomposed into components with definite transformation properties under (10.67):

$$O \longrightarrow \pm O \quad \text{with} \quad \Delta O \longrightarrow \mp \Delta O\,. \tag{10.68}$$

The terms of the effective action transform as

$$S_0 \longrightarrow S_0\,, \quad S_1 \longrightarrow -S_1\,. \tag{10.69}$$

Using (10.68) and (10.69) we can analyze the contributions to the expectation values (10.66). For chirally even operators we find the symmetries

$$\langle O\, S_1 \rangle_0 = -\langle O\, S_1 \rangle_0 = 0\,, \quad \langle \Delta O \rangle_0 = -\langle \Delta O \rangle_0 = 0\,, \tag{10.70}$$

while for odd operators

$$\langle O \rangle_0 = -\langle O \rangle_0 = 0\,, \quad \langle O\, S_1 \rangle_0 = \langle O\, S_1 \rangle_0\,, \quad \langle \Delta O \rangle_0 = \langle \Delta O \rangle_0\,. \tag{10.71}$$

Putting things together we conclude

$$\langle O \rangle = \langle O \rangle_0 + \mathcal{O}(a^2) \qquad\qquad \text{for } O \text{ even} , \qquad (10.72)$$
$$\langle O \rangle = a \left( \langle \Delta O \rangle_0 - \langle O\, S_1 \rangle_0 \right) + \mathcal{O}(a^2) \qquad \text{for } O \text{ odd} .$$

This implies that either only $\mathcal{O}(a^2)$ terms survive (for even operators) or the expectation value $\langle O \rangle$ vanishes in the continuum limit (odd operators). Thus we can conclude that the vacuum expectation values of nonvanishing operators are $\mathcal{O}(a)$ improved at tmQCD with maximal twist. The only parameter that has to be tuned to achieve this is the bare quark mass which has to be set to its critical value, $m = m_c$.

The ease with which $\mathcal{O}(a)$ improvement is obtained and the technical advantages which come with the removal of exceptional configurations make tmQCD an attractive lattice formulation. From a numerical point of view no new techniques are required, which, for example, are needed in the case of dynamical overlap calculations. A rich program of numerical simulations with tmQCD was conducted in recent years and for a comprehensive recent overview we refer to [33].

## 10.4 Effective theories for heavy quarks

For lattice calculations involving heavy quarks, such as the charm and the bottom quark, special techniques are needed. The central idea is to remove the dominant scale, the mass $m_h$ of the heavy quark, and to work with an effective Lagrangian. This procedure gives rise to two formulations known as *nonrelativistic QCD* (NRQCD) and *heavy quark effective theory* (HQET).

### 10.4.1 The need for an effective theory

We begin with a short discussion of the scales involved when working with heavy quarks. The masses of the charm and bottom quarks in the $\overline{\text{MS}}$ scheme [38] are

$$m_c \approx 1.27(9) \text{ GeV} , \quad m_b \approx 4.20(12) \text{ GeV} . \qquad (10.73)$$

The masses for typical charmonium or bottomium states are, e.g.,

$$M_{J/\psi} = 3.096 \text{ GeV} , \quad M_\Upsilon = 9.460 \text{ GeV} . \qquad (10.74)$$

If we want to reliably describe these states on the lattice the cutoff $1/a$ should be larger than these masses. Using (6.71) for converting MeV to fm, one finds that the lattice spacing has to obey

$$a < 0.064 \text{ fm for charm} , \quad a < 0.021 \text{ fm for bottom} . \qquad (10.75)$$

On the other hand we need a spatial volume that is sufficiently large to accommodate the hadrons we want to study and a long enough temporal extent to be able to track the Euclidean propagators when we determine their masses.

A typical value would be a spatial extent of about 2 fm. While for the charm quark lattices of sufficient size will be in reach soon, for the bottom quark a naive approach to simulations with heavy quarks would require very large lattices with typical sizes of at least $\mathcal{O}(100)^4$.

A large part of the mass of the $J/\psi$ and $\Upsilon$ mesons comes from the valence quark masses themselves. Thus if one is able to remove this trivial contribution to the hadron mass from the theory, the typical energy scales of the problem are reduced considerably. Then one can work again with a lattice spacing which is large enough such that the required number of lattice points is sufficiently small for a numerical simulation.

### 10.4.2 Lattice action for heavy quarks

The first ideas for treating heavy quarks on the lattice were put forward in [39–43]. As for the case of heavy quarks in the continuum (see [44] for a standard text), the central idea is to scale out the mass $m_h$ of the heavy quark. This is done in a systematic way by the Foldy–Wouthuysen transformation well known from quantum mechanics. In the system where the heavy quark is at rest it provides an expansion of the action for the heavy flavors in terms of $1/m_h$. Following a standard textbook such as [45] (see also [42, 44]), the Euclidean continuum action density $L = \bar{q}\left(\gamma_\mu D_\mu + m_h\right)q$ with $D_\mu = \partial_\mu + iA_\mu$ is expanded in the form (4 is the time direction)

$$\bar{q}\left(\gamma_\mu D_\mu + m_h\right)q \longrightarrow \bar{\psi}_h\left(m_h + D_4 - \frac{\boldsymbol{D}^2}{2m_h} - \frac{\boldsymbol{\sigma}\cdot\boldsymbol{B}}{2m_h}\right)\psi_h + \mathcal{O}\left(1/m_h^2\right).$$
(10.76)

Here, $\boldsymbol{B}$ denotes the chromomagnetic fields and $\boldsymbol{D}^2 = \sum_{j=1}^3 D_j^2$ is the spatial covariant Laplace operator. On the right-hand side the original 4-spinors $q, \bar{q}$ for the relativistic quarks are replaced by projected, nonrelativistic spinors $\psi_h, \bar{\psi}_h$ which obey

$$P_+\psi_h = \psi_h \,, \quad \bar{\psi}_h P_+ = \bar{\psi}_h \,, \quad P_+ = \tfrac{1}{2}(\mathbb{1} + \gamma_4) \,.$$
(10.77)

It has to be remarked that in expansion (10.76) the usual renormalizability of QCD is obtained only by including all orders in $1/m_h$. How vacuum expectation values with only a finite number of terms in (10.76) can be defined such that they have a finite continuum limit will be described in the next section.

Let us now invoke the lattice. Guided by expansion (10.76) of the continuum action we write the lattice action up to a given order $N$ in the expansion as

$$S_{h,N} = S_{\text{stat}} + \sum_{\nu=1}^{N} S^{(\nu)} \,,$$

$$S_{\text{stat}} = \sum_{n\in\Lambda} \bar{\psi}(n)\left(\tilde{\nabla}_4 + \delta m_h\right)\psi(n) \,, \quad S^{(\nu)} = \sum_i \omega_i^{(\nu)} S_i^{(\nu)} \,. \quad (10.78)$$

Here we have written explicitly the static contribution $S_{\text{stat}}$ which does not give rise to spatial propagation. In this term the mass $m_h$ was dropped to get rid of the unwanted scale $m_h$. For dimensional reasons a mass type of term has to be included, which is taken into account by a residual mass parameter $\delta m_h$ which is also needed for matching and renormalization purposes. By $\widehat{\nabla}_4$ we denote the temporal backward derivative on the lattice, $\widehat{\nabla}_4 f(n) \equiv f(n) - U_4(n - \hat{4}) f(n - \hat{4})$.

The nonstatic terms are organized in powers of $1/m_h$ labeled by the index $\nu$. The leading nonstatic terms are (compare (10.76))

$$S_1^{(1)} = -\frac{1}{2} \sum_{n \in \Lambda} \overline{\psi}_h(n) \, \boldsymbol{\sigma} \cdot \boldsymbol{B} \, \psi_h(n) \, ,$$

$$S_2^{(1)} = -\frac{1}{2} \sum_{n \in \Lambda} \overline{\psi}_h(n) \, \boldsymbol{D}^2 \, \psi_h(n) \, ,$$

$$S_1^{(2)} = \frac{1}{8} \sum_{n \in \Lambda} \overline{\psi}_h(n) \, \boldsymbol{\nabla} \cdot \boldsymbol{E} \, \psi_h(n) \, , \qquad \text{etc} \, . \tag{10.79}$$

On the lattice, the chromomagnetic field $\boldsymbol{B}$ and the chromoelectric field $\boldsymbol{E}$ are implemented by a suitable discretization which expresses these fields through the plaquettes $U_{\mu\nu}$, e.g.,

$$E_j = F_{j4} \, , \quad B_j = \frac{\epsilon_{j4\mu\nu}}{2} F_{\mu\nu} \, , \quad F_{\mu\nu} = i \, \text{Im} \, (\mathbb{1} - U_{\mu\nu}) \, . \tag{10.80}$$

This discretization of the field strength tensor $F_{\mu\nu}$ follows from expansion (2.53) of the plaquette $U_{\mu\nu}$ used for the construction of the gauge action.

The covariant (spatial) Laplace operator $\boldsymbol{D}^2$ is discretized as (we set $a \equiv 1$)

$$\boldsymbol{D}^2 f(n) \longrightarrow \sum_{j=1}^{3} \left( U_j(n) f(n + \hat{\jmath}) - 2f(n) + U_j(n - \hat{\jmath})^\dagger f(n - \hat{\jmath}) \right) \, . \tag{10.81}$$

In the classical theory the expansion coefficients $w_i^{(\nu)}$ would be given by $w_i^{(\nu)} = 1/m_h^\nu$, as can be seen from comparing (10.76), (10.78), and (10.79). Here we are using HQET as an effective theory for fully quantized QCD and have to keep the coefficients $w_i^{(\nu)}$ as free parameters which will be fixed later when matching HQET and QCD. It is, however, important to keep in mind the power counting of the coefficients in expansion (10.78):

$$w_i^{(\nu)} = \mathcal{O}\left(m_h^{-\nu}\right) \qquad (\text{and } \mathcal{O}(a) \equiv \mathcal{O}(1/m)) \, . \tag{10.82}$$

We remark that ultimately we are interested in QCD with both light and heavy flavors. The action for the light flavors is included in a form discussed in the previous chapters and will be added in the next section. We stress that also the observables one is interested in may mix heavy and light quarks. A typical example would be the study of the heavy–light D and B mesons.

### 10.4.3 General framework and expansion coefficients

Having discussed the expansion of the action and its discretization, we need to address the quantization and renormalization of our effective theory. In other words, the coefficients of the effective action have to be determined. Several approaches to this problem can be found in the literature. In our presentation we follow the nonperturbative strategy outlined [46] for HQET and reviewed in great detail in [47]. For a more general presentation of recent results in NRQCD and HQET we refer the reader to [48, 49].

We have already pointed out that expanding the action in $1/m_h$ up to a fixed-order $N$ spoils the usual renormalizability of QCD. However, it can be shown that the static term $S_{\text{stat}}$ of the effective action alone gives rise to a renormalizable theory. Thus the Boltzmann factor for the heavy quark part of the action is expanded as

$$ e^{-S_h} = e^{-S_{\text{stat}}} \left( 1 - S^{(1)} - S^{(2)} + \frac{1}{2} \left( S^{(1)} \right)^2 + \dots \right) , \qquad (10.83) $$

and the nonstatic terms appear as insertions in expectation values computed with the static action. The expansion is terminated after all terms up to the desired order $N$ are included. In (10.83) all terms up to order $N = 2$ are displayed.

The form (10.83) of the expansion leads to the formulation usually referred to as HQET. An alternative is to keep in the exponent also the $D^2$ contribution, the $S_2^{(1)}$ term in (10.79), together with $S_{\text{stat}}$. This leads to the NRQCD formulation (see, e.g., [50]).

For our final expression we still have to add the action for the light quarks, $S_L$, and the action for gauge fields, $S_G$. We obtain for the expectation value of some observable $O$ in HQET on the lattice

$$ \langle O \rangle_h = \frac{1}{Z_h} \int \mathcal{D}[\Phi] \, e^{-S_G - S_L - S_{\text{stat}}} \left( 1 - S^{(1)} - S^{(2)} + \frac{1}{2} \left( S^{(1)} \right)^2 + \dots \right) O , $$

$$ Z_h = \int \mathcal{D}[\Phi] \, e^{-S_G - S_L - S_{\text{stat}}} \left( 1 - S^{(1)} - S^{(2)} + \frac{1}{2} \left( S^{(1)} \right)^2 + \dots \right) . (10.84) $$

The integration is over all fields collectively denoted by $\Phi$, i.e., the light and heavy quark fields and the gauge variables. We remark that for many applications also the observable $O$ is expanded in $1/m_h$. It can be shown that the expanded static vacuum expectation values in (10.84) can be renormalized when all local operators with the proper symmetries and dimension $\leq N$ are included.

The theory defined by (10.84) is described by the following parameters: From the gauge and light quark sectors we have the gauge coupling, the masses of the light quarks, and maybe some coefficients of improvement terms. The heavy quark part contains $\delta m_h$ and the coefficients $\omega_i^{(\nu)}, \nu = 1, \dots N$, where $N$ is the desired order of the expansion.

The remaining problem is how to set the parameters of HQET such that expectation values match their QCD counterparts. In principle it would be possible to evaluate a set of observables $\Phi_k(m_h)$ in both theories and to require

$$\Phi_k^{\text{HQET}}(m_h) = \Phi_k^{\text{QCD}}(m_h) + \mathcal{O}\left(m_h^{-(N+1)}\right) , \qquad (10.85)$$

where the number of observables has to equal the number of parameters. However, such a brute force procedure faces exactly the difficulties which we outlined in Sect. 10.4.1, i.e., the need for large, very fine lattices.

A nonperturbative strategy to overcome this obstacle was presented in [46]. The idea is to start with a sufficiently fine lattice such that the boundaries for the cutoff given in (10.75) are obeyed. In order to have a lattice which in lattice units is small enough to be accessible to a numerical simulation, one is restricted to lattices with a small physical size of $L = 0.2-0.5$ fm. On the small lattice the matching conditions

$$\Phi_k^{\text{HQET}}(m_h, L) = \Phi_k^{\text{QCD}}(m_h, L) + \mathcal{O}\left(m_h^{-(N+1)}\right) \qquad (10.86)$$

are imposed. The HQET part of the calculation is repeated now using a lattice of extent $2L$. The two results can be used to compute the *step-scaling functions* $F_k$ (see [51]) defined as

$$\Phi_k^{\text{HQET}}(m_h, 2L) = F_k\left(\Phi_k^{\text{HQET}}(m_h, L)\right) . \qquad (10.87)$$

The dimensionless functions $F_k$ describe the change of the complete set of observables under a volume change $L \to 2L$. For the step-scaling functions the continuum limit can be performed and subsequently they are used to transport the physical observables to sufficiently large volumes where contact with experimental results can be made.

We conclude with emphasizing again that, although for the matching we here mainly follow the proposal [46], several variants of obtaining effective theories for heavy quarks have been discussed. Various applications of HQET, e.g., the determination of the mass of the b-quark [46, 52], have been proposed and more general reviews of the corresponding results can be found in [48–50, 54, 55].

# References

1. J. Kogut and L. Susskind: Phys. Rev. D **11**, 395 (1975)
2. G. Kilcup and S. Sharpe: Nucl. Phys. B **283**, 493 (1987)
3. F. Gliozzi: Nucl. Phys. B **204**, 419 (1982)
4. H. Kluberg-Stern, A. Morel, O. Napoly, and B. Petersson: Nucl. Phys. B **220**, 447 (1983)
5. T. Takaishi: Phys. Rev. D **54**, 1052 (1996)

6. K. Orginos, D. Touissant, and R. Sugar: Phys. Rev. D **60**, 054503 (1999)
7. G. P. Lepage: Phys. Rev. D **59**, 074502 (1999)
8. M. Albanese et al.: Phys. Lett. B **192**, 163 (1987)
9. A. Hasenfratz and F. Knechtli: Phys. Rev. D **64**, 034504 (2001)
10. E. Follana, A. Hart, and C. Davies: Phys. Rev. Lett. **93**, 241601 (2004)
11. S. Dürr, C. Hoelbling, and U. Wenger: Phys. Rev. D **70**, 094502 (2004)
12. S. Dürr: PoS **LAT2005**, 021 (2005)
13. S. R. Sharpe: Phys. Rev. D **74**, 014512 (2006)
14. M. Creutz: Phys. Lett. B **649**, 230 (2007)
15. A. Kronfeld: PoS **LATTICE2007**, 016 (2007)
16. M. F. L. Golterman (unpublished) arXiv:0812.3110 [hep-lat] (2008)
17. C. T. H. Davies et al.: Phys. Rev. Lett. **92**, 022001 (2004)
18. C. Aubin et al.: Phys. Rev. D **70**, 114501 (2004)
19. C. W. Bernard et al.: Phys. Rev. D **64**, 054506 (2001)
20. A. Bazavov et al. (unpublished) arXiv:0903.3598 [hep-lat] (2009)
21. D. B. Kaplan: Phys. Lett. B **288**, 342 (1992)
22. Y. Shamir: Nucl. Phys. B **406**, 90 (1993)
23. Y. Shamir: Phys. Rev. Lett. **71**, 2691 (1993)
24. V. Furman and Y. Shamir: Nucl. Phys. B **439**, 54 (1995)
25. P. M. Vranas: Phys. Rev. D **57**, 1415 (1998)
26. H. Neuberger: Phys. Rev. D **57**, 5417 (1998)
27. Y. Kikukawa and T. Noguchi (unpublished) arXiv:hep-lat/9902022 (1999)
28. C. R. Allton et al.: Phys. Rev. D **78**, 114509 (2008)
29. R. Frezzotti, P. Grassi, S. Sint, and P. Weisz: Nucl. Phys. (Proc. Suppl.) **83**, 941 (2000)
30. R. Frezzotti, P. Grassi, S. Sint, and P. Weisz: JHEP **0108**, 058 (2001)
31. R. Frezzotti, S. Sint, and P. Weisz: JHEP **0107**, 048 (2001)
32. S. Sint: in *Perspectives in Lattice QCD, Proceedings of the Workshop in Nara, Japan, 31 Oct–11 Nov 2005*, edited by Y. Kuramashi, p. 169 (World Scientific, Singapore 2007)
33. A. Shindler: Phys. Rep. **461**, 37 (2008)
34. R. Frezzotti and G. C. Rossi: Nucl. Phys. (Proc. Suppl.) **128**, 193 (2004)
35. C. Pena, S. Sint, and A. Vladikas: JHEP **09**, 069 (2004)
36. C. Gattringer and S. Solbrig: Phys. Lett. B **621**, 195 (2005)
37. R. Frezzotti and G. C. Rossi: JHEP **0408**, 007 (2004)
38. C. Amsler et al. [Particle Data Group]: Phys. Lett. B **667**, 1 (2008)
39. E. Eichten and B. Hill: Phys. Lett. B **234**, 511 (1990)
40. E. Eichten and B. Hill: Phys. Lett. B **240**, 193 (1990)
41. B. A. Thacker and G. P. Lepage: Phys. Rev. D **43**, 196 (1991)
42. G. P. Lepage et al.: Phys. Rev. D **46**, 4052 (1992)
43. M. Neubert: Phys. Rept. **245**, 259 (1994)
44. A. V. Manohar and M. B. Wise: *Heavy Quark Physics* (Cambridge University Press, Cambridge 2000)
45. J. D. Bjorken and S. D. Drell: *Relativistic Quantum Mechanics* (McGraw-Hill, New York 1964)
46. J. Heitger and R. Sommer: JHEP **0402**, 022 (2004)
47. R. Sommer: in *Perspectives in Lattice QCD, Proceedings of the Workshop in Nara, Japan, 31 Oct–11 Nov 2005*, edited by Y. Kuramashi, p. 209 (World Scientific, Singapore 2007)

48. A. Kronfeld: Nucl. Phys. B (Proc. Suppl.) **129**, 46 (2004)

49. T. Onogi: PoS **LAT2006**, 017 (2006)

50. A. X. El-Khadra, A. S. Kronfeld, and P. B. Mackenzie: Phys. Rev. D **55**, 3933 (1997)

51. M. Lüscher, P. Weisz, and U. Wolff: Nucl. Phys. B **359**, 221 (1991)

52. M. Della Morte, N. Garron, M. Papinutto, and R. Sommer: JHEP **01**, 007 (2007)

53. M. Della Morte: PoS **LATTICE2007**, 008 (2007)

54. J. Heitger: Nucl. Phys. B (Proc. Suppl.) **181–182**, 156 (2008)

55. E. Gamiz: PoS **LATTICE2008**, 014 (2009)

# Hadron structure

In this section we address an important type of nonspectrum calculations. Hadron structure is explored by matrix elements of suitable operators between hadronic states or the vacuum. The simplest such properties are the coefficients of the hadron propagators, related to matrix elements between the hadronic state and the vacuum, the so-called *decay constants*. They describe the weak decay properties of the hadron. We discuss in some detail the lattice calculation of decay constants as a representative of many technically similar problems.

The matrix elements of vector or axial vector currents between single hadron states lead to the *electromagnetic* and the *weak form factors*. Further matrix elements provide information on semileptonic decays, quark- and gluon structure functions, and other information for effective descriptions like the operator product expansion.

All these involve field values and as such have to be related to continuum quantities by renormalization prescriptions. Lattice perturbation theory provides such relations. The results, however, are often not applicable in the nonperturbative regime at the moderately small lattice spacing where one usually works. One therefore relies also on nonperturbative determinations of the renormalization constants that allow contact with observables given in a continuum renormalization scheme such as $\overline{\text{MS}}$.

In this chapter we will give an introductory survey of the related methods to obtain matrix elements and renormalization factors in lattice calculations.

## 11.1 Low-energy parameters

In the early years of hadron theory several hypotheses were introduced to describe the observed features. The current algebra hypothesis related the (flavor $SU(3)$) vector and the axial vector currents in a triplet of current commutation relations. Since the vector current is normalized due to its relation to the electric charge, the commutation relations also lead to a normalization of the

Gattringer, C., Lang, C.B.: *Hadron Structure*. Lect. Notes Phys. **788**, 267–299 (2010)
DOI 10.1007/978-3-642-01850-3_11    © Springer-Verlag Berlin Heidelberg 2010

axial vector current. Coupled to the weak interaction this current describes, among other features, the decay of the pion. The *partially conserved axial current* (PCAC) hypothesis then relates the divergence of the axial vector current to the pion field, thus defining the pion decay constant as a proportionality factor. Another factor in the relation is the pion mass squared, indicating that the axial charge would be conserved for exactly massless pions, i.e., Goldstone bosons. From these assumptions many experimental results in hadron physics can be explained. Meanwhile QCD turned out to be the underlying theory, and what were hypotheses are now properties of that theory. In the lattice approach we want to verify these properties and measure the corresponding physical parameters.

In this section we consider only $u$ and $d$ quarks. These are light enough such that the explicit breaking of chiral flavor SU(2) is small.

### 11.1.1 Operator definitions

In calculations of matrix elements one has to specify the relation between the fields and the observed particle states. In quantum field theory the pion field couples to all states with the quantum numbers $J^{PC} = 0^{-+}$ of the pion. The pseudoscalar interpolator

$$P^a = \tfrac{1}{2}\,\overline{\psi}\,\gamma_5\,\tau^a\,\psi \tag{11.1}$$

has these quantum numbers. We here write the fermion fields as flavor doublets $\psi = (u, d)$ and $\tau^a$, $a = 1, 2, 3$, are the Pauli matrices.[1] Charged quark bilinear operators are defined through

$$P^+ \equiv P^1 - \mathrm{i}\,P^2 = \tfrac{1}{2}\,\overline{\psi}\,\gamma_5\,(\tau^1 - \mathrm{i}\,\tau^2)\,\psi = \overline{d}\,\gamma_5\,u\,,$$
$$P^- \equiv P^1 + \mathrm{i}\,P^2 = \tfrac{1}{2}\,\overline{\psi}\,\gamma_5\,(\tau^1 + \mathrm{i}\,\tau^2)\,\psi = \overline{u}\,\gamma_5\,d\,. \tag{11.2}$$

Analogously we introduce the vector and axial vector fields

$$V_\mu = \tfrac{1}{2}\,\overline{\psi}\,\gamma_\mu\,\psi\,, \qquad V_\mu^a = \tfrac{1}{2}\,\overline{\psi}\,\gamma_\mu\,\tau^a\,\psi\,, \qquad V_\mu^\pm = V_\mu^1 \mp \mathrm{i}\,V_\mu^2\,,$$
$$A_\mu = \tfrac{1}{2}\,\overline{\psi}\,\gamma_\mu\,\gamma_5\,\psi\,, \qquad A_\mu^a = \tfrac{1}{2}\,\overline{\psi}\,\gamma_\mu\,\gamma_5\,\tau^a\,\psi\,, \qquad A_\mu^\pm = A_\mu^1 \mp \mathrm{i}\,A_\mu^2\,. \tag{11.3}$$

The axial vector current is related to the vector current by commutation relations in current algebra. For conserved vector currents the normalization is therefore fixed for both $A_\mu$ and $V_\mu$.

One may obtain equivalent relations also directly in Euclidean space in terms of identities of integrals over expectation values of products of field operators [1]. In our presentation we translate the Minkowskian results of the continuum quantum field theory to Euclidean form by replacing i$t$ with $t$. In order to facilitate switching between continuum and lattice notation we

---

[1] In our discussion we only consider the flavor SU(2) symmetry group.

introduce the abbreviation $\omega$ for the normalization factor of the spatial Fourier transform, with

$$\omega \equiv \begin{cases} (2\pi)^3 & \text{for the continuum,} \\ |\Lambda_3| & \text{for the lattice,} \end{cases} \tag{11.4}$$

where $|\Lambda_3| = N^3$ denotes the spatial lattice volume.

Results of a quantum field theory calculation need to be renormalized before they can be compared to experimental numbers (see Sect. 11.2). We denote quantities renormalized according to a continuum renormalization scheme with a superscript $(r)$. As our reference scheme we use the $\overline{\text{MS}}$ scheme at a scale of $\mu = 2\,\text{GeV}$. We define the renormalization factors $Z_P$, $Z_V$, and $Z_A$ through

$$P^{(r)a} = Z_P\, P^a\,, \quad V^{(r)a} = Z_V\, V^a\,, \quad A^{(r)a} = Z_A\, A^a \tag{11.5}$$

and note that they may depend on details like the lattice action and the scale. In Sect. 11.2 we discuss how one determines the renormalization factors relating the bare lattice fields to the renormalized fields.

Also the quark mass renormalizes multiplicatively, as does the condensate

$$m^{(r)} = Z_m\, m\,, \quad \langle \overline{\psi}\psi \rangle^{(r)} = Z_S\, \langle \overline{\psi}\psi \rangle\,, \tag{11.6}$$

which introduces further renormalization factors $Z_m$ and $Z_S$. These are related via $Z_S = 1/Z_m$ since the combination $m\,\overline{\psi}\psi$ is a renormalization invariant. If chiral symmetry holds, one has the additional relations $Z_S = Z_P$ and $Z_A = Z_V$.

If the fermionic lattice action does not obey chiral symmetry, the unrenormalized bare quark mass may have an additive contribution, called residual mass, such that $m = m_{\text{bare}} + m_{\text{res}}$.

Free bosonic fields in quantum field theory obey a standard normalization. Let us assume that the asymptotic, physical pion is described by such a renormalized triplet field with components $\phi(x)^{(r)a}$. Then in Euclidean space–time the matrix element with the single pion state at rest $|\pi^b(\boldsymbol{p} = 0)\rangle$ is

$$\langle 0|\,\phi(x)^{(r)a}\,|\pi^b(\boldsymbol{p} = 0)\rangle = \delta_{ab}\, e^{-M_\pi t}\,, \tag{11.7}$$

where $x = (\boldsymbol{x}, t)$. The propagator of this field reads

$$\langle 0|\phi^{(r)a}(\boldsymbol{p} = 0, t)\phi^{(r)b}(0)|0\rangle = \frac{1}{2\,M_\pi\,\sqrt{\omega}}\,\delta_{ab}\, e^{-M_\pi t}\,, \tag{11.8}$$

where $(\boldsymbol{p} = \boldsymbol{0}, t)$ denotes the projection to zero spatial momentum in the time slice $t$ (see Sect. 6.1.4). However, we still have to relate this field to the pseudoscalar lattice field operator $P$. This will be done in the next section.

In the standard model of electroweak interaction the decay of hadrons (like the pion or the kaon) into leptons is mediated by a – compared to hadron masses – very heavy intermediate boson. The effective interaction is then described by a Lagrangian coupling the left-handed weak to the hadronic current, e.g., for the decay $\pi \to e\,\nu_e$ one has the coupling term

$$\frac{G_F}{\sqrt{2}} \cos\theta_c \, (\,\overline{u}\gamma_\mu(1 - \gamma_5)d\,) \, (\,\overline{e}\gamma_\mu(1 - \gamma_5)\nu_e\,) \ . \tag{11.9}$$

This is an example of the general operator product expansion of the weak Hamiltonian, which will be discussed in more detail in Sect. 11.4.2.

The pion decay according to (11.9) leads to the matrix element of the axial vector current between the $\pi$ state and the vacuum. The relation to the physical (renormalized) isovector pion field defines the *pion decay constant* $F_\pi$, i.e.,

$$\partial_\mu A_\mu^{(r)a} = M_\pi^2 \, F_\pi \, \phi^{(r)a} \ . \tag{11.10}$$

(There are also other conventions differing by, e.g., a factor of $\sqrt{2}$. Our definition corresponds to an experimental value of $F_\pi = 92.4(3)$ MeV [2]). The divergence of the axial vector current acts as interpolating field operator of the pion. This is an operator identity in Minkowski space; in Euclidean space it holds in expectation values.

Various equivalent expressions for expectation values may be derived. With (11.7) we get the so-called *partially conserved axial vector current* or *PCAC* relation:

$$\partial_\mu \langle 0|\, A_\mu^{(r)a}(x)\, |\pi^b(\boldsymbol{p}=0)\rangle = \delta_{ab}\, M_\pi^2\, F_\pi\, e^{-M_\pi t} \ . \tag{11.11}$$

These relations allow us to derive the decay constant $F_\pi$ from pair correlators between the operators $A_4$ and $\partial_t A_4$. As an example, the $\langle A_4\, A_4\rangle$ correlation function is dominated by the single pion at large $t$ and therefore has the asymptotic behavior

$$\langle\, A_4^{(r)+}(\boldsymbol{p}=0,t)\, A_4^{(r)-}(0)\,\rangle \sim \frac{M_\pi\, F_\pi^2}{\sqrt{\omega}}\, e^{-M_\pi t} \ , \tag{11.12}$$

where the symbol $\sim$ denotes the asymptotic behavior at large $t$. The kaon decay constant is computed analogously, with one of the quarks replaced by the strange quark. More details will be discussed in Sect. 11.1.4.

Decay constants are but one of the quantities that may be obtained. Other low-energy parameters include the quark masses and the condensate. For these we introduce a new tool in the next section: Ward–Takahashi identities.

## 11.1.2 Ward identities

In the path integral expression for the partition function

$$Z = \int \mathcal{D}[\psi,\overline{\psi},U]\, e^{-S[\psi,\overline{\psi},U]} = \int \mathcal{D}[\psi',\overline{\psi}',U']\, e^{-S[\psi',\overline{\psi}',U']}, \tag{11.13}$$

the field variables $\psi, \overline{\psi}, U$ are integrated over. In the second step we have used transformed variables $\psi', \overline{\psi}', U'$ for the path integral, which in general gives the same result. The exceptions are anomalous transformations which affect the integration measure. This strategy of comparing the path integrations using transformed variables was already applied successfully in Chaps. 7 and 9. *Ward identities* express the invariance of the partition function (and expectation values) under such a transformation of the field variables.

Let us study the effect of a local, unitary transformation. The expectation value of an arbitrary interpolator $O$ is

$$\langle 0|O|0 \rangle = \frac{1}{Z} \int \mathcal{D}[\psi, \overline{\psi}, U] \; O[\psi, \overline{\psi}, U] \; e^{-S[\psi, \overline{\psi}, U]} \; . \tag{11.14}$$

We consider an infinitesimal symmetry transformation $\psi \to \psi + \delta\psi$, $\overline{\psi} \to \overline{\psi} + \delta\overline{\psi}$ of the fermion fields in this path integral. The result has to be invariant under such a transformation of integration variables. In case of a non-anomalous transformation the integration measure is invariant and one gets (Ward-) identities of the form

$$0 = \langle 0|\delta O|0 \rangle - \langle 0|O \delta S|0 \rangle \; , \tag{11.15}$$

where $\delta O$ and $\delta S$ denote the linear change of the operator $O$ and the action $S$ under the transformation. The simplest case for $O = 1$ is

$$\langle 0|\delta S|0 \rangle = 0 \; , \tag{11.16}$$

which leads to relations analogous to the classical Noether conservation laws. For some transformations the functional interaction measure is not invariant. This then gives rise to the so-called anomalous contributions (cf., Chaps. 7 and 9).

Let us first derive such relations in the familiar continuum form and for $O = 1$. Later we will point out differences to the lattice notation. We study the transformation of the action ($\mathcal{M} = \text{diag}(m_u, m_d)$ denotes the mass matrix in flavor space)

$$\int d^4x \; \overline{\psi}(\gamma_\mu D_\mu + \mathcal{M})\psi \; , \quad D_\mu = \partial_\mu + iA_\mu \; , \tag{11.17}$$

with regard to infinitesimal SU(2) flavor transformations, corresponding to the finite transformations (7.13), (7.14), (7.15), and (7.16). The linearized, infinitesimal transformations may be written as

$$\begin{aligned}
\psi(x) &\to \psi'(x) = \left(1 + i\,\varepsilon(x)\,\lambda\right)\psi(x) \; , \\
\overline{\psi}(x) &\to \overline{\psi}'(x) = \overline{\psi}(x)\left(1 + i\,\varepsilon(x)\,\hat{\lambda}\right) \; ,
\end{aligned} \tag{11.18}$$

where $\lambda$ and $\hat{\lambda}$ are products of matrices in Dirac and flavor spaces, e.g., $\lambda = 1$, $\tau^a$, $\gamma_5$, $\gamma_5\tau^a$ and $\hat{\lambda} = -1$, $-\tau^a$, $\gamma_5$, $\gamma_5\tau^a$. We assume that $\varepsilon(x)$ is a function that vanishes smoothly outside some bounded region. This will be of importance since we use partial integration and the boundary terms should not contribute.

To $\mathcal{O}(\varepsilon)$ the change of action (11.17) under such an infinitesimal transformation then is

$$\delta S = \mathrm{i} \int \mathrm{d}^4x \, \overline{\psi} \left( \varepsilon \hat{\lambda} \gamma_\mu \partial_\mu + \gamma_\mu \lambda \, \partial_\mu \varepsilon + \mathrm{i} \varepsilon A_\mu (\hat{\lambda} \gamma_\mu + \gamma_\mu \lambda) + \varepsilon (\hat{\lambda} \mathcal{M} + \mathcal{M} \lambda) \right) \psi \, . \tag{11.19}$$

Here we study only transformations that have the property

$$\hat{\lambda} \gamma_\mu + \gamma_\mu \lambda = 0 \, , \tag{11.20}$$

and since the gauge field is flavor and Dirac blind, we get rid of the term involving $A_\mu$. We have furthermore

$$\partial_\mu \left( \varepsilon \, \psi \right) = \left( \partial_\mu \varepsilon \right) \psi + \varepsilon \, \partial_\mu \psi \, , \tag{11.21}$$

leading to

$$\delta S = \mathrm{i} \int \mathrm{d}^4x \, \left( \left( \partial_\mu \varepsilon(x) \right) \overline{\psi} \gamma_\mu \lambda \psi + \varepsilon(x) \, \overline{\psi} \left( \hat{\lambda} \mathcal{M} + \mathcal{M} \lambda \right) \psi \right)$$
$$= \mathrm{i} \int \mathrm{d}^4x \, \varepsilon(x) \left( -\partial_\mu \left( \overline{\psi} \gamma_\mu \lambda \psi \right) + \overline{\psi} \left( \hat{\lambda} \mathcal{M} + \mathcal{M} \lambda \right) \psi \right) \, . \tag{11.22}$$

We have used integration by part in the last step, assuming that the boundary terms do not contribute, as guaranteed by our choice of $\varepsilon(x)$. Since this choice is arbitrary, the condition (11.16) has to hold for any $x$. From this the operator identity results

$$\partial_\mu \left( \overline{\psi} \gamma_\mu \lambda \psi \right) = \overline{\psi} (\hat{\lambda} \mathcal{M} + \mathcal{M} \lambda) \psi \, . \tag{11.23}$$

Then, depending on the choice of $\lambda$ and $\hat{\lambda}$, but respecting (11.20), we obtain the relations

$$\lambda = \mathbb{1}, \hat{\lambda} = -\mathbb{1} \quad \Rightarrow \quad \partial_\mu \left( \overline{\psi} \gamma_\mu \psi \right) = 0 \, , \tag{11.24}$$

$$\lambda = \tau^a, \hat{\lambda} = -\tau^a \quad \Rightarrow \quad \partial_\mu (\overline{\psi} \, \gamma_\mu \, \tau^a \, \psi) = \overline{\psi} \, [\mathcal{M}, \tau^a] \, \psi \, , \tag{11.25}$$

$$\lambda = \hat{\lambda} = \gamma_5 \quad \Rightarrow \quad \partial_\mu (\overline{\psi} \, \gamma_\mu \, \gamma_5 \, \psi) = 2 \, \overline{\psi} \, \mathcal{M} \, \gamma_5 \, \psi \quad (\text{+anomaly}) \, , \tag{11.26}$$

$$\lambda = \hat{\lambda} = \gamma_5 \, \tau^a \quad \Rightarrow \quad \partial_\mu (\overline{\psi} \, \gamma_\mu \, \gamma_5 \, \tau^a \, \psi) = \overline{\psi} \, \{\mathcal{M}, \tau^a\} \, \gamma_5 \, \psi \, , \tag{11.27}$$

where we used that the mass matrix $\mathcal{M}$ commutes with $\gamma_5$: $\mathcal{M} \gamma_5 = \gamma_5 \mathcal{M}$. With the shorter notation (11.1) and (11.3) for the flavor singlet and nonsinglet vector and axial vector currents and the pseudoscalar field the relations then assume the form

$$\partial_\mu V_\mu = 0 \, , \tag{11.28}$$

$$\partial_\mu V_\mu^a = \tfrac{1}{2} \overline{\psi} \, [\mathcal{M}, \tau^a] \, \psi \, , \tag{11.29}$$

$$\partial_\mu A_\mu = \overline{\psi} \, \mathcal{M} \, \gamma_5 \, \psi \quad (\text{+anomaly}) \, , \tag{11.30}$$

$$\partial_\mu A_\mu^a = \tfrac{1}{2} \overline{\psi} \, \{\mathcal{M}, \tau^a\} \, \gamma_5 \, \psi \, . \tag{11.31}$$

Equation (11.28) for the flavor singlet vector current is the conservation law for the total baryon number. In the other three relations we find that the Noether symmetry may be broken by the mass term. For mass-degenerate quarks one has isospin symmetry and therefore $[\mathcal{M}, \tau^a] = 0$. Consequently (11.29) defines the conservation of the flavor vector current. In (11.30), when written for the quantized theory, the anomaly has to be added due to a noninvariance of the integration measure in the functional integration (cf., Sects. 7.1, 7.3, and 9.4).

Equation (11.31) is called the *nonsinglet axial Ward identity* (AWI): The divergence of the flavor nonsinglet axial vector current is related to the pseudoscalar field.

We stress again that due to (11.16) these relations hold as expectation values. Since the choice of $O$ is arbitrary we can choose the operator to be nonvanishing and constant inside some compact, arbitrarily small space–time region, and zero outside. Therefore (11.28), (11.29), (11.30), and (11.31) may be considered to be valid locally. The relations are preserved under quantization and renormalization (up to the anomaly) and therefore hold for the renormalized quantities as well.

For degenerate fermion masses $m$ the AWI (11.31) has the form

$$\partial_\mu A_\mu^{(r)a} = 2\, m^{(r)}\, P^{(r)a} \;. \tag{11.32}$$

This relation is useful in many aspects. Comparing it with (11.10) and evaluating it between the vacuum and the pion we find

$$M_\pi^2\, F_\pi \, \langle 0 | \phi^{(r)a} | \pi \rangle = \langle 0 | \partial_\mu A_\mu^{(r)a} | \pi \rangle = 2\, m^{(r)} \langle 0 | P^{(r)a} | \pi \rangle \;. \tag{11.33}$$

This allows to determine the renormalized quark mass $m^{(r)}$ and the pion decay constant from the asymptotic behavior of correlators of $P^{(r)a}$ and $A_\mu^{(r)a}$ (see Sect. 11.1.4).

Further exploitation of (11.32) leads to another interesting relation. We utilize (11.22) for transformation (11.31) by multiplying it with a local pseudoscalar operator and taking its expectation value over the Grassmann fields. Specifically we use (11.18) with

$$\lambda = \hat{\lambda} = \gamma_5 \tfrac{1}{2}(\tau^1 + i\tau^2) \;,$$
$$O(0) = \overline{\psi}(0)\, \gamma_5 \tfrac{1}{2}(\tau^1 - i\tau^2)\, \psi(0) = \overline{d}(0)\, \gamma_5\, u(0) = P^+(0) \;. \tag{11.34}$$

This operator is just the pseudoscalar field. In the chiral limit $\mathcal{M} = 0$ Eq. (11.15) with (11.22) now leads to

$$-\int \mathrm{d}^4 x \, \langle \partial_\mu A_\mu^{(r)-}(x)\, P^+(0) \rangle = \langle \overline{u}(0)u(0) + \overline{d}(0)d(0) \rangle \;. \tag{11.35}$$

Inserting on the left-hand side a complete set of states, the 2-point function is saturated by the pion and the expression becomes proportional to $\langle P^-(x)P^+(0)\rangle$. The left-hand side turns into

$$-\frac{F_\pi^2 M_\pi^4}{2m^{(r)}} \int \mathrm{d}^4x \, \langle \phi^-(x)\phi^+(0)\rangle = -\frac{F_\pi^2 M_\pi^2}{m^{(r)}} \,, \tag{11.36}$$

where we use that the integral is twice the momentum space propagator $(p^2 + M_\pi^2)^{-1}$ at vanishing momentum.

The right-hand side of (11.35) is the condensate $\Sigma^{(r)}$

$$\langle \overline{u}(0)u(0) + \overline{d}(0)d(0)\rangle \equiv N_f \, \Sigma^{(r)} \,, \tag{11.37}$$

where in our case $N_f = 2$.

We finally obtain a relation that holds for the renormalized quark mass and the renormalized condensate $\Sigma^{(r)}$:

$$F_\pi^2 \, M_\pi^2 = -m^{(r)} \, N_f \, \Sigma^{(r)} \,. \tag{11.38}$$

The pion decay constant $F_\pi$ and the pion mass $M_\pi$ do not depend on the renormalization scheme; they are physical, measurable quantities. In our normalization the experimental value of $F_\pi$ is close to 93 MeV.

Equation (11.38) is known as *Gell–Mann–Oakes–Renner* (GMOR) relation due to the seminal paper [3], where the mechanism of symmetry breaking by the quark mass term was discussed. Note that this equation has possible correction terms of $\mathcal{O}(m^2)$. This is the lowest order of a systematic expansion leading to an effective field theory known as *chiral perturbation theory* [4, 5].

When one attempts to derive the lattice analogies [6, 7] to the continuum identities one runs into a problem. For actions that are not chirally symmetric, like the Wilson action, the chiral variation of the lattice action cannot be expressed in terms of the axial current and density alone [6] and additional terms appear. One then argues, however, that the correlation functions of the renormalized lattice fields should converge to the corresponding continuum ones. This allows to derive information from expectation values like (11.58) for nonzero distances of propagation.

### 11.1.3 Naive currents and conserved currents on the lattice

We now change to the lattice formulation and consider the construction of lattice currents corresponding to symmetry transformations of the form (11.18), i.e.,

$$\delta\psi(n) = \mathrm{i}\,\varepsilon(n)\,\lambda\,\psi(n)\,, \quad \delta\overline{\psi}(n) = \mathrm{i}\,\overline{\psi}(n)\,\varepsilon(n)\,\hat{\lambda}\,. \tag{11.39}$$

Under this transformation the action

$$S = \sum_{n,m} \overline{\psi}(n)\, D(U;n|m)\, \psi(m) \tag{11.40}$$

varies, with the linear term of the variation given by

$$\delta S = \mathrm{i}\sum_{n,m} \overline{\psi}(n)\left(D(U;n|m)\,\varepsilon(m)\,\lambda + \varepsilon(n)\,\hat{\lambda}\,D(U;n|m)\right)\psi(m)\,. \tag{11.41}$$

Let us first consider Wilson fermions (5.51) and $\lambda = \tau^a$ and $\hat{\lambda} = -\tau^a$. The term $\mathcal{O}(\varepsilon)$ in the transformation of the Wilson action has to vanish for arbitrary $\varepsilon(n)$. After shifting the index for the summation over the lattice sites this can be cast into a local form. We end up with an equation equivalent to (11.29), with the replacement of the continuum derivatives by a nearest neighbor difference

$$\sum_{\mu=1}^{4} \frac{1}{a} \left( V_\mu^a(n + \hat{\mu}) - V_\mu^a(n) \right) = \frac{1}{2} \overline{\psi}(n)[M, \tau^a]\psi(n) , \tag{11.42}$$

with the lattice vector current

$$V_\mu^a(n) = \frac{1}{4} \sum_{\mu=1}^{4} \left( \overline{\psi}(n + \mu)(\mathbb{1} + \gamma_\mu)U_\mu(n)^\dagger \tau^a \psi(n) \right. \tag{11.43}$$
$$\left. - \overline{\psi}(n)(\mathbb{1} - \gamma_\mu)U_\mu(n)\tau^a \psi(n + \mu) \right) .$$

We can write (11.42) as

$$\Delta_\mu V_\mu^a(n) = \frac{1}{2} \overline{\psi}(n)[M, \tau^a]\psi(n) , \tag{11.44}$$

where $\Delta_\mu$ is the forward lattice derivative $\Delta_\mu f(n) \equiv (f(n + \hat{\mu}) - f(n))/a$. This agrees (up to higher orders in the lattice spacing) with the corresponding continuum expression (11.29). For $u$ and $d$ quarks with identical masses the current is conserved like in the continuum.

For the axial transformations, however, this simple equivalence does not work any more for fermion actions which violate chiral symmetry (like the Wilson action). There are symmetry-breaking terms divergent with $1/a$ and it is not possible to define an exactly conserved axial vector current for vanishing quark masses.

In Chap. 7 we discussed Dirac operators which obey the Ginsparg–Wilson relation and thus chiral symmetry on the lattice. The corresponding operators, e.g., the overlap operator, are more complicated. It therefore pays off to consider the problem of constructing Noether currents for a general action without specifying a particular form explicitly. For this we follow [8] and discuss as an example a simple $n$-dependent scalar transformation like in (11.18) and (11.24) with $\lambda = 1$, $\hat{\lambda} = -1$.

The variation (11.41) assumes the form ($k$ is a position index here)

$$\delta S = i\, a \sum_{k,\mu} (\Delta_\mu \varepsilon(k))\, J_\mu(k) = i \sum_{k,\mu} (\varepsilon(k + \hat{\mu}) - \varepsilon(k))\, J_\mu(k)$$
$$= i \sum_{k,\mu} \varepsilon(k)\, (J_\mu(k - \hat{\mu}) - J_\mu(k)) = -i\, a \sum_{k,\mu} \varepsilon(k)\, \Delta_\mu^* J_\mu(k) , \tag{11.45}$$

where $\Delta_\mu$ and $\Delta_\mu^*$ denote forward and backward lattice derivatives. In case the action is invariant under such a change of variables, i.e., $\delta S = 0$, we could identify $J_\mu(k)$ with the conserved Noether current on the lattice.

For this end let us consider a transformation of each individual link variable:

$$U_\mu(n) \quad \rightarrow \quad \widetilde{U}_\mu(n) = U_\mu^{(\alpha)}(n) \equiv e^{i\,\alpha_\mu(n)}\,U_\mu(n) \equiv e^{i\,\varepsilon(n)}\,U_\mu(n)\,e^{-i\,\varepsilon(n+\hat{\mu})}\,, \tag{11.46}$$

where $\alpha_\mu(n) = -\Delta_\mu\varepsilon(n)$. We have rewritten this transformation in order to utilize the transformation properties of the Dirac operator under gauge transformations, hence

$$D(U;n|m) \quad \rightarrow \quad D(\widetilde{U};n|m) = e^{i\,\varepsilon(n)}\,D(U;n|m)\,e^{-i\,\varepsilon(m)}\,. \tag{11.47}$$

Taking the difference between transformed and untransformed Dirac operators we find at $\mathcal{O}(\varepsilon)$:

$$-\Big(D(\widetilde{U};n|m) - D(U;n|m)\Big) = i\,(D(U;n|m)\,\varepsilon(m) - \varepsilon(n)\,D(U;n|m))\,, \tag{11.48}$$

where we have taken into account just the leading order in $\varepsilon$. This, however, is the kernel of (11.41). Alternatively, we can determine this leading order difference also by a derivative with regard to $\alpha_\mu$:

$$-\Big(D(\widetilde{U};n|m) - D(U;n|m)\Big) = -\sum_{k,\mu}\alpha_\mu(k)\left(\frac{\partial D\left(U_\mu^{(\alpha)}(k);n|m\right)}{\partial\alpha_\mu(k)}\right)_{\alpha=0}$$

$$= \sum_{k,\mu}(\varepsilon(k+\hat{\mu}) - \varepsilon(k))\left(\frac{\partial D(U_\mu^{(\alpha)}(k);n|m)}{\partial\alpha_\mu(k)}\right)_{\alpha=0}. \tag{11.49}$$

Comparison with (11.45) leads to the identification of the Noether current

$$J_\mu(k) = \sum_{n,m}\overline{\psi}(n)K_\mu(k;n|m)\psi(m)\,, \quad \text{with}$$

$$K_\mu(k;n|m) \equiv -i\left(\frac{\partial D\left(U_\mu^{(\alpha)}(k);n|m\right)}{\partial\alpha_\mu(k)}\right)_{\alpha=0}. \tag{11.50}$$

We find that if we imagine the Dirac operator as a sum over paths connecting $n$ with $m$, then all paths which run through the link $U_\mu(k)$ contribute to $J_\mu(k)$. From the construction the gauge invariance of $J_\mu(k)$ is obvious. In practical applications the derivatives have to be performed numerically by taking differences. Applying this construction to the Wilson Dirac operator we recover the simple local currents as, e.g., in (11.43).

For chirally symmetric actions it is possible to construct conserved and covariantly transforming vector currents as well as covariant scalar densities. These then obey Ward identities like in the continuum.

Following the strategy outlined in our example, in [8, 9] chiral covariant and conserved currents have been constructed for general, exactly chiral-invariant actions like the overlap action. We refer to the original papers for the explicit derivation and only quote the results. With the abbreviation

$$\hat{\gamma}_5 = \gamma_5(\mathbb{1} - D), \tag{11.51}$$

we have

$$V_\mu^a(k) = \frac{1}{4} \sum_{n,m} \overline{\psi}(n) \left( K_\mu(k;n|m) - \gamma_5 \sum_j K_\mu(k;n|j) \, (\hat{\gamma}_5)_{jm} \right) \tau^a \, \psi(m) \,,$$

$$A_\mu^a(k) = \frac{1}{4} \sum_{n,m} \overline{\psi}(n) \left( -\gamma_5 K_\mu(k;n|m) + \sum_j K_\mu(k;n|j) \, (\hat{\gamma}_5)_{jm} \right) \tau^a \, \psi(m) \,,$$

$$S^a(k) = \frac{1}{4} \sum_{n,m} \overline{\psi}(n) \left( \delta_{nk} \left( \mathbb{1} - \tfrac{1}{2}D \right)_{km} \right.$$
$$\left. + \frac{1}{2} \sum_j \left( \mathbb{1} - \tfrac{1}{2}D \right)_{nj} \left( \delta_{jk}\delta_{mk} + (\hat{\gamma}_5)_{jk} \, (\hat{\gamma}_5)_{km} \right) \right) \tau^a \, \psi(m) \,,$$

$$P^a(k) = \frac{1}{4} \sum_{n,m} \overline{\psi}(n) \left( \gamma_5 \delta_{nk} \left( \mathbb{1} - \tfrac{1}{2}D \right)_{km} \right.$$
$$\left. + \frac{1}{2} \sum_j \left( \mathbb{1} - \tfrac{1}{2}D \right)_{nj} \left( (\hat{\gamma}_5)_{jk} \, \delta_{mk} + \delta_{jk} \, (\hat{\gamma}_5)_{km} \right) \right) \tau^a \, \psi(m) \,.$$
$$\tag{11.52}$$

These conserved currents and equivalently constructed scalar and pseudoscalar densities transform covariantly under the chiral transformations, like in the continuum. For conserved vector currents the renormalization is trivial (the charge is conserved and the renormalization constants obey $Z_A = Z_V = 1$). An important advantage is that the currents and densities (dimension-3 operators) are automatically $\mathcal{O}(a)$ improved and do not mix with dimension-4 or dimension-3 operators, respectively.

Since the implementation of prescription (11.50) is rather involved, the simpler expressions

$$\widetilde{V}_\mu^a(k) = \frac{1}{2} \sum_m \overline{\psi}(k)\gamma_\mu \left( \mathbb{1} - \tfrac{1}{2}D \right)_{km} \tau^a \, \psi(m) \,,$$

$$\widetilde{A}_\mu^a(k) = \frac{1}{2} \sum_m \overline{\psi}(k)\gamma_\mu\gamma_5 \left( \mathbb{1} - \tfrac{1}{2}D \right)_{km} \tau^a \, \psi(m) \,,$$

$$\widetilde{S}^a(k) = \frac{1}{2} \sum_m \overline{\psi}(k) \left( \mathbb{1} - \tfrac{1}{2}D \right)_{km} \tau^a \, \psi(m) \,,$$

$$\widetilde{P}^a(k) = \frac{1}{2} \sum_m \overline{\psi}(k)\gamma_5 \left( \mathbb{1} - \tfrac{1}{2}D \right)_{km} \tau^a \, \psi(m)$$
$$\tag{11.53}$$

have been used. They are also covariant but not conserved and thus not related by Ward identities [8]. For direct constructions based on Ginsparg–Wilson fermions see also [10, 11].

In correlation functions over distances larger than the range of the Dirac operator the covariant densities in (11.52) can be replaced by the even simpler, point-like expressions

$$S^a(k) \approx \frac{1}{2} \frac{1}{1 - m/2} \overline{\psi}(k)\, \tau^a\, \psi(k) \,,$$

$$P^a(k) \approx \frac{1}{2} \frac{1}{(1 - m/2)(1 - m^2/4)} \overline{\psi}(k)\, \gamma_5\, \tau^a\, \psi(k) \,. \tag{11.54}$$

This comes about since the propagator involves the inverse massive Dirac operator $D_m^{-1}$ and from (7.41) and (7.42) we know that

$$\mathbb{1} - \tfrac{1}{2}D = \frac{1}{1 - m/2}\left(\mathbb{1} - \tfrac{1}{2}D_m\right) \,. \tag{11.55}$$

The insertions $\mathbb{1} - \tfrac{1}{2}D$ in (11.53) thus can be expressed in terms of the massive Dirac operator $D_m$. When considering correlation functions of the interpolators (11.53) the insertions give rise to products $D_m D_m^{-1}$ which just contribute locally ("contact term") and may be omitted [8].

In the continuum limit the currents renormalize multiplicatively with factors like $Z_V$ and $Z_A$ and we will discuss the determination of these factors in Sect. 11.2.

### 11.1.4 Low-energy parameters from correlation functions

In the last section we have shown that the continuum identities for correlation functions may be translated to the lattice, although the identification of the currents and densities on the lattice is not straightforward. We can expect that the asymptotic behavior of correlation functions may be compared. Thus we may utilize equations like (11.12) to determine low-energy constants from the asymptotic (long time distance) behavior of certain correlation functions and ratios thereof.

The renormalized quark mass may be determined utilizing the axial Ward identities (11.31), (11.33) and considering the (large $t$) asymptotic behavior of

$$\frac{1}{2} \frac{\langle 0|\partial_t A_4^{-(r)}(\boldsymbol{p} = \mathbf{0}, t)X(0)|0\rangle}{\langle 0|P^{-(r)}(\boldsymbol{p} = \mathbf{0}, t)X(0)|0\rangle} = \frac{1}{2} \frac{Z_A}{Z_P} \frac{\langle 0|\partial_t A_4^{-}(\boldsymbol{p} = \mathbf{0}, t)X(0)|0\rangle}{\langle 0|P^{-}(\boldsymbol{p} = \mathbf{0}, t)X(0)|0\rangle} \sim m^{(r)} \,. \tag{11.56}$$

For $X$ one chooses an interpolator coupling to the pion and often one takes $X = P$. Numerator and denominator are both asymptotically dominated by the pion intermediate state and the asymptotic $t$-dependence cancels. In the calculations one can identify a plateau as soon as the asymptotic behavior has set in. In this relation one needs to know the renormalization constants $Z_A$ and $Z_P$ relating the lattice interpolators with the continuum normalization. These can be determined separately, as will be discussed later.

Neglecting the renormalization constants one often just determines the asymptotic ratio of the unrenormalized correlators:

$$\frac{1}{2} \frac{\langle 0|\partial_t A_4^{-}(\boldsymbol{p} = \mathbf{0}, t)X(0)|0\rangle}{\langle 0|P^{-}(\boldsymbol{p} = \mathbf{0}, t)X(0)|0\rangle} \sim m_{\mathrm{AWI}} \,. \tag{11.57}$$

Due to its origin this ratio is called the *AWI mass* $m_{\mathrm{AWI}}$ or *PCAC mass*. We have already used (11.57) as a tool for determining the coefficients $c_{\mathrm{sw}}$ and $c_A$ in the Symanzik improvement program in Sect. 9.1.4.

The quark mass of the lattice action may undergo two renormalizations. In case the action is not chirally symmetric there may be an additive renormalization called *residual mass* $m_{\mathrm{res}}$ and $m_{\mathrm{AWI}} = m_{\mathrm{bare}} + m_{\mathrm{res}}$. We then have $m^{(r)} = (Z_A/Z_P)\, m_{\mathrm{AWI}}$. In relation to, e.g., the continuum $\overline{\mathrm{MS}}$ scheme one has a multiplicative renormalization factor $m^{(r)} = Z_m\, m$. For chirally symmetric actions $Z_m\, Z_P = 1$ and thus the bare mass in lattice units is $m = Z_A\, m_{\mathrm{AWI}}$.

The AWI mass vanishes at the points in the space of couplings, where chiral symmetry holds. This is particularly useful in cases with additive mass renormalization, since it allows to identify the residual mass (see below). For the Wilson action one identifies the critical value of the hopping parameter $\kappa_c(\beta)$ with its help: It is defined as the value where $m_{\mathrm{AWI}}$ vanishes and the pions are massless (see also Sect. 9.1.4).

The pion decay constant may be obtained, e.g., from the axial current correlation function (11.12):

$$Z_A^2 \, \langle\, A_4^+(\boldsymbol{p}=\boldsymbol{0},t)\, A_4^-(0)\,\rangle \sim \frac{M_\pi\, F_\pi^2}{\sqrt{\omega}}\, \mathrm{e}^{-M_\pi t}\,. \tag{11.58}$$

Due to (11.33) we also find

$$\langle 0| P^+(\boldsymbol{p}=\boldsymbol{0},t) P^-(0)|0\rangle \sim \frac{A}{\sqrt{\omega}}\, \mathrm{e}^{-M_\pi t} \quad \text{with} \quad A = \frac{M_\pi^3\, F_\pi^2}{4\, m^{(r)2} Z_P^2}\,. \tag{11.59}$$

From the coefficient to the asymptotic behavior we get

$$F_\pi = 2\, Z_P\, m^{(r)} \sqrt{\frac{A}{M_\pi^3}} = 2\, Z_A\, m_{\mathrm{AWI}} \sqrt{\frac{A}{M_\pi^3}}\,. \tag{11.60}$$

The GMOR equation (11.38) may be used to compute the quark condensate, once $F_\pi$, $M_\pi$, and $m^{(r)}$ are known. One may also use correlation functions like

$$Z_A\, Z_P\, |\langle 0| A_4^+(\boldsymbol{p}=\boldsymbol{0},t) P^-(0)|0\rangle| \sim \frac{|\Sigma^{(r)}|}{\sqrt{\omega}}\, \mathrm{e}^{-M_\pi t} \tag{11.61}$$

for its determination (cf. [12] for more such relations).

## 11.2 Renormalization

### 11.2.1 Why do we need renormalization?

When quantizing a field theory one has to regularize it, and the lattice formulation is one possible ultraviolet regulator. Identification of computed quantities with physical observables then completes the renormalization process. Different regulators and actions require different renormalization parameters.

Some observables like hadron masses can be measured directly as dimensionless numbers $a\,m_{\text{phys}}$ from the exponential decay of the correlation functions. By comparing one such number with the physical mass $m_{\text{phys}}$ of that particle one may determine the lattice spacing $a$ and establish the scale (see Sect. 6.3). Other quantities like decay constants or form factors involve fields and their relationship to experiments and continuum normalization is more involved.

Matrix elements and low-energy parameters like quark masses or the condensate play an important role in QCD phenomenology. However, their values depend on the definitions in some renormalization scheme. Since one wants to compare with physical, experimentally measured quantities one has to relate these to the parameters of the underlying formulation of the quantized theory. For that purpose one has to determine the scaling properties, i.e., the dependence on the scale $a$, of the observables and their behavior under renormalization. Many quantities are intrinsically scale dependent and are even divergent when the cutoff parameter $a$ is removed. We can distinguish the following scenarios:

- Finite operators, e.g., vector currents and axial vector currents or the ratios of scalar and pseudoscalar densities: The multiplicative renormalization factors $Z_V$, $Z_A$ and the ratio $Z_S/Z_P$ assume finite values for vanishing regularization parameter $a$, thus they should be scale independent. For conserved currents and chirally symmetric actions $Z_V = Z_A = 1$. Depending on the Dirac operator, it may be complicated to use the conserved currents (see Sect. 11.1.3) and one therefore relies on simpler, e.g., point-like currents where the $Z$-factors are not known a priori.
- Logarithmically divergent operators (which are, e.g., scalar and pseudoscalar densities or flavor $\Delta F = 2$ changing 4-fermion operators) are scale dependent.
- Power divergent operators: These occur in regularizations with intrinsic mass scale like in the lattice regularizations. Then mixing with lower dimensional operators becomes possible, leading to power divergences $\mathcal{O}(1/a^n)$ [13]. Examples are $\mathcal{O}(1/m)$ terms in *heavy quark effective theory (HQET)* discussed in Sect. 10.4.

In the regularization some symmetries of the original theory are lost. In the lattice approach these are, e.g., the continuous space–time transformations and (for some actions) the chiral symmetry. Exact GW fermions are protected by their chiral symmetry and renormalization of operators constructed from them is simpler than for, e.g., Wilson fermions. Chiral symmetry implies several relations between renormalization constants, e.g., $Z_A = Z_V$ and $Z_S = Z_P$.

In order to compute renormalization constants nonperturbatively on the lattice one needs a renormalization scheme which can be implemented in lattice Monte Carlo simulations and in continuum perturbation theory. The latter property is necessary to enable the conversion of the lattice results to the continuum scheme. The most favored continuum scheme is the modified minimal subtraction $\overline{\text{MS}}$ scheme. In such a mass-independent scheme the renormalization factors depend only on a normalization mass $\mu$ and the coupling constant, but not on quark masses.

## 11.2.2 Renormalization with the Rome–Southampton method

Since one compares the lattice results with results of the continuum one could utilize lattice perturbation theory (see [13] and references therein). This has been done but in many cases the convergence is too bad to lead to reliable numbers. One therefore often relies on nonperturbative methods [14].

The idea is to compare bare lattice correlation functions, couplings, and masses determined in nonperturbative lattice calculations with quantities in the so-called *Regularization-independent* (RI) scheme (often also called RI/MOM scheme). The connection between the RI quantities and those defined in the $\overline{\text{MS}}$ scheme can then be done in continuum perturbation theory [15].

As an example we discuss the renormalization factors of quark bilinear operators. To be more explicit, let us consider the local, flavor nonsinglet quark field bilinear operators

$$O_\Gamma \equiv \overline{u}\,\Gamma\,d \,. \tag{11.62}$$

Here $\Gamma$ denotes a Clifford algebra matrix. According to their Lorentz symmetry we denote the five types of $\Gamma$ by S, V, A, T, and P corresponding to scalar, vector, axial vector, tensor, and pseudoscalar (corresponding to $\mathbb{1}$, $\gamma_\mu$, $\gamma_\mu\gamma_5$, $\frac{i}{2}[\gamma_\mu,\gamma_\nu]$, and $\gamma_5$).

One studies expectation values $\langle p \,|\, O_\Gamma \,|\, p \rangle$ (i.e., at zero momentum transfer) of the bilinear quark operators between quark fields $|p\rangle$ at a specific momentum value $p^2 = \mu^2$ and matches them to the corresponding tree-level matrix element:

$$Z_\Gamma \left.\langle p \,|\, O_\Gamma \,|\, p \rangle\right|_{p^2=\mu^2} = \left.\langle p \,|\, O_\Gamma \,|\, p \rangle_0\right|_{p^2=\mu^2} \,. \tag{11.63}$$

The renormalization constant $Z_\Gamma$ is the proportionality factor between the interacting and the free case.

This procedure is expected to work in a window

$$\Lambda^2_{\text{QCD}} \ll \mu^2 \ll 1/a^2 \,, \tag{11.64}$$

where discretization effects can be neglected, because the renormalization scale $\mu$ is small compared with the lattice cutoff $1/a$. Then (few-loop) continuum perturbation theory can be used to connect different schemes, because $\mu$ is much larger than the QCD scale parameter $\Lambda_{\text{QCD}}$. For comparing with the $\overline{\text{MS}}$ scheme a typical value is $\mu = 2\,\text{GeV}$. In most calculations one has $(a\,\mu) \approx 1$ (or even somewhat larger) and the upper limit is not strictly obeyed. On the other hand, the limit also depends on the scaling properties of the actions involved.

Since (11.63) is gauge variant, one has to work in a fixed gauge and must compare the gauge dependent lattice matrix elements with the continuum results in the same gauge. Landau gauge fixing is a suitable choice, but one has to keep in mind that the Gribov copies uncertainty could spoil the comparison. In the lattice calculations one finds little, if any, signal of such an effect [16–18].

We summarize the method following [14] in the modification of [19]. Note that in (11.63) one compares matrices in color and Dirac spaces. Taking the trace one obtains for the renormalization condition

$$Z_\Gamma \frac{1}{12} \, \mathrm{tr} \left[ \langle p \mid O_\Gamma \mid p \rangle \, \langle p \mid O_\Gamma \mid p \rangle_0^{-1} \right] \Big|_{p^2 = \mu^2} = 1 \, . \tag{11.65}$$

The matrix element

$$\langle p \mid O_\Gamma \mid p \rangle = \frac{1}{Z_q} \Lambda_\Gamma(p) \tag{11.66}$$

is proportional to the amputated Green function

$$\Lambda_\Gamma(p) = S^{-1}(p) \, G_\Gamma(p) \, S^{-1}(p) \, , \tag{11.67}$$

and $Z_q$ is the quark field renormalization constant to be discussed below. The Green function $G_\Gamma(p)$ is determined as the expectation value:

$$G_\Gamma(p)_{\substack{\alpha\beta \\ ab}} = \frac{1}{V} \sum_{x,y} e^{-ip(x-y)} \left\langle u_\alpha(x) \sum_z O_\Gamma(z) \, \overline{d}_\beta(y) \right\rangle \, . \tag{11.68}$$

The indices $\alpha, \beta$ and $a, b$ run over Dirac and color indices, respectively, and $V$ denotes the lattice volume. The quark propagator is

$$S_{\substack{\alpha\beta \\ ab}}(x,y) = \langle u_\alpha(x) \, \overline{u}_\beta(y) \rangle = \langle d_\alpha(x) \, \overline{d}_\beta(y) \rangle \tag{11.69}$$

(assuming that $u$ and $d$ have equal masses and using the Landau gauge for the expectation value).

So we have to compute $G_\Gamma(p)$ and $S(p)$. This is done in the following way. For the quark propagator $S^{(n)}$ evaluated on a single gauge configuration $n$ we define

$$S^{(n)}(x|p) = \sum_y e^{ipy} S^{(n)}(x,y) \, . \tag{11.70}$$

Taking into account $\gamma_5$-hermiticity of the propagator we may, for quark bilinear operators $O_\Gamma$ as defined in (11.62), rewrite $G_\Gamma(p)$ in terms of the quantities (11.70):

$$G_\Gamma(p) = \frac{1}{V} \sum_{x,y,z} e^{-ip(x-y)} \langle u(x) \overline{u}(z) \, \Gamma \, d(z) \overline{d}(y) \rangle \tag{11.71}$$

$$\approx \frac{1}{V N} \sum_{n=1}^{N} \sum_z \gamma_5 \, S^{(n)}(z|p)^\dagger \, \gamma_5 \, \Gamma \, S^{(n)}(z|p) \, , \tag{11.72}$$

where we approximate the expectation value by averaging over $N$ gauge configurations. Similarly we find the quark propagator in momentum space

$$S(p) \approx \frac{1}{VN} \sum_{n=1}^{N} \sum_{x} \mathrm{e}^{-\mathrm{i}px} \, S^{(n)}(x|p) \,. \tag{11.73}$$

$S^{(n)}(y|p)$ is computed by solving the lattice Dirac equation ($D$ denotes the Dirac operator)

$$\sum_{y} D(z,y) \, S^{(n)}(y|p) = \mathrm{e}^{\mathrm{i}pz} \tag{11.74}$$

with a momentum source (cf. [19]). This has the disadvantage that one has to determine the quark propagators for several momentum sources, whereas in the original method [14] one uses point sources (i.e., taking into account just $z = 0$ instead of summing over all $z$) and projects the quark sink to the desired momentum values. However, using momentum sources has the big advantage of a significantly better signal.

The quark field renormalization constant is obtained by comparing the quark propagator to the free (lattice) propagator. Using the so-called RI′ scheme we take

$$Z_q' = \frac{1}{12} \, \mathrm{tr} \left[ S^{-1}(p) \, \frac{\mathbb{1} \, R(p) - \mathrm{i}\gamma_\nu v_\nu(p)}{R(p)^2 + \sum_\nu v_\nu(p)^2} \right] \Bigg|_{p^2 = \mu^2} \,. \tag{11.75}$$

The RI′ scheme differs from the RI scheme only by the definition of the quark field renormalization constant. In (11.75), $R$ and $v_\nu$ are the scalar and vector terms appearing in the free Dirac operator, which in momentum space reads $D(p) = \mathrm{i}\gamma_\nu v_\nu(p) + \mathbb{1} \, R(p)$. Using Fourier transformation one can compute $R$ and $v_\nu$ from the definition of $D$. They are normalized such that one finds

$$v_\nu(p) = \mathrm{i} p_\nu + \mathcal{O}\left(a p^2\right) \quad \text{and} \quad R(p) = \mathcal{O}\left(a p^2\right) \,. \tag{11.76}$$

Landau gauge fixing $\partial_\nu A_\nu = 0$ is implemented as discussed in [17, 20] by iteratively minimizing a functional of the link variables with stochastic overrelaxation [21] (cf., Sect. 3.2.2). As is well known this type of gauge fixing still allows for Gribov copies corresponding to further local minima. This gives rise to an uncertainty which one has to check, e.g., by studying the effect of random gauge transformations.

It is easy to check that at tree level one finds $\langle p \mid O_\Gamma \mid p \rangle_0 = \Gamma$. Putting things together we obtain the final formula for $Z_\Gamma$ in the RI′ scheme:

$$Z_\Gamma^{\mathrm{RI}'} = \frac{12 \, Z_q'}{\mathrm{tr}\left[\Lambda_\Gamma(p) \, \Gamma^{-1}\right]} \Bigg|_{p^2 = \mu^2} \,. \tag{11.77}$$

For $\Gamma = \gamma_\nu, \gamma_\nu\gamma_5$ averaging over the index $\nu$ under the trace is implied.

The connection between different renormalization schemes is established using continuum perturbation theory. In the two schemes $\overline{\mathrm{MS}}$ and RI′ the renormalization factors are related, and we obtain the final result for the lattice numbers $Z_\Gamma^{RI'}$ as

$$Z_\Gamma^{\overline{MS}}(\mu^2) = R_\Gamma(\mu^2)\, Z_\Gamma^{RI'}(\mu^2)\,, \qquad (11.78)$$

where the ratio $R_\Gamma(\mu^2)$ may be computed in perturbation theory. Both, $Z_\Gamma^{RI'}(\mu^2)$ and $Z_\Gamma^{\overline{MS}}(\mu^2)$ may be divergent in the continuum limit, the ratio stays finite, though.

Conversion factors between RI, RI', and $\overline{MS}$ schemes have been determined in Landau gauge and 3-loop order in [15, 22, 23]. Integrating the renormalization group differential equations one can also define scale-independent quantities $Z^{RGI}$ (RGI = renormalization group invariant) [19, 24].

Here we have only discussed the simple case of renormalizing quark bilinears of the form (11.62). Operators with derivatives, as they are used for matrix elements discussed in Sect. 11.4, may be treated with similar methods.

In the RI scheme one has to extrapolate to the chiral limit for comparison with the mass-independent $\overline{MS}$ scheme. An alternative approach is the Schrödinger functional method (see Sect. 9.1), where one may work directly at the chiral point [13, 25].

## 11.3 Hadronic decays and scattering

### 11.3.1 Threshold region

In finite volumes the observables show finite size effects. Let us first discuss the propagation of a single hadron. We assume that the time extent is sufficiently large and thus not important for finite size effects. We also assume periodic boundary conditions for the spatial directions of the volume. Then there are two ways the single particle propagator is affected by the finite volume: On the one hand the propagator between two points separated by some distance in Euclidean time is actually a sum over propagators between the source and all spatial mirror images. On the other hand mass renormalizing interactions (like loops) will also have paths running around a periodic spatial direction before coming back to the vertex, as in Fig. 11.1. One could interpret this as a squeezing effect of the finite box on the virtual polarization cloud of the propagating particle. The leading effect comes from the lightest interacting particles in the system, i.e., the pions in QCD. The control parameter is the ratio of the box size $L$ to the light particle correlation length, or, equivalently, the product $M_\pi L$.

This has been discussed in [26] for the continuum quantum field theory of hadrons. In QCD we have quarks and gluons as the basic objects. However, due to confinement the quark loops are suppressed and the leading effect is expected to be due to pions. Thus the derivation should be applicable to QCD.

A single hadron does not change its momentum when emitting and absorbing a pion, as in Fig. 11.1. Thus the interaction is related to the forward

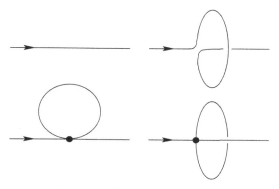

**Fig. 11.1.** Propagation of a pion. *Left:* free (*above*) and with interaction loop (*below*). *Right:* additional terms in a finite, periodic spatial box, where the pion also can run around the torus due to periodic boundary conditions

scattering amplitude. If, in an approximation, this is just a constant $\alpha$, then the mass change is proportional to the (divergent) loop contribution

$$m^2 = m_0^2 + \alpha \int \frac{\mathrm{d}^4 p}{(2\pi)^4} \, \widehat{G}(p) = m_0^2 + \alpha \, G(0) \, , \qquad (11.79)$$

where $\widehat{G}(p)$ is the propagator in momentum space. The integral gives the real space propagator at the origin $G(0)$, as made explicit in the second step of (11.79). The divergence is absorbed in the definition of the bare mass (renormalization).

We now consider a finite, periodic, spatial box of size $L^3$ and temporal extent large enough to neglect modifications due to its finiteness. In such a finite volume we have to replace the real space propagator by a sum

$$G(x) \quad \Rightarrow \quad \sum_{s \in \mathbb{Z}^3} G(x + L s) \, , \qquad (11.80)$$

where $s \in \mathbb{Z}^3$ gives the shift between the original lattice and its mirror images. The mass difference between finite and infinite box due to the interaction term, i.e., due to the contributions from propagation from the mirror images, is given by

$$m^2(L) - m^2 = \alpha \sum_{s \neq 0} G(L s) \, . \qquad (11.81)$$

In order to estimate these contributions we now need the large distance behavior of the free boson propagator. There are various ways to determine the Fourier transform

$$G(x) = \int \frac{\mathrm{d}^4 p}{(2\pi)^4} \frac{\mathrm{e}^{\mathrm{i} p x}}{p^2 + m^2} \sim \frac{m^2 \sqrt{8\pi}}{(4\pi)^2} \frac{\mathrm{e}^{-m \, |x|}}{(m \, |x|)^{3/2}} \, . \qquad (11.82)$$

We therefore can expect a leading size dependence $\propto \exp(-m\,L)/L^{3/2}$. The coefficient can be determined, e.g., from ChPT [27–29].

Lüscher [26] sums all orders of perturbation theory of the loop contributions on a periodic, finite lattice and obtains the expression

$$m(L) - m = -\frac{3}{(4\pi)^2}\frac{1}{mL}\int_{-\infty}^{\infty} dy\, e^{-\sqrt{y^2+m^2}\,L} F(iy) + \mathcal{O}(e^{-\overline{m}L}), \quad (11.83)$$

where the second term is suppressed since $\overline{m} > \sqrt{3/2}\,m$ can be shown. $F(iy)$ denotes the $X\pi \to X\pi$ forward-scattering amplitude $F(\nu = s - u) = A(s, t = 0, u)$, analytically continued to imaginary argument, where $s$, $t$, and $u$ are the relativistically invariant Mandelstam variables.

We will not reproduce the derivation here but just make the result plausible. For this we integrate (11.82) by choosing $x$ along the time axis and write

$$\begin{aligned}
G(x) &= \int_{-\infty}^{\infty}\frac{dE}{2\pi}\, e^{iEx}\int_{-\infty}^{\infty}\frac{d^3p}{(2\pi)^3}\frac{1}{E^2 + p^2 + m^2} \\
&= \int_{-\infty}^{\infty}\frac{dE}{2\pi}\, e^{iEx}\int_0^{\infty}\frac{dy}{2\pi^2}\frac{y^2}{E^2 + y^2 + m^2}
\end{aligned} \quad (11.84)$$

(here $y = |\boldsymbol{p}|$). The integral over $E$ may be solved using Cauchy's theorem, i.e., completing the contour by a half-circle at infinity and picking up the pole at $E = i\sqrt{y^2 + m^2}$ in the upper half-plane,

$$\begin{aligned}
G(x) &= \frac{1}{4\pi^2}\int_0^{\infty} dy\, e^{-\sqrt{y^2+m^2}\,x}\frac{y^2}{\sqrt{y^2 + m^2}} \\
&= -\frac{1}{4\pi^2}\int_0^{\infty} dp\,\frac{p}{x}\frac{d}{dp}\left(e^{-\sqrt{p^2+m^2}\,x}\right) = \frac{1}{8\pi^2 x}\int_{-\infty}^{\infty} dp\, e^{-\sqrt{p^2+m^2}\,x}.
\end{aligned} \quad (11.85)$$

In the second step we did a partial integration and then doubled the integration interval to get a form resembling (11.83). We now add up just the leading contributions of the sum (11.81) due to the six neighbor images and get

$$m^2(L) - m^2 = \frac{6\,\alpha}{8\pi^2 L}\int_{-\infty}^{\infty} dp\, e^{-\sqrt{p^2+m^2}\,L}, \quad (11.86)$$

and from this

$$m(L) - m \approx \frac{m^2(L) - m^2}{2\,m} = -\frac{3}{(4\pi)^2}\frac{(-2\alpha)}{m\,L}\int_{-\infty}^{\infty} dp\, e^{-\sqrt{p^2+m^2}\,L}. \quad (11.87)$$

This is just (11.82) with the scattering amplitude replaced by the constant coupling $F \to -2\alpha$. Thus we reproduce the simple leading behavior discussed earlier.

In the second paper of the series [30] Lüscher then relates the energy spectrum of two stable particles (at rest) in a finite, periodic box to their elastic scattering amplitude. Whereas the finite volume effects for the single particle are exponentially small, now the interaction between the two particles leads to only polynomial suppression $\mathcal{O}(1/L^3)$. The physical picture is that the localized particles need to be close to each other to interact and the probability to find one particle close to another is inversely proportional to the spatial volume.

Near threshold the scattering amplitude is dominated by the s-wave scattering length $a_0 = \lim_{k \to 0} \delta_0(k)/k$ (where $k$ is the 2-particle momentum and $\delta_0$ the s-wave scattering phase shift). For a 2-particle system the expansion of the ground state energy in $1/L$ is

$$W = 2M_\pi - \frac{4\pi a_0}{M_\pi L^3} \left( 1 + c_1 \frac{a_0}{L} + c_2 \frac{a_0^2}{L^2} + \mathcal{O}\left(\frac{1}{L^3}\right) \right) . \tag{11.88}$$

The coefficients $c_1 = -2.837297$ and $c_2 = 6.375183$ depend on the lattice shape and are related to the generalized zeta function of the momentum lattice. Similar relations for nonzero momenta involve other phase shifts $\delta_l(k)$ as well.

Relation (11.88) can be generalized to particles with unequal mass and nonzero total momentum. The leading term always is proportional to the inverse volume [30].

## 11.3.2 Beyond the threshold region

Asymptotically only stable states can be observed. Thus in full QCD simulations resonances usually will have to be identified by their impact on intermediate states. Utilizing equations like (11.88), Lüscher [26, 30–32] derived a relation between the two-particle eigenstates energies on the finite lattice and the scattering phase shift, valid below the inelastic threshold.

Let us discuss the idea in a simple situation. Consider the scattering of two bosons in a 1+1 dimensional system of finite spatial extension $L$, but infinite time extension. This situation is most conveniently described by using a wave function for the relative motion of the scattering partners. The basic assumption is that the interaction range is finite and smaller than $L$. Thus the wave function outside the interaction range is a plane wave. The effect of the interaction is taken into account by a momentum-dependent phase shift $\delta(k)$, acquired in the interaction region as illustrated in Fig. 11.2.

Imposing periodic boundary conditions, the matching of the plane wave with momentum $k$ at $x = L$ implies

$$e^{ikL+2i\delta(k)} = e^{ik0} = 1 . \tag{11.89}$$

This gives rise to the quantization condition for the relative momenta $k_n$:

$$k_n L + 2\delta(k_n) = 2n\pi , \tag{11.90}$$

where $n$ is an integer. For vanishing $\delta$ one recovers the free case where the momenta are $k_n = 2\pi n/L$. For the interacting situation nontrivial momenta

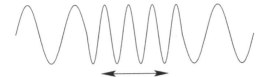

**Fig. 11.2.** This figure illustrates the behavior of the wave function: Outside the interaction region it is an unperturbed plane wave which picks up an extra phase shift in the interaction region (indicated by the *arrow*)

may be deduced from (11.90). If the functional form $\delta(k_n)$ were known, one could use this relation to find the quantized values of the momentum in this finite volume. On the other hand, given the momentum spectrum from some measurement, (11.90) allows the determination of the phase shift $\delta(k_n)$ for each $k_n$.

The momenta can be obtained from the energy values of the two particle states which are accessible in the simulation. For given $L$ one computes the discrete levels $W_0, W_1, W_2, \dots$ and from these the values of $k_n$ using the dispersion relation

$$W_n = 2\sqrt{m^2 + k_n^2} \, . \tag{11.91}$$

The technical problem lies in the precise determination of the single-particle mass and of the energy levels $W_n$.

For the determination of the energy spectrum one has to use techniques like the variational method (discussed in Sect. 6.3.3) considering correlation functions of a sufficiently large number of interpolators with the correct quantum numbers, capable of representing the space of scattering states [33, 34], including the coupled single-particle channels. Usually several of the lowest energy eigenmodes can be determined with sufficient reliability. Varying the spatial size $L$ of the system allows one to cover different values of the momentum. In [34] a simple 2D system was studied which couples a heavier and two lighter bosons on the lattice, with mass and coupling parameters allowing for a decay like in the $\rho \to \pi\pi$ system. Figure 11.3 demonstrates the expected phenomenon of level-crossing avoidance, which leads to a resonating phase shift.

One has to respect carefully the limitations of the approach: The interaction region and the single-particle correlation length ought to be smaller than the spatial volume, in particular $m\,L \gg 1$. The relation is applicable only below the first inelastic threshold. Polarization effects due to virtual particles running around the torus should be under control. Lattice artifacts will turn up for large values of $k$.

In the physical 4D situation the relationship between phase shift, lattice size, and momentum becomes somewhat more complicated:

$$\delta(k) = \phi\left(\frac{kL}{2\pi}\right) \bmod \pi \quad \text{with} \quad \tan(-\phi(q)) = \frac{q\pi^{3/2}}{\mathcal{Z}_{00}(1; q^2)} \, , \quad \phi(0) = 0 \, . \tag{11.92}$$

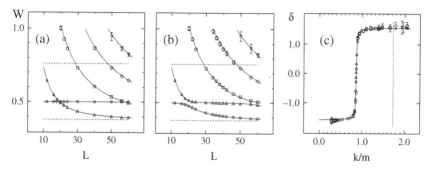

**Fig. 11.3.** Demonstration of the relation between scattering phase shifts and the finite volume energy spectrum in a simple toy model coupling a heavy particle to two light ones (figures from [34]). We show the energy levels observed without (**a**) and with (**b**) interaction coupling the single heavy particle to the two light ones. The *horizontal dashed lines* in (**a**) and (**b**) indicate the 2- and 4-particle thresholds, the *horizontal full line* in (**a**) denotes the single-(heavy)particle mass. Figure (**c**) shows the resulting resonance phase shift for (**b**); the *vertical dashed line* denotes the 4-particle threshold. For the 2D model the phase shift starts at $-\pi/2$ (no interaction). In all plots the symbols show the numerical data and the *full curves* represent models for the free (**a**) or resonating interacting case (**b**), (**c**)

Here $\mathcal{Z}_{00}$ denotes the Riemann zeta function. A 4D bosonic model has been studied in [35, 36] and a quenched QCD simulation in [37].

The $\rho$-meson is the resonance in the $\pi\pi$ p-wave channel and therefore the pions in the decay have nonzero momentum. In [38] (cf., also [39]) the relation was extended to nonrest frame situations. This has the advantage that on the same set of configurations both, zero and nonzero momentum cases, can be studied simultaneously, resulting in more data points. Smaller volumes and a smaller volume range can be utilized. A dynamical fermion study of the $\rho$-decay has used this method [40, 41].

In [42] the decay is studied by analyzing the energy levels in a situation when the $\rho$ mass lies below the two-pion mass. From the $N_T$-dependence the transition amplitude $\rho \rightarrow \pi\pi$ was determined, and from this the effective coupling.

## 11.4 Matrix elements

When exploring the internal structure of a hadron one wants to know the spatial distribution of charge, matter, and spin. One also needs to know, however, the momentum distribution. Both quantities are intrinsically nonperturbative and therefore a challenge for lattice QCD.

The spatial distribution may be probed by elastic scattering processes, giving rise to the concept of form factors and charge distributions. The momentum distribution can be explored experimentally in hard (high-energy)

**Fig. 11.4.** Matrix elements of operators (*squares*) between mesonic states (*circles*)

collisions like deep-inelastic scattering or Drell–Yan processes. In the parton model [43] quarks, antiquarks, and gluons are all partons and considered in a hadron at very high momentum. The point-like probes identify the *parton probability distributions* to find the partons in the infinite momentum frame with some fraction $x$ of the hadron's longitudinal momentum and a momentum transfer $Q^2$.

A quark distribution function in a hadron $h$ is defined through a matrix element in continuum light-cone notation

$$q(x) = \frac{1}{2p^+} \int \frac{d\lambda}{2\pi} e^{i\lambda x} \langle h(p)|\overline{\Psi}(0)\gamma^+\Psi(\lambda n)|h(p)\rangle , \qquad (11.93)$$

with $p^+ = (p^0 + p^3)/\sqrt{2}$, $n^\mu = (1,0,0,-1)$ and $n \cdot p = 1$. The quark field $\Psi$ is connected from its position to infinity with a gauge transporter to make the expression gauge-invariant.

The *generalized parton distributions* (GPDs) are an extension of this concept, introducing further variables like the spin and the transverse momentum and correlating spatial and momentum values. Via this extension one hopes to describe also scattering processes at lower energies [44, 45].

Wilson's *operator product expansion* (OPE) [46–49] relates the light-cone matrix elements (11.93) or GPDs to matrix elements of local operators $\langle h|O_i|h'\rangle$. One also needs renormalization factors $Z_{O_i}(a\mu)$ converting the lattice result to, e.g., the $\overline{\text{MS}}$ scheme. The local operators can be simple quark bilinears like $\overline{u}\gamma_\nu d$ (for form factors), they may include derivatives like $\overline{u}\gamma_\nu D_\mu d$ (for GPDs), or more quarks like in $\overline{s}\gamma_\mu u\,\overline{u}\gamma_\mu d$ (for weak interaction processes). Figure 11.4 gives examples of the matrix elements that one has to determine.

### 11.4.1 Pion form factor

Meson form factors are among the simplest matrix elements. As a prototype example we discuss the *electromagnetic form factor of the pion* $F^{(\pi)}$ that describes the coupling of the pion to the photon. It is defined by the matrix element

$$\langle\pi^+(\mathbf{p}_f)|\,V_\mu\,|\pi^+(\mathbf{p}_i)\rangle_{\text{cont}} = (p_f + p_i)_\mu\, F^{(\pi)}(Q^2) , \qquad (11.94)$$

where $p_i$ and $p_f$ are the 4-momenta of the initial and final pions, $Q^2 = (p_f - p_i)^2 \equiv -t$ is the space-like invariant momentum transfer squared, and

$$V_\mu = \frac{2}{3}\,\overline{u}\,\gamma_\mu\,u - \frac{1}{3}\,\overline{d}\,\gamma_\mu\,d \qquad (11.95)$$

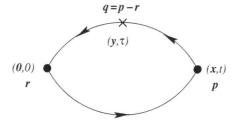

**Fig. 11.5.** Schematic diagram for the matrix element (11.98)

is the isovector vector current normalized to unit electric charge at zero momentum transfer (i.e., the on-shell photon coupling): $F^{(\pi)}(0) = 1$. Its space-like values are determined from experiment.

The connection to the continuum normalization gives another factor,

$$\langle \pi^+(\mathbf{p}_f)| V_\mu |\pi^+(\mathbf{p}_i)\rangle_{\text{latt}} = \frac{1}{2\sqrt{E_i E_f}} \langle \pi^+(\mathbf{p}_f)| V_\mu |\pi^+(\mathbf{p}_i)\rangle_{\text{cont}} . \qquad (11.96)$$

The pion form factor is an analytic function of $t$. Its space-like values can be determined from the values on the boundary of its analyticity domain, i.e., the cut along the positive real $t$-axis, the time-like region. Although the cut starts at $t = 4\,m_e^2$, significant contributions only come from the hadronic resonance region, starting with the two-pion threshold at $4\,M_\pi^2$. Due to unitarity along the first part of the cut, until inelastic channels become important, the phase shift is essentially the p-wave phase shift of the elastic two-pion channel. That region is dominated by the $\rho$-meson resonance. This feature has been exploited by various dispersion relation representations [50–53].

Although there are further inelastic contributions from the four-pion channel, these remain tiny until the $\pi\omega$ channel opens. There is also a small contribution from the isoscalar $\omega$ coupling through higher-order electromagnetic interactions. All these contributions show small effects on the near space-like region. The dominance of the vector meson motivates a simple pole parameterization for this region.

The mean charge radius squared $\langle r^2\rangle_V$ is defined through

$$F^{(\pi)}(Q^2) = 1 - \frac{1}{6}\langle r^2\rangle_V Q^2 + \mathcal{O}\left(Q^4\right) \quad \Rightarrow \quad \langle r^2\rangle_V \equiv 6\,dF^{(\pi)}(t)/dt\big|_{t=0} .$$
$$(11.97)$$

The current PDG average for its value is $0.45(1)$ fm$^2$ [2].

With present lattice tools one studies the pion form factor in the space-like region. For this we need to evaluate off-forward matrix elements (preferably at several transferred momenta), i.e., the expectation values of

$$\text{tr}\left[\sum_{y;\,y_0=\tau} e^{i\,\mathbf{q}\cdot\mathbf{y}} S(0,y)\,O(y) \sum_{x;\,x_0=t} e^{-i\,\mathbf{p}\cdot\mathbf{x}} S(y,x)\gamma_5 S(x,0)\gamma_5\right] , \qquad (11.98)$$

where $S(y,x)$ is the quark propagator from $x$ to $y$ and $O(y)$ denotes the operator inserted at $y$. We use the notations $\mathbf{p} \equiv \mathbf{p}_f$, $\mathbf{r} \equiv \mathbf{p}_i$ and denote the momentum transfer as $\mathbf{q} = \mathbf{p} - \mathbf{r}$ (cf. Fig. 11.5). Ideally the time distances between source and operator insertion and between the operator insertion and the sink are long enough, such that the higher excitations in the pion channel may be disregarded and the pion ground state dominates the insertion region. Smearing of source and sink operators also helps to get a better ground state signal.

In evaluating matrix elements one has to compute expectation values of combinations of quark propagators as in Figs. 11.4 and 11.5. Each calculation of a quark propagator starting from a given source requires an inversion of the Dirac operator and thus a significant amount of computer time. One therefore wants to reduce the number of propagator calculations. This is achieved by using the so-called *sequential source method* [54, 55]. We first calculate the quark propagator from the source at the origin to all other points of the lattice as discussed in Chap. 6. This provides us with both the propagator from the source origin to $x$ and, due to the $\gamma_5$-hermiticity property, also from $y$ to the origin. All we need is another propagator from $x$ to $y$. The matrix element is then written as

$$\mathrm{tr}\left[\sum_{\mathbf{y};y_0=\tau} e^{i\,\mathbf{q}\cdot\mathbf{y}}\, S(0,y)\, O(y)\, \Sigma(y,0)\, \gamma_5\right], \tag{11.99}$$

where the sequential propagator is defined as

$$\Sigma(y,0) = \sum_{\mathbf{x};x_0=t} e^{-i\,\mathbf{p}\cdot\mathbf{x}}\, S(y,x)\, \gamma_5\, S(x,0)\,. \tag{11.100}$$

It can be easily computed by an additional inversion of the Dirac operator $D$ for each choice of the final momentum $\mathbf{p}$:

$$\sum_y D(z,y)\, \Sigma(y,0) = e^{-i\,\mathbf{p}\cdot\mathbf{z}}\, \gamma_5\, S(z,0)\Big|_{z_0=t}\,. \tag{11.101}$$

Changing the properties of the sink requires the computation of new sequential propagators, and so simulating several final momenta, different field interpolators, or a different smearing for the sink rapidly becomes rather expensive. For this reason one often works only at one value of the final momentum. This approach then allows not only the determination of the mentioned matrix element but, with no more effort, also the consideration of other operator insertions, e.g., for the scalar form factor.

One extracts the physical matrix elements by computing ratios of 3-point and 2-point correlators. To see this, let us look at the 3-point correlator for in- and outgoing pseudoscalar interpolating field operators $P$ and an operator $O$ at insertion time $\tau$. At $\tau$ we insert a complete set of states. For large enough separation distance this can be saturated by the ground state in the channel, the pion, and we find

$$\langle 0|P(t;\mathbf{p})\,O(\tau)\,\overline{P}(\mathbf{r})|0\rangle \tag{11.102}$$

$$= \langle 0|P(t;\mathbf{p})|\pi(\mathbf{p})\rangle \frac{e^{-E_p(t-\tau)}}{2E_p}\langle \pi(\mathbf{p})|\,O(\tau)|\pi(\mathbf{r})\rangle \frac{e^{-E_r\tau}}{2E_r}\langle \pi(\mathbf{r})|\overline{P}(0;\mathbf{r})|0\rangle\ ,$$

and for the pseudoscalar propagator,

$$\langle 0|P(t;\mathbf{p})\overline{P}(\mathbf{p})|0\rangle = \langle 0|P(t;\mathbf{p})|\pi(\mathbf{p})\rangle \frac{e^{-E_p t}}{2E_p}\langle \pi(\mathbf{p})|\overline{P}(0;\mathbf{p})|0\rangle\ . \tag{11.103}$$

Here we neglect the lattice periodicity in time direction, otherwise there will be further terms from the backward propagation. This has been discussed, e.g., in [56].

One now constructs a suitable combination of these functions, such that the unknown factors and the exponentials cancel:

$$R(t,\tau;\mathbf{p},\mathbf{q}\equiv \mathbf{p}-\mathbf{r}) = \frac{\langle \pi(\mathbf{p})|\,O(\tau;\mathbf{q}))|\pi(\mathbf{r})\rangle}{2\sqrt{E_i E_f}} = \frac{\langle P(t;\mathbf{p})\,O(\tau;\mathbf{q})\,\overline{P}(0;\mathbf{r})\rangle}{\langle P(t;\mathbf{p})\,\overline{P}(0;\mathbf{p})\rangle}$$

$$\times \sqrt{\frac{\langle P(t;\mathbf{p})\,\overline{P}(0;\mathbf{p})\rangle\,\langle P(\tau;\mathbf{p})\,\overline{P}(0;\mathbf{p})\rangle\,\langle P(t-\tau;\mathbf{r})\,\overline{P}(0;\mathbf{r})\rangle}{\langle P(t;\mathbf{r})\,\overline{P}(0;\mathbf{r})\rangle\,\langle P(\tau;\mathbf{r})\,\overline{P}(0;\mathbf{r})\rangle\,\langle P(t-\tau;\mathbf{p})\,\overline{P}(0;\mathbf{p})\rangle}}\ . \tag{11.104}$$

One can keep the sink fixed at time $t$ and vary the time slice $\tau$ where the operator $O$ sits (scanning a range of time slices). As a consequence, $R(t,\tau;\mathbf{p},\mathbf{q})$ exhibits two plateaus in $\tau$: $0 \ll \tau \ll t$ and $t \ll \tau \ll N_T$. In the case in which the sink is put at $t = N_T/2$ the ratio is (in our case) anti-symmetric in $\tau - N_T/2$. The value of the form factor can be read off from the plateau values. Further normalization factors due to smearing of source and sink operators cancel also in the ratio.

The resulting plateau value for $Q^2 = 0$ will not necessarily be 1, which is the electric charge in our example. This is because the naive lattice current $V_\mu$ of (11.95) is not conserved. One could replace it by the conserved current, which, however, is numerically demanding for chiral actions, as discussed in Sect. 11.1.3. For other actions even the conserved current will need $\mathcal{O}(a)$ correction terms to obtain improved currents (see Sect. 9.1). Taking the simple point-like current one has to determine the renormalization factor $Z_V$ relating it to the renormalized current $V_\mu^{(r)}$.

Source, insertion, and sink are located on three time slices. The source operator usually is localized on or around a site. In a Fourier decomposition it therefore contributes to all possible 3-momenta. By projecting the sink to a 3-momentum $\mathbf{p}$ and the insertion to $\mathbf{q}$, the relevant component of the source is $\mathbf{r} = \mathbf{p}-\mathbf{q}$. All other contributions cancel in the sum over all field configurations due to translation invariance. In a realistic situation with a large but finite number of gauge configurations the cancellation is only approximate.

For the electromagnetic pion form factor a suitable choice of the 3-momenta is $|\mathbf{p}_f| = |\mathbf{p}_i|$ such that $E_i = E_f$. In that situation $Q^2 = |\mathbf{q}|^2$ and combining (11.94) and (11.96) the kinematical factors simplify due to

$$\frac{E_i + E_f}{2\sqrt{E_i E_f}} = 1 . \tag{11.105}$$

This removes another possible source of statistical errors. A possible way to minimize the computational effort is to fix the momentum of the sink to $\boldsymbol{p} = (1, 1, 0)\Delta p$ for $\Delta p = 2\pi/(aN)$, the spatial momentum gap. Thus one has to compute one further (sequential) propagator for each configuration. The 3-point function is then evaluated for $\boldsymbol{q}$ values such that $|\boldsymbol{r}| = |\boldsymbol{p} - \boldsymbol{q}| = \sqrt{2} = |\boldsymbol{p}|$. These are the 12 combinations $(0, 0, 0)$, $(2, 2, 0)$, $(2, 0, 0)$, $(0, 2, 0)$, $(0, 1, \pm 1)$, $(1, 0, \pm 1)$, $(2, 1, \pm 1)$, $(1, 2, \pm 1)$. This leads to results for $Q^2 = 2n(\Delta p)^2$ with $n = 0, 1, 2, 3, 4$.

The momentum projection for the 2-point and 3-point functions gives rise to significant statistical noise, in particular close to the symmetry point $t = N_T/2$. One therefore chooses the position of the sink at time slices somewhat closer to the source. The matrix elements are then obtained by combining the plateau values on either side of the sink $R_{\mathrm{lhs}} \pm R_{\mathrm{rhs}}$ where the relative sign depends on the parity properties of the interpolator. The sign is negative for the vector form factor.

The earliest lattice studies of the pion form factor were [57, 58]. The space-like form factor usually is well approximated by the time-like $\rho$ pole, which in quenched calculations [59, 60] at unphysical large quark masses is larger than the experimental value. For simulations with dynamical quarks, extrapolating to physical quark masses, agreement with the physical $\rho$ mass and the measured pion form factor improves [56, 61, 62]. The pion is at the heart of chiral symmetry breaking and therefore ChPT provides several constraints [53, 63–65].

### 11.4.2 Weak matrix elements

Weak interactions couple almost point-like to quarks, but the quarks live in the QCD-dominated environment of the hadron. Nonperturbative effects define the hadron structure and therefore are important to analyze the way the weak interactions "see" the hadron.

In the standard model (SM) the electroweak interactions are mediated by the photon and the heavy vector bosons $W^\pm$ and $Z^0$. These bosons couple to bilinear combinations of leptons and bilinear combinations of quarks, thus allowing for quark flavor changing processes. The six flavors of quarks are arranged in three families: $(u, d)$, $(c, s)$, and $(t, b)$. The weak interactions "see" the quark flavors slightly differently than the strong interactions. The corresponding flavor eigenstates are denoted by $d'$, $s'$, and $b'$ and are related to the mass eigenstates $d$, $s$, and $b$ via the *Cabibbo–Kobayashi–Maskawa (CKM) matrix* [66, 67]

$$\begin{pmatrix} d' \\ s' \\ b' \end{pmatrix} = V_{\mathrm{CKM}} \begin{pmatrix} d \\ s \\ b \end{pmatrix} = \begin{pmatrix} V_{ud} & V_{us} & V_{ub} \\ V_{cd} & V_{cs} & V_{cb} \\ V_{td} & V_{ts} & V_{tb} \end{pmatrix} \begin{pmatrix} d \\ s \\ b \end{pmatrix} . \tag{11.106}$$

The other three quarks may be left unchanged without loss of generality. The CKM matrix is complex (allowing for CP violation) and unitary (enforcing absence of flavor changing neutral current transitions at tree level). For a review on these aspects, see [68, 69].

With certain further phenomenologically motivated approximations the matrix is often formulated in terms of four so-called generalized Wolfenstein parameters: $\lambda = V_{us} = \sin\theta_c$, $A$, $\bar\rho$, and $\bar\eta$. These parameters may be related to various experiments. A nonvanishing $\bar\eta$ is responsible for CP violation in the standard model.

Unitarity of the CKM matrix gives relations like

$$V_{ud}V_{ub}^* + V_{cd}V_{cb}^* + V_{td}V_{tb}^* = 0 .\tag{11.107}$$

This relation can be represented as the "unitarity triangle" in the $(\bar\rho, \bar\eta)$ plane. The endpoints of the triangle's base are $(0,0)$ and $(1,0)$ and the third corner is then constrained by experiments, in particular by $K$ and $B$ physics results. Lattice calculations may help in determining the QCD input relevant for the computation of the CKM matrix elements [70–72].

### 11.4.3 OPE expansion and effective weak Hamiltonian

Weak decays of hadrons involve two mass scales: the heavy intermediate vector bosons $W^\pm$ and $Z$ and the QCD low-energy scale dominating the hadron structure. One separates the problem into distinct parts by introducing an effective weak Hamiltonian

$$H_{\text{eff}} = \frac{G_F}{\sqrt{2}} \sum_i V_{\text{CKM}}^i C_i(\mu) Q_i(\mu) .\tag{11.108}$$

Here $G_F$ is the Fermi constant and the $Q_i$ denote local operators, built from quark and lepton fields. The CKM factors and the *Wilson coefficients* $C_i$ govern the weight of the operators. This is the so-called *operator product expansion (OPE)* [46–49]. One of the local operators is just the well-known product of $(V - A)$ currents from the Fermi theory for $\beta$-decay. Expectation values of these terms between hadronic states then are related to the experimental observables.

The Wilson coefficients (short-distance scale) can be computed in perturbation theory (cf., the NLO calculation [68]). The long-distance (nonperturbative) scale is encoded in the $Q_i$ evaluated at a renormalization scale $\mu$. Typically $\mu$ is of the order of 1–2 GeV or higher; it is, however, much smaller than the weak interaction scale $M_W$ and therefore corrections $\mathcal{O}(\mu^2/M_W^2)$ are negligible. In lattice calculations one computes the hadronic matrix elements $\langle h|Q_i|h'\rangle$, i.e., matrix elements of four-quark operators between hadronic states (cf., Fig. 11.4).

When one uses lattice Dirac operators that do not obey the Ginsparg–Wilson equation, the lack of chiral symmetry leads to significant complications [73]:

- For $K - \overline{K}$ mixing and computation of $B_K$ one needs operators like $O_{LL} = (\overline{s}\gamma_\mu(\mathbb{1} - \gamma_5)d)^2$. If chiral symmetry is not guaranteed, upon renormalization these operators mix with other operators of different chirality like $O_{PP} = (\overline{s}\gamma_5 d)^2$. Whereas the matrix element $\langle \overline{K}|O_{LL}|K\rangle$ vanishes in the chiral limit (proportional to the quark mass), the element $\langle \overline{K}|O_{PP}|K\rangle$ approaches a constant and therefore high accuracy is necessary to separate the physically interesting part.
- Operators of the effective Hamiltonian mix with lower dimensional operators. An example is the $\Delta S = 1$ Hamiltonian relevant to $K \to \pi\pi$. The dimension-6 operator $\overline{s}\gamma_\mu(\mathbb{1} - \gamma_5)u\,\overline{u}\gamma_\mu(\mathbb{1} - \gamma_5)d$ mixes with operators like $\overline{s}d$, $\overline{s}\gamma_5 d$, and $\overline{s}\gamma_\mu\gamma_\nu G_{\mu\nu}d$ (where $G$ denotes the gluon field). The mixing coefficients diverge in powers of $1/a$ and precise nonperturbative methods have to be employed to deal with these problems.

With actions like the Wilson action, that violate chiral symmetry, computations are therefore cumbersome due to the need to subtract divergent terms. Using Ginsparg–Wilson fermions, on the other hand, improves that situation significantly. For the $\Delta I = \frac{1}{2}$ problem one does not have any power divergent subtractions [74] and for the computation of $\varepsilon'/\varepsilon$ subtraction is needed (for a review see, e.g., [13]).

*Nonleptonic weak decays*, where the final state consists exclusively out of hadrons, like $K \to \pi\pi$, are a challenging task. There are two main groups of problems. The ultra-violet problem is to construct finite matrix elements of renormalized operators from the bare lattice operators; this is discussed in [6, 75–77]. The infrared problem has to do with re-scattering of the outgoing two-particle system, the continuation from Euclidean to Minkowski space–time and the effects of the spatial volume.

The Maiani–Testa theorem [78] states that in Euclidean calculations in very large spatial volumes the matrix element is dominated by the pair of zero momentum single-particle states. In infinite volume in Euclidean correlation functions one obtains the average of matrix elements into in- and out two-pion states. This situation improves, however, when dealing with finite volumes. Lellouch and Lüscher [79] derived a relation between the $K \to \pi\pi$ matrix elements in finite volume and the physical kaon-decay amplitudes. In accessible finite volumes the two-pion energy spectrum is far from being continuous and a kaon at rest cannot decay into two pions (unless one of the two-particle energy levels is close to its mass). A simple expression then relates the transition amplitude in finite volume to the decay rates in infinite volume. The work has been adapted [80] to include all elastic states below the inelastic threshold. The methods allow to extract the decay amplitudes also when the kaon mass is not equal to the two-pion energy, i.e., when an inserted (weak Hamiltonian) operator carries nonvanishing energy momentum. For a review on lattice results for kaon physics see [81] and, more generally, for what lattice QCD can do for the standard model, see [82].

# References

1. M. Lüscher: in *Probing the Standard Model of Particle Interactions, Proceedings Les Houches 1997,* edited by R. Gupta, A. Morel, E. de Rafael, and F. David, Vol. 1 (Elsevier, Amsterdam 1998)
2. C. Amsler et al. [Particle Data Group]: Phys. Lett. B **667**, 1 (2008)
3. M. Gell-Mann, R. J. Oakes, and B. Renner: Phys. Rev. **175**, 2195 (1968)
4. S. Weinberg: Physica A **96**, 327 (1979)
5. J. Gasser and H. Leutwyler: Nucl. Phys. B **250**, 465 (1985)
6. M. Bochicchio et al.: Nucl. Phys. B **262**, 331 (1985)
7. P. Hasenfratz: Nucl. Phys. B **525**, 401 (1998)
8. P. Hasenfratz et al.: Nucl. Phys. B **643**, 280 (2002)
9. Y. Kikukawa and A. Yamada: Nucl. Phys. B **547**, 413 (1999)
10. S. Capitani et al.: Phys. Lett. B **468**, 150 (1999)
11. L. Giusti, C. Hoelbling, and C. Rebbi: Phys. Rev. D **64**, 114508 (2001)
12. C. Gattringer, P. Huber, and C. B. Lang: Phys. Rev. D **72**, 094510 (2005)
13. S. Capitani: Phys. Rept. **382**, 113 (2003)
14. G. Martinelli et al.: Nucl. Phys. B **445**, 81 (1995)
15. E. Franco and V. Lubicz: Nucl. Phys. B **531**, 641 (1998)
16. M. L. Paciello, S. Petrarca, B. Taglienti, and A. Vladikas: Phys. Lett. B **341**, 187 (1994)
17. L. Giusti et al.: Int. J. Mod. Phys. A **16**, 3487 (2001)
18. L. Giusti, S. Petrarca, B. Taglienti, and N. Tantalo: Phys. Lett. B **541**, 350 (2002)
19. M. Göckeler et al.: Nucl. Phys. B **544**, 699 (1999)
20. H. Suman and K. Schilling (unpublished) arXiv:hep-lat/9306018 (1993)
21. J. E. Mandula and M. Ogilvie: Phys. Lett. B **B48**, 156 (1990)
22. K. G. Chetyrkin and A. Retey: Nucl. Phys. B **583**, 3 (2000)
23. J. A. Gracey: Nucl. Phys. B **662**, 247 (2003)
24. V. Gimenez, L. Giusti, F. Rapuano, and M. Talevi: Nucl. Phys. B **531**, 429 (1998)
25. S. Capitani, M. Lüscher, R. Sommer, and H. Wittig: Nucl. Phys. B **544**, 669 (1999)
26. M. Lüscher: Commun. Math. Phys. **104**, 177 (1986)
27. P. Hasenfratz and H. Leutwyler: Nucl. Phys. B **343**, 241 (1990)
28. A. Hasenfratz et al.: Z. Physik C **46**, 257 (1990)
29. A. Hasenfratz et al.: Nucl. Phys. B **356**, 332 (1991)
30. M. Lüscher: Commun. Math. Phys. **105**, 153 (1986)
31. M. Lüscher: Nucl. Phys. B **354**, 531 (1991)
32. M. Lüscher: Nucl. Phys. B **364**, 237 (1991)
33. M. Lüscher and U. Wolff: Nucl. Phys. B **339**, 222 (1990)
34. C. R. Gattringer and C. B. Lang: Nucl. Phys. B **391**, 463 (1993)
35. J. Nishimura: Phys. Lett. B **294**, 375 (1992)
36. M. Göckeler, H. Kastrup, J. Westphalen, and F. Zimmermann: Nucl. Phys. B **425**, 413 (1994)
37. S. Aoki et al.: Phys. Rev. D **67**, 014502 (2003)
38. K. Rummukainen and S. Gottlieb: Nucl. Phys. B **450**, 397 (1995)
39. C. Kim, C. T. Sachrajda, and S. R. Sharpe: Nucl. Phys. B **727**, 218 (2005)
40. S. Aoki et al.: PoS **LAT2006**, 110 (2006)

41. S. Aoki et al.: Phys. Rev. D **76**, 094506 (2007)
42. C. McNeile and C. Michael: Phys. Lett. B **556**, 177 (2006
43. R. P. Feynman: *Photon-Hadron Interactions* (Benjamin, New York 1972)
44. D. Diakonov: Prog. Part. Nucl. Phys. **51**, 173 (2003)
45. X. Ji: Ann. Rev. Nucl. Part. Sci. **54**, 413 (2004)
46. K. G. Wilson: Phys. Rev. **179**, 1499 (1969)
47. K. G. Wilson and W. Zimmermann: Commun. Math. Phys. **24**, 87 (1972)
48. W. Zimmermann: Ann. Phys. **77**, 570 (1973)
49. E. Witten: Nucl. Phys. B **120**, 189 (1977)
50. J. Gounaris and J. J. Sakurai: Phys. Rev. Lett. **21**, 244 (1968)
51. M. F. Heyn and C. B. Lang: Z. Phys. **C7**, 169 (1981)
52. J. F. de Trocóniz and F. J. Ynduráin: Phys. Rev. D **65**, 093001 (2002)
53. H. Leutwyler: in *Continuous Advances in QCD 2002: Conference Proceedings Minneapolis, Minnesota, 2002*, edited by K. A. Olive, M. A. Shifman, and M. B. Voloshin (World Scientific, Singapore 2002)
54. C. W. Bernard, T. Draper, G. Hockney, and A. Soni: in *Lattice Gauge Theory: A Challenge in Large Scale Computing*, edited by B. Bunk, K. H. Mütter, and K. Schilling (Plenum, New York 1986)
55. G. W. Kilcup et al.: Phys. Lett. B **164**, 347 (1985)
56. D. Brömmel et al.: Eur. Phys. J. C **51**, 335 (2007)
57. G. Martinelli and C. T. Sachrajda: Nucl. Phys. B **306**, 865 (1988)
58. T. Draper, R. Woloshyn, W. Wilcox, and K.-F. Liu: Nucl. Phys. B **318**, 319 (1989)
59. J. van der Heide, J. Koch, and E. Laermann: Phys. Rev. D **69**, 094511 (2004)
60. S. Capitani, C. Gattringer, and C. B. Lang: Phys. Rev. D **73**, 034505 (2005)
61. F. D. R. Bonnet et al.: Phys. Rev. D **72**, 054506 (2005)
62. S. Hashimoto et al.: PoS **LAT2005**, 336 (2005)
63. H. Gausterer and C. B. Lang: Z. Phys. C **28**, 475 (1985)
64. G. Colangelo, J. Gasser, and H. Leutwyler: Nucl. Phys. B **603**, 125 (2001)
65. J. Bijnens: PoS **LATTICE2007**, 004 (2007)
66. N. Cabibbo: Phys. Rev. Lett. **10**, 531 (1963)
67. M. Kobayashi and T. Maskawa: Prog. Theor. Phys. **49**, 652 (1973)
68. J. Buras: in *Proccedings of the International School of Subnuclear Physics: 38th Course: Theory and Experiment Heading for New Physics, Erice, Italy, 27 Aug–5 Sep 2000*, edited by A. Zichichi (World Scientific, Singapore 2001)
69. J. Buras: in *Lectures given at the European CERN School, Saint Feliu de Guixols, 2004* (CERN, Geneva, Switzerland 2005)
70. Z. Ligeti: PoS **LAT2005**, 012 (2005)
71. M. Okamoto: PoS **LAT2005**, 013 (2005)
72. 5th International Workshop on the CKM Unitarity Triangle: http://ckm2008.roma1.infn.it (2008)
73. A. Soni: Pramana **62**, 415 (2004)
74. S. Capitani and L. Giusti: Phys. Rev. D **64**, 014506 (2001)
75. L. Maiani, G. Martinelli, G. C. Rossi, and M. Testa: Phys. Lett. B **176**, 445 (1986)
76. L. Maiani, G. Martinelli, G. C. Rossi, and M. Testa: Nucl. Phys. B **289**, 505 (1987)
77. C. Bernard, T. Draper, G. Hockney, and A. Soni: Nucl. Phys. B (Proc. Suppl.) **4**, 483 (1988)

78. L. Maiani and M. Testa: Phys. Lett. B **245**, 585 (1990)
79. L. Lellouch and M. Lüscher: Commun. Math. Phys. **219**, 31 (2001)
80. C.-J. Lin, G. Martinelli, C. Sachrajda, and M. Testa: Nucl. Phys. B **619**, 467 (2001)
81. L. Lellouch: PoS **LATTICE2008**, 015 (2009)
82. T. DeGrand and C. DeTar: *Lattice Methods for Quantum Chromodynamics* (World Scientific, Singapore 2006)

# 12

# Temperature and chemical potential

Since the big bang the universe has cooled down substantially. Still, understanding QCD at high temperature and at high matter density is important for various reasons. One of them is obviously a better understanding of quark- and gluon matter shortly after the big bang and of the condensation of hadrons in the cooling process. Many objects in the universe, like neutron stars, supposedly have high enough density or temperature to expect that hadronic matter behaves differently compared to usual atoms. On earth very energetic collisions of heavy ions provide also possibilities to study matter under extreme conditions. Thus studying QCD at high temperature and density from first principles is a challenging task for the lattice approach.

Intuitively one may expect that at a density higher than in the nucleons and at a temperature above the pion mass the hadrons loose their individuality as bound states and new structural phases emerge. One anticipates phase transitions and the lattice formulation is a natural setting for the study of such collective phenomena. The studies should reveal the existence and the kind of transitions to different states of matter and their properties, as well as the behavior of the possible other states of matter in an unusual environment, thus helping us to understand experiments as well as astrophysical observations. Reviews for progress in lattice thermodynamics are, e.g., found in [1–5].

In today's lattice approach we always study a system in equilibrium and not the evolution and decay dynamics of the system. This challenge is left for the future.

## 12.1 Introduction of temperature

In Sect. 1.4.3 we have discussed the relation of a Euclidean quantum field theory to the partition function (1.13) of a classical statistical mechanics system in a heat bath with temperature $T$. For the quantum mechanical case the partition function is given by

Gattringer, C., Lang, C.B.: *Temperature and Chemical Potential*. Lect. Notes Phys. **788**, 301–326 (2010)
DOI 10.1007/978-3-642-01850-3_12    © Springer-Verlag Berlin Heidelberg 2010

$$Z(T) = \text{tr} \left[ e^{-\widehat{H}/(k_B T)} \right] = \text{tr} \left[ e^{-\beta \widehat{H}} \right] , \qquad (12.1)$$

where $\widehat{H}$ is the Hamiltonian operator and $\beta$ now denotes the inverse temperature $\beta = 1/(k_B T)$, with the Boltzmann constant $k_B$. This is the standard notation, which, unfortunately, often leads to confusing $\beta$ with the inverse gauge coupling. Unless stated otherwise, in this chapter $\beta$ exclusively denotes the inverse temperature. In our subsequent discussion we follow the usual choice of units setting $k_B = 1$. Then the temperature $T$ is given in units of energy or mass and $\beta = 1/T$.

Due to the trace in (12.1) we are restricted to fields that are periodic (bosons) or anti-periodic (fermions) in time. Like in Chap. 1 the partition function can be transformed to a path integral over such field configurations. However, contrary to our discussion there, we now do not assume that the time extent becomes infinite. Instead we obtain the functional integral ($\Phi$ being some generic field)

$$Z(T) = \int \mathcal{D}[\Phi] \, e^{-S_E[\Phi]} , \qquad (12.2)$$

where the integration is over fields, (anti-)periodic in the finite time direction. The Euclidean action results from an integral over all space but finite time extent:

$$S_E[\Phi] = \int_0^\beta dt \int_{\mathbb{R}^3} d^3x \, L_E \left( \Phi(t, \boldsymbol{x}), \partial_\mu \Phi(t, \boldsymbol{x}) \right) . \qquad (12.3)$$

The measure $\mathcal{D}[\Phi]$ and the action $S_E[\Phi]$ are discretized on a lattice as usual. In the discretization we find that only our point of view changes. Until now, working at "zero temperature" we were interested in results in the infinite space–time volume limit and we discussed deviations as finite size effects. Space and time extent were considered as much larger than the largest correlation length in the system – that of the pion. Now space is still considered in that limit, but the physical extent of time is limited to $\beta$. For a finite lattice the space extent is $a \, N$ and the time extent is $a \, N_T$. We thus have

$$\beta = a \, N_T = \frac{1}{T} \qquad (12.4)$$

and find that the limit $\beta \to \infty$ corresponds to $T \to 0$. We interpret this as a system with finite spatial volume and fixed *temperature* $T$. The continuum limit of such a system corresponds to $a \to 0$ while holding $a \, N$ and $a \, N_T$ fixed. Finite volume effects become smaller when $N/N_T$ becomes larger.

Arguing again with physical intuition we may expect to observe finite temperature effects when the correlation length of the pion $1/M_\pi$ reaches the time extent $a \, N_T$, and it has been found that $T$ is indeed not far from the value $M_\pi$.

The fact that the temporal extent $\beta$ is kept finite has important physical consequences. If Fourier transformation is applied to the finite time direction one finds that only discrete energy levels with spacing $\Delta = 2\pi/\beta$ are allowed. Integer multiples of this value $\Delta$ are often referred to as *Matsubara frequencies*.

Due to the lattice structure the Matsubara frequencies are not only discrete but also limited to the range $(-\pi/a, \pi/a]$, taking for boson states values $k\Delta$ (for $-N_T/2 + 1 \leq k \leq N_T/2$) with $\Delta = 2\pi T = 2\pi/(aN_T)$. For fermions, which obey anti-periodic boundary conditions in time direction, the values are shifted by $\Delta/2$ (cf. Appendix A) and the smallest quark Matsubara mode thus has energy $\Delta/2$.

### 12.1.1 Analysis of pure gauge theory

Let us first discuss pure gauge theory at finite temperature. For this case we introduced in Sect. 3.3.5 an important observable, the Polyakov loop. It is defined as the trace of the ordered product of gauge link variables in time direction, closed due to periodicity:

$$P(\boldsymbol{m}) = \mathrm{tr}\left[\prod_{j=0}^{N_T-1} U_4(\boldsymbol{m}, j)\right] . \tag{12.5}$$

The correlator $\langle P(\boldsymbol{m})P(\boldsymbol{n})^\dagger\rangle$ is related to the potential between a pair of a static quark and antiquark (compare (3.61)), i.e., the free energy $F_{\bar{q}q}$ of such a pair at a given temperature:

$$\langle P(\boldsymbol{m})P(\boldsymbol{n})^\dagger\rangle = \mathrm{e}^{-a\,N_T\,F_{\bar{q}q}(a|\boldsymbol{m}-\boldsymbol{n}|)} = \mathrm{e}^{-F_{\bar{q}q}(r)/T} . \tag{12.6}$$

As usual the energy has to be normalized at some distance. For large distances we expect factorization

$$\lim_{a|\boldsymbol{m}-\boldsymbol{n}|\to\infty} \langle P(\boldsymbol{m})P(\boldsymbol{n})^\dagger\rangle = \langle P(\boldsymbol{m})\rangle\langle P(\boldsymbol{n})^\dagger\rangle = |\langle P\rangle|^2 . \tag{12.7}$$

Due to translational invariance the spatial position of the Polyakov loop is irrelevant and we have therefore omitted it on the right-hand side of (12.7) and instead use the spatial average

$$P = \frac{1}{N^3}\sum_{\boldsymbol{m}} P(\boldsymbol{m}) . \tag{12.8}$$

We observe that for static potentials that grow indefinitely with separation distance, such as the confining potential (3.62) at $\sigma > 0$, $|\langle P\rangle|$ has to vanish. Therefore, we conclude

$$\begin{aligned} \langle P\rangle = 0 &\iff \text{confinement} , \\ \langle P\rangle \neq 0 &\iff \text{no confinement} . \end{aligned} \tag{12.9}$$

At low temperatures QCD is confining. As the temperature is increased, pure gauge theory undergoes a phase transition [6, 7] at a *critical temperature* $T_c$ of about 270 MeV. There the system deconfines and $\langle P \rangle$ acquires a nonvanishing value (cf., the left-hand side of Fig. 12.1).

An individual Polyakov loop $P(\boldsymbol{m})$ is nothing but a Wilson loop at fixed spatial position extending in the temporal direction and closed due to periodicity. We may interpret its expectation value as the probability to observe a single static charge. With

$$|\langle P \rangle| \sim \mathrm{e}^{-F_q/T} \tag{12.10}$$

we relate it to the free energy of a single color charge. From (12.9) we conclude that $F_q \to \infty$ for confinement and $F_q$ finite in the *deconfined phase*.

The deconfinement transition in quenched QCD has another interesting interpretation in terms of the so-called *center symmetry* or $Z_3$ *symmetry*. From the definition of the Polyakov loop in (3.60) it is obvious that it is a gauge-invariant quantity and cannot be made unity by a gauge transformation. However, on a finite lattice any given gauge configuration has the same statistical weight as the one obtained by a *center transformation*. This transformation consists of multiplying all temporal links in a given time slice $n_4 = t_0$ with the same element $z$ of the center group $Z_3$ of gauge group SU(3):

$$U_4(\boldsymbol{n}, t_0) \to z\, U_4(\boldsymbol{n}, t_0) . \tag{12.11}$$

The center elements of SU(3) are the cubic roots $(\mathbb{1}, \mathbb{1}\,\mathrm{e}^{2\pi i/3}, \mathbb{1}\,\mathrm{e}^{-2\pi i/3})$. The reason for this symmetry is that the gauge action is constructed from products of link variables along trivially closed loops. Each trivially closed loop has as many elements in one direction as in the opposite. Center elements commute with all group elements and therefore cancel in the closed loop products. For example, a plaquette in the $(\mu, 4)$-plane at $n_4 = t_0$ transforms according to

$$
\begin{aligned}
&\mathrm{tr}\left[ U_\mu(\boldsymbol{n}, t_0)\, U_4(\boldsymbol{n} + \hat{\mu}, t_0)\, U_\mu(\boldsymbol{n}, t_0 + 1)^\dagger\, U_4(\boldsymbol{n}, t_0)^\dagger \right] \\
&\longrightarrow\quad \mathrm{tr}\left[ U_\mu(\boldsymbol{n}, t_0)\, z\, U_4(\boldsymbol{n} + \hat{\mu}, t_0)\, U_\mu(\boldsymbol{n}, t_0 + 1)^\dagger\, U_4(\boldsymbol{n}, t_0)^\dagger\, z^\dagger \right] \\
&=\quad \mathrm{tr}\left[ z\, z^\dagger\, U_\mu(\boldsymbol{n}, t_0)\, U_4(\boldsymbol{n} + \hat{\mu}, t_0)\, U_\mu(\boldsymbol{n}, t_0 + 1)^\dagger\, U_4(\boldsymbol{n}, t_0)^\dagger \right] .
\end{aligned}
\tag{12.12}
$$

Since $z\,z^\dagger = \mathbb{1}$, the gauge action is invariant. Plaquettes in spatial planes or with $n_4 \neq t_0$ are not affected. This global symmetry is the heart of the proof of confinement of static charges in the strong coupling limit in Sect. 3.4.

The Polyakov loop, however, does not close in a topologically trivial way, but instead winds once around the compact time direction. Thus it is not invariant under transformation (12.11). Since we multiply the temporal links in only one time slice with a center element $z$, the Polyakov loop transforms as

$$P \quad \longrightarrow \quad z\, P . \tag{12.13}$$

Due to the symmetry of the action the expectation value of the Polyakov loop may be written as

$$\langle P \rangle = \frac{1}{3} \langle P + z\,P + z^2\,P \rangle = \frac{1}{3} \left( 1 + \mathrm{e}^{\mathrm{i}2\pi/3} + \mathrm{e}^{-\mathrm{i}2\pi/3} \right) \langle P \rangle = 0 \,. \qquad (12.14)$$

The right-hand side vanishes since the sum over the center elements is zero. The analysis (12.14) breaks down, when the center symmetry is broken. In particular this symmetry is broken spontaneously above the critical value of the temperature. Thus the finite temperature transition of quenched QCD signals a spontaneous breaking of the center symmetry [8]. The Polyakov loop (also called *thermal Wilson line*) is an order parameter distinguishing between a confinement phase, where free charges cannot be found, and a deconfinement phase, where single charges are screened and may be observed.

Combining several of such charges leads to nonvanishing expectation values as long as the charges combine to a color singlet. This allows the definition of, e.g., the static potential between static charges. Also the evolution of flux tubes (concentration of gluonic energy) in situations of two or more static charges has been studied [9–13].

The center symmetry property resembles that of a three-state spin model with spin variables $\in Z_3$ living on the sites of the 3D spatial lattice volume [14, 15]. Such a model exhibits a phase transition at some value of its nearest neighbor coupling – but only for an infinite volume. In a finite volume the symmetry cannot be broken in a strict sense. However, computer simulations at coupling values, where one would expect symmetry breaking in the infinite volume situation, may stay for long persistence periods in one of the three sectors, before a tunneling to another sector occurs.

The definition and measurement of the order parameter are a delicate issue. In spin systems one introduces an external field, lets the volume go to infinity and, then the external field to zero. This order of limits is important, since the spontaneous magnetization vanishes for any finite volume. For the gauge theory we have a similar situation preventing the correct determination of $\langle P \rangle$ in a finite volume. Technically, larger lattices lead to long periods in one of the sectors and thus one may expect good approximate determinations of this would-be order parameter. However, the systematic error due to tunneling is not under control. In the updating procedure one can even explicitly enforce the center symmetry by regularly multiplying all gauge links in a time slice with a nontrivial center element. This does not change the gauge action. Then it becomes obvious that $\langle P \rangle$ is bound to vanish for sufficiently long runs. Instead of $\langle P \rangle$ one therefore often plots and analyses $\langle |P| \rangle$, which agrees with $|\langle P \rangle|$ in the infinite volume limit.

In order to simulate at finite temperature $1/(aN_T)$ one either changes $N_T$ or the gauge coupling and therefore the lattice spacing $a$. Changing $a$ allows for a continuous scan over $T$. In Fig. 12.1 we show the result of a simulation of pure gauge theory on a lattice of size $16^3 \times 4$ for a range of coupling values such that $T$ is close to the transition point from confinement to deconfinement. In the left-hand side plot we show $\langle |P| \rangle$ as a function of $T$. The right-hand side is a scatter plot of $P$ for individual configurations at a fixed value of $T > T_c$. In the simulation also the center symmetry was updated and the values of $P$

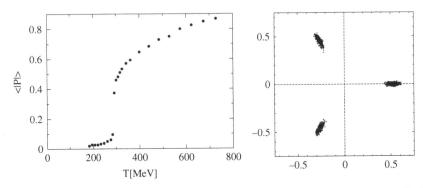

**Fig. 12.1.** *L.h.s.*: The expectation value $\langle |P| \rangle$ as function of the temperature $T$. In computing $T = 1/(N_T\, a) = 1/(4a)$ the lattice spacing has been determined using the static potential as discussed in Sect. 3.5, cf. (3.83) (for $r_0 = 0.5$ fm). The results are from a simulation on lattice size $16^3 \times 4$ averaging over 100 measured configurations for each value of the gauge coupling (i.e., the temperature); for a realistic determination better statistics and an extrapolation to infinite volume should be performed. *R.h.s.*: Scatter plot of Polyakov loop results in the complex plane for 1500 individual gauge configurations. In the broken phase the values of $P$ concentrate around the three center phases. The data are from a run with $\langle |P| \rangle = 0.53$

concentrate in the directions of the center elements. The results in general will be affected by the finiteness of the spatial volume and by scaling corrections. Comparing different parameter settings allows one to study both.

Monte Carlo calculations for the quenched theory have been performed already very early [16–18] and improved considerably since then [19, 20]. Careful studies revealed that the phase transition is weakly first order. This is in agreement with the analogy to the $Z_3$-spin model [14].

In the determination of the transition temperature the lattice spacing enters and the scale has to be fixed, as discussed in Sect. 3.5. The currently accepted value is $T_c \approx 270$ MeV for the quenched situation. Results from different improved gauge actions, extrapolated to the continuum limit, agree within a few percent.

In studies of the transition various standard tools of statistical physics have been employed. An informative quantity is the Polyakov loop susceptibility

$$\chi_P = N^3 \left( \langle P^2 \rangle - \langle P \rangle^2 \right) , \qquad (12.15)$$

which peaks at the transition. The value and position of this peak follow finite size scaling laws [21] and lead to information on the infinite volume values from finite volume measurements.

Typical computer simulations at finite $T$ should have spatial extents much larger than the temporal size in order to reduce finite size effects. Thus $N_T$ is relatively small, which limits the number of different Matsubara frequencies. It also complicates attempts to determine exponential decay properties in time direction of observables like particle propagators since at large $T$, i.e., small $N_T$, they can be followed only over very few time steps. For that reason one also has introduced different lattice spacings $a$ and $a_T$ in spatial and temporal directions, respectively, with $a > a_T$. This may be obtained by choosing different gauge couplings multiplying the time-directed plaquettes and the purely spatial plaquettes. Keeping the ratio $\xi = a/a_T$ large allows one to have large $N_T$ while keeping the temperature constant. This method introduced in [22, 23] is discussed in more detail in [24].

### 12.1.2 Switching on dynamical fermions

Dynamical fermions break the $Z_3$ symmetry explicitly and this affects the Polyakov loop as well. A simple way to see this comes from the hopping expansion of the fermion determinant introduced in Sect. 5.3.2. The exponent in (5.65) may be considered as an effective fermionic action. It is a sum over closed loops, like the gauge action itself. However, these are now all possible loops on the lattice, including the ones that wind nontrivially around the compactified time direction. In particular this sum will also include Polyakov loops and therefore terms proportional to

$$\kappa^{N_T} (P + P^*) . \tag{12.16}$$

Due to the transformation property (12.13) such a term in the effective action is no longer invariant under an overall multiplication of all gauge links in a time slice with an element of the group's center and therefore the fermions break the center symmetry. In full QCD one finds that the Polyakov loop settles around real positive values. In some sense one can consider the fermions' role like that of an external, symmetry-breaking magnetic field in a spin model.

Another, more intuitive, picture is that the dynamically generated vacuum loops screen the individual charges, similar to the Debye screening of electric charges in a polarizable medium. The potential computed from the correlation function (12.6) will saturate at some distance and may be approximated by a screened Coulomb potential $\propto \exp(-\mu|\boldsymbol{r}|)/|\boldsymbol{r}|$. Eventually the notion "potential" looses its meaning, since it becomes energetically more favorable to produce a quark–antiquark pair from the vacuum, saturating the test charges: the "string breaks." The Polyakov loop and the Wilson loop are no longer true-order parameters. Confinement now has to be defined differently, namely through the spectrum of the asymptotic states of the system, i.e., the color neutral hadrons. Based on these considerations an alternative definition of the order parameter was given in [25, 26].

In our theoretical world we can change the quark mass parameter $m$ continuously. Let us first discuss the case of two flavors with equal mass $m_{u,d}$ (cf. Fig. 12.2). If we let the mass approach infinity, the quarks decouple and we are in the quenched situation with a first-order temperature phase transition. Decreasing the quark mass this transition persists but the latent heat decreases and the transition becomes weaker. Eventually the first-order transition line

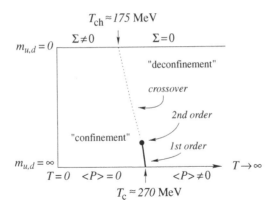

**Fig. 12.2.** The plot elaborates the phase structure in the $(T, m)$-plane for the $N_f = 2$ sub-manifold (*front side* of the 3D plot in Fig. 12.3)

in the $(g, m)$ plane ends with a second-order point. The regions neighboring to the confinement and the deconfinement domains are therefore analytically connected. Such a *crossover* is not unusual. Just think of water in its three phases: ice, liquid, and vapor. Depending on the pressure and temperature there may be phase transitions but also continuous crossover-type transitions.

However, we are not yet done. Decreasing the quark mass further we finally arrive at the situation of massless quarks, where the action is chirally invariant (if a suitable discretization is used). As we have discussed in Chap. 7, the full theory breaks the chiral symmetry spontaneously at zero temperature. The corresponding order parameter is the quark condensate $\Sigma$ defined in (7.63), which, however, is an order parameter only for the massless theory. As one increases $T$ the condensate shows a transition. Its value vanishes at some critical temperature $T_{\mathrm{ch}}$ and stays zero for higher temperatures. Thus the condensate may be considered a valid order parameter indicating the chiral temperature transition (cf. Fig. 12.2).

In Sect. 7.3.5 we have related the chiral condensate to the spectral density of the Dirac operator eigenvalues near the origin via the Banks–Casher relation (7.77). This formula describes the situation also above the transition: At $T_{\mathrm{ch}}$ a gap opens up in the spectrum and the chiral condensate vanishes.

A highly nontrivial observation is that confinement also seems to go away at about the same temperature $T_{\mathrm{ch}}$ which suggests a strong connection between confinement and chiral symmetry breaking. This equality of $T_{\mathrm{deconf}}$ and $T_{\mathrm{ch}}$ has been disputed recently [1, 2, 27]. Possible reasons for the controversy may be related to the definition of the pseudo-critical transition point(s) and to the fermion species and masses used in the simulations.

The value of $T_{\mathrm{ch}}$ is not easy to determine, since it requires a simulation at vanishing quark masses. Also, it depends on the number of quark species considered. Typical values obtained by various simulations and extrapolations range between $T_{\mathrm{ch}} \approx 150$ MeV [27] and $T_{\mathrm{ch}} \approx 190$ MeV [28]. This – compared

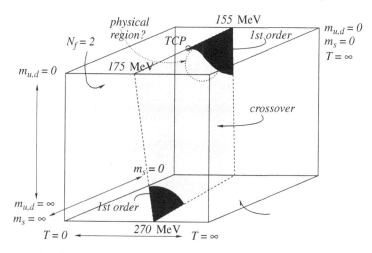

**Fig. 12.3.** Sketch of the phase structure in the $(T, m_{u,d}, m_s)$-space for two mass-degenerate flavors (mass $m_{u,d}$) and another third flavor ($m_s$) and vanishing quark number density/chemical potential $\mu = 0$. The *skewed shaded area* indicates the position of the crossover transition and, where the shading is dark, the region where the transition is of first order. At the boundaries of the first-order region one has a second-order phase transition. The boundary lying in the $m_{u,d} = 0$ plane may have a second-order part, joining the first-order domain at a tricritical point. The figure is based on information discussed and summarized in [29]

to pure gauge theory – smaller value may be understood from the observation that in the quenched case the only bound states are rather heavy glueballs. Therefore, larger temperatures are needed to generate a glueball gas dense enough to enable deconfinement. There is some caution advisable, though, as mentioned above.

Figure 12.3 illustrates our present knowledge on the phase diagram in a space of two mass-degenerate quarks with mass $m_{u,d}$ and a third quark with mass $m_s$. It is presently uncertain, whether for the physical quark masses (indicated by a dotted circle in the plot) the transition is located in the crossover or in the first-order domain.

Lattice results with two light and with or without a heavier strange quark near their physical mass values find a rapid but smooth crossover from lower to higher temperature, but no convincing evidence for a universal critical behavior has been found. Working at smaller lattice spacing, closer to the continuum limit, and with improved fermion actions with smaller cutoff effects may help to further clarify that situation [30, 31].

The order of the chiral temperature transition is less established than that for the pure gauge theory and depends on the number of quark flavors. Pisarski and Wilczek [32] have studied an effective 3D model field theory for the order parameter, assuming the underlying spontaneously broken chiral symmetry of QCD. A renormalization group analysis then led to the suggestion that the chiral transition is second order for $N_f = 2$ but weakly first order for $N_f = 3$. In the latter case the transition again extends toward $m > 0$ with a second-order endpoint. This endpoint and the endpoint of the transition line starting at the pure gauge theory are connected by a so-called crossover curve. This is a maximum of some observables (like the specific heat) but not the locus of nonanalyticity.

Like for the Polyakov loop, also the derivative of the condensate, the chiral susceptibility

$$\chi_{\text{ch}} = \frac{\partial \Sigma}{\partial m} \tag{12.17}$$

is a useful quantity to locate the transition point, and it also has definite finite size scaling behavior. Although neither $\Sigma$ nor $\langle P \rangle$ are strict order parameters for $0 < m < \infty$ their values can still be used to study the position of the phase transition lines.

### 12.1.3 Properties of QCD in the deconfinement phase

Formally the partition function depends on several parameters: the volume, the temperature, the quark masses, and the gauge coupling. One defines *thermodynamic quantities* like the *energy density* $\varepsilon$ or the *pressure* $p$ by derivatives of the free energy $(F = -\ln(Z)/\beta)$:

$$\varepsilon \equiv -\frac{1}{V}\frac{\partial \ln Z}{\partial \beta} = \frac{T^2}{V}\frac{\partial \ln Z}{\partial T} \ , \quad p \equiv T\frac{\partial \ln Z}{\partial V} \ , \quad \Sigma_i = \frac{\partial \ln Z}{\partial m_i} \ . \tag{12.18}$$

Here $V$ denotes the spatial volume and, as usual, $\beta = 1/T$.

As already addressed, the confinement and the deconfinement regions are analytically connected. Still we may expect quite different properties of matter in these regions, although the transition may be a gradual one. For large temperature the scale for the running coupling is set by $T$ and this brings one to the small gauge coupling perturbative regime. Asymptotic freedom then lets one expect a gas of weakly interacting quarks and gluons: a plasma. This is why the high-temperature phase is usually called the *quark-gluon plasma* phase [33].

A relativistic ideal gas of (noninteracting) QCD gluons and quarks at high temperature follows the Stefan–Boltzmann law and has the pressure and energy density [34],

$$p = \frac{\pi^2}{45}\left(8 + \frac{21}{4}N_f\right)T^4 \ , \quad \varepsilon = 3p \ (\text{equation of state}) \ . \tag{12.19}$$

Due to the larger number of degrees of freedom of quarks and gluons this is much larger than what one expects from a gas of, say, the three light pions $\pi^0$ and $\pi^\pm$. It is not clear how quickly hadrons dissociate and such a behavior is obtained. Figure 12.4 demonstrates that this may happen only at quite high values of $T$.

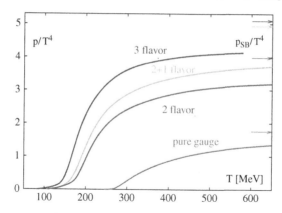

**Fig. 12.4.** Pressure $p$ determined from simulations for pure SU(3) gauge theory and QCD with two or three light quarks or two light and one heavy quark (2+1) with $m_s \approx T$ [35] (figure from [36]). The *arrows* on the *right-hand side* of the plot indicate the corresponding infinite temperature limits from the Stefan–Boltzmann law (12.19). (From: Karsch [36]. Reprinted with permission from AIP)

Due to the crossover nature of the transition the particle spectrum may change gradually from the (for zero temperature well-known) spectrum of mesons and baryons to some intermediate state before asymptotic behavior sets in and a plasma of quarks and gluons defines the ground state. The restoration of chiral symmetry above $T_{ch}$ may be realized by degenerate parity partners, but still, e.g., mesonic bound states. The region presently accessible in heavy ion experiments is typically close to the transition and the observations thus may be dominated and obscured by the intermediate phase-changing properties.

As mentioned, because of the small Euclidean time extent (less than 1 fm) it is quite problematic to use the standard approach for determining the particle spectrum by studying correlation functions in time direction, although this has been tried [37–39]. One can, however, analyze the spatial correlation functions, which leads to so-called *screening masses* [40–43]. Integrating the propagators over all space and time gives the thermal susceptibilities involving a contact term, which has to be subtracted. The integration involves a sum over all states in the corresponding quantum channel, though. Following these through the transition indicates that the masses of $\pi(0^{-+})$ and the isoscalar $f_0(0^{++})$ become degenerate quickly above $T_{ch}$, indicating a restoration of SU(2) flavor symmetry. The mass of the isovector $a_0(0^{++})$ is still different from that of the pion, though, indicating persistent $U_A(1)$ symmetry breaking.

## 12.2 Introduction of the chemical potential

The net number of baryons or quarks in the vacuum is zero, but it is not in matter. In extreme situations, like heavy ion collisions or ultra-dense matter in neutron stars, one has to consider the effect of nonvanishing quark number density $n_q$. In this case we have to extend the partition function of the grand canonical ensemble (12.1):

$$Z(T, \mu) = \mathrm{tr} \left[ e^{-(\hat{H} - \mu \hat{N}_q)/T} \right] . \tag{12.20}$$

We have added to the total action a term introducing a *quark chemical potential* $\mu$ which multiplies the quark number operator $\hat{N}_q$. Sometimes the baryon number operator $\hat{N}_B = \hat{N}_q/3$ and the *baryon chemical potential* $\mu_B = 3\mu$ are used instead.

Since the quark number operator $\hat{N}_q$ assumes integer values, one may expand the grand canonical partition function $Z(T, \mu)$ in a power series of the *fugacity variable*

$$z = e^{\mu/T} . \tag{12.21}$$

The result of this *fugacity expansion* is a sum over canonical partition functions $Z_n(T)$ with a fixed quark number $n \in \mathbb{Z}$

$$Z(T, \mu) = \sum_n \left( e^{\mu/T} \right)^n Z_n(T) , \tag{12.22}$$

where negative values of $n$ correspond to a net surplus of antiquarks. On a finite lattice the sum over the quark number $n$ is finite, i.e., $|n| \le n_{max}$. For a spatial volume with $N^3$ points each flavor of Wilson fermions contributes to $n_{max}$ the number $6N^3$.

With the help of the grand canonical partition function (12.20) we can compute new observables such as the expectation value of the *quark number density*

$$n_q \equiv \frac{1}{V} \langle \hat{N}_q \rangle = \frac{T}{V} \frac{\partial \ln Z(T, \mu)}{\partial \mu} . \tag{12.23}$$

The quark number density of the physical environment will affect the phase structure and the location and properties of the finite temperature phase transition.

In addition to the gauge coupling and the quark mass parameters, familiar from previous chapters, and the temperature $T$ introduced in the last section, we now have the additional parameter $\mu$. The phase diagram will become increasingly complicated, as we will see in the discussion in Sect. 12.2.2.

### 12.2.1 The chemical potential on the lattice

How can we introduce the quark number density in the lattice action? The Euclidean continuum quark number operator is given by the spatial volume integral over the temporal component

$$\overline{\psi}(x)\,\gamma_4\,\psi(x) \tag{12.24}$$

of the conserved vector current $\overline{\psi}\gamma_\mu\psi$. In a lattice implementation one is tempted to add to the Dirac operator (2.36) such a term multiplied with $\mu$, since the factor $1/T$ comes from integrating over the finite time direction.

That simplistic ansatz runs into problems, though. In order to see this let us discuss the free case. We will show that there the energy density $\varepsilon(\mu)$ becomes divergent in the continuum limit and that this unphysical divergence is a consequence of the naive introduction of the chemical potential according to (12.24).

The free energy density is given by

$$\varepsilon(\mu) \;=\; \frac{1}{V_3}\langle\widehat{H}\rangle \;=\; -\frac{1}{V_3 Z}\frac{\partial}{\partial\beta}\,\mathrm{tr}\left[\mathrm{e}^{-\beta\widehat{H}}\right] \;=\; -\frac{1}{V_3 Z}\frac{\partial}{\partial\beta}Z \;=\; -\frac{1}{V_3}\frac{\partial}{\partial\beta}\ln Z\;, \tag{12.25}$$

where $V_3$ is the spatial volume and $Z = \mathrm{tr}\left[\exp(-\beta\widehat{H})\right]$. When evaluating the energy density on the lattice it is convenient to introduce a different lattice spacing $a_T$ for the temporal direction, which in the end of the calculation can be set equal to the lattice spacing $a$ used for the spatial direction. Due to the fact that for the free case the partition function is $Z = \det[D]$, we find for the energy density (using $\beta = N_T a_T$)

$$\varepsilon(\mu) \;=\; -\frac{1}{(aN)^3}\frac{\partial}{\partial(N_T a_T)}\ln\left(\det[D]\right)\Big|_{\mu a_T=\mathrm{const}}$$

$$=\; -\frac{1}{(aN)^3 N_T}\frac{\partial}{\partial(a_T)}\ln\left(\det[D]\right)\Big|_{\mu a_T=\mathrm{const}}. \tag{12.26}$$

For the evaluation of $\varepsilon(\mu)$ we need to compute the determinant of the lattice Dirac operator $D$. For free fermions with a chemical potential introduced in the naive way (12.24) it reads

$$D(n|m) \;=\; \sum_{j=1}^{3}\gamma_j\,\frac{\delta_{n+\hat{j},m}-\delta_{n-\hat{j},m}}{2a} + \gamma_4\,\frac{\delta_{n+\hat{4},m}-\delta_{n-\hat{4},m}}{2a_T}$$

$$+\,m\,\delta_{nm} + \mu\,\gamma_4\,\delta_{nm}\,. \tag{12.27}$$

For the evaluation of the determinant we Fourier transform $D$ to momentum space (cf. Appendix A.3) and obtain

$$\widetilde{D}(p|q) \;=\; \frac{1}{|A|}\sum_{n,m\in A}\mathrm{e}^{-\mathrm{i}p\cdot n}\,D(n|m)\,\mathrm{e}^{\mathrm{i}q\cdot m} \;=\; \delta(p-q)\widetilde{D}(p)\,, \tag{12.28}$$

with the Dirac operator in momentum space $\widetilde{D}$ given by

$$\widetilde{D}(p) \;=\; \frac{1}{a_T}\left(\mathrm{i}\frac{a_T}{a}\sum_{j=1}^{3}\gamma_j\,\sin(p_j a) + \mathrm{i}\gamma_4\,\sin(p_4 a_T) + a_T\,m\,\mathbb{1} + a_T\,\mu\,\gamma_4\right).$$

$$\tag{12.29}$$

For the logarithm of the determinant needed in (12.26) we find (use (A.54))

$$\ln\left(\det[D]\right) = \ln\left(\prod_p \det_d\left[\tilde{D}(p)\right]\right) = \sum_p \ln\left(\det_d\left[\tilde{D}(p)\right]\right)$$
$$= \sum_p \ln\left(e^{\mathrm{tr}_d[\ln(\tilde{D}(p))]}\right) = \sum_p \mathrm{tr}_d\left[\ln\left(\tilde{D}(p)\right)\right] . \quad (12.30)$$

We use the subscript $d$ attached to the determinant and the trace to indicate that both are computed only for the remaining $4 \times 4$ matrices in Dirac space.

Using (12.26), (12.27), (12.28), (12.29), and (12.30) we obtain for the derivative with respect to $a_T$,

$$\frac{\partial}{\partial a_T} \ln\left(\det[D]\right)\bigg|_{\mu a_T = \mathrm{const}} = \sum_p \mathrm{tr}_d\left[\frac{\partial}{\partial a_T} \ln\left(\tilde{D}(p)\right)\right]\bigg|_{\mu a_T = \mathrm{const}}$$

$$= C + \sum_p \mathrm{tr}_d\left[\left(a_T \tilde{D}(q)\right)^{-1}\left(\frac{i}{a}\sum_{j=1}^{3} \gamma_j \sin(p_j a) + m\,\mathbb{1}\right)\right] , \quad (12.31)$$

where the constant $C$ is given by $C = -4 N_T N^3 / a_T$. Evaluating the remaining traces in Dirac space and inserting the result into (12.26) we find after setting $a_T = a$,

$$\varepsilon(\mu) = C - \frac{4}{N^3 N_T a^4} \sum_p F(ap, am, a\mu) , \quad (12.32)$$

with

$$F(ap, am, a\mu) = \frac{\sum_{j=1}^{3} \sin^2(a\,p_j) + (a\,m)^2}{\sum_{k=1}^{3} \sin^2(a\,p_j) + (a\,m)^2 + (\sin(a\,p_4) - i\,a\,\mu)^2} . \quad (12.33)$$

When computing the final result for the energy density we need to normalize by removing the value for $\mu = 0$, i.e., we consider $\varepsilon(\mu) - \varepsilon(0)$. We return to zero temperature and, letting $N = N_T \to \infty$, we change to continuous integration over momenta $p$:

$$\frac{1}{N^3 N_T a^4} \sum_p \quad \Rightarrow \quad \frac{1}{(2\pi)^4} \int_{-\pi}^{\pi} \mathrm{d}^4 p \quad (12.34)$$

giving rise to

$$\varepsilon(\mu) - \varepsilon(0) = -\frac{4}{(2\pi)^4} \int_{-\pi}^{\pi} \mathrm{d}^4 p \,\left(F(ap, am, a\mu) - F(ap, am, 0)\right) , \quad (12.35)$$

where in $F$ we have still kept $a$ finite. Using a contour integral in the complex $p_4$-plane, the leading term in the limit $a \to 0$ may be extracted. From the $p_4 \to 0$ contribution one obtains the behavior

$$\lim_{a \to 0} \left(\varepsilon(\mu) - \varepsilon(0)\right) \propto \left(\frac{\mu}{a}\right)^2 . \quad (12.36)$$

This is divergent in the continuum limit and thus simply adding the term (12.24) in the action does not give rise to a proper discretization of the chemical potential.

Closer inspection of the continuum situation (cf. [44–46]) clarifies the problem. The quark number is the conserved charge of the U(1) global symmetry. Determining the Noether current for the lattice action gives the current expressed by nearest neighbor terms. The space integral then produces a suitable form of the chemical potential term. One therefore implements the chemical potential by replacing the temporal hopping term in (5.51) with

$$
-\frac{1}{2a} \sum_{n \in \Lambda} \Big( f(a\mu)(\mathbb{1} - \gamma_4)_{\alpha\beta} U_4(n)_{ab} \delta_{n+\hat{4},m}
$$
$$
+ f(a\mu)^{-1}(\mathbb{1} + \gamma_4)_{\alpha\beta} U_4(n - \hat{4})^\dagger_{ab} \delta_{n-\hat{4},m} \Big) , \tag{12.37}
$$

where $f(a\mu)$ is a real function not yet specified.

For $\mu = 0$ the original action should be recovered, thus we request $f(0) = 1$. In an expansion in $\mu$ the next term should be linear in the chemical potential in order to reproduce the density term. We therefore have $f(a\mu) = 1 + a\mu + \mathcal{O}(a\mu)^2$. Time reflection invariance (cf. Sect. 5.4) requires $f(a\mu) = 1/f(-a\mu)$. The simplest choice fulfilling all these conditions is

$$
f(a\mu) = \exp(a\mu) . \tag{12.38}
$$

In this formulation the propagation forward in time is favored by a factor of $\exp(a\mu)$, whereas propagation backward in time is disfavored by $\exp(-a\mu)$. This introduces the desired particle–antiparticle asymmetry.

In Sect. 5.3.1 we have shown that the hopping expansion expresses the fermion determinant as a sum over closed loops. In such a loop the forward hopping factors $f(a\mu)$ and the backward hopping factors $f(a\mu)^{-1}$ cancel unless the loop winds nontrivially around the compact time direction. The total contribution of the chemical potential then is $f(a\mu)^{wN_T}$, where $w \in \mathbb{Z}$ denotes the number of windings of the loop.

Based on that observation, we note that in a simulation instead of modifying all link terms in time direction one may also modify just all forward time-directed hopping terms in only a single time slice with the factor

$$
f(a\mu)^{N_T} = \exp(a\mu N_T) = \exp(\mu/T) , \tag{12.39}
$$

and the corresponding backward-oriented terms with the inverse factor. Like in (12.20) we find that the chemical potential always enters in the form $\mu/T$.

The introduction of the chemical potential comes with a serious technical drawback: For $a\mu \neq 0$ the Dirac operator is no longer $\gamma_5$-hermitian. It is easy to see that multiplying the hopping term (12.37) with $\gamma_5$ from the left and the right changes the sign of $\gamma_4$; hermitian conjugation then exchanges $U$ and $U^\dagger$ and we find instead of the $\gamma_5$-hermiticity relation (5.76) the modified equation

$$\gamma_5 D(\mu)\gamma_5 = D^\dagger(-\mu) \,. \tag{12.40}$$

This, however, is no longer a symmetry transformation but leads to a term with $f$ replaced by $1/f^*$. We find

$$\gamma_5\, D(f)\,\gamma_5 = D^\dagger\,(1/f^*) \quad \Rightarrow \quad \det[D(f)] = \det\left[D\,(1/f^*)\right]^* \,. \tag{12.41}$$

This is an invariance operation only for $f = 1/f^*$. For real $f$ one therefore has $f = 1/f = 1$ which implies $a\,\mu = 0$. Consequently for nonvanishing real $\mu$ the determinant of the Dirac operator then is complex and one cannot obtain a real Boltzmann weight by doubling the number of fermion flavors. A nonvanishing real $\mu$ creates a particle–antiparticle asymmetry which destroys the reality of the determinant and a straightforward application of importance sampling.

It is interesting to notice, however, that for *imaginary chemical potential* $\mu = i\eta$, $\eta \in \mathbb{R}$ we have

$$f(i\,a\,\eta) = 1/f(i\,a\,\eta)^* = 1/f(-i\,a\,\eta) \,. \tag{12.42}$$

Then the Dirac operator is $\gamma_5$-hermitian in the standard way and the determinant is real [47–49]. Based on this observation one might try to work at imaginary chemical potential and analytically continue to real values (cf. Sect. 12.3.3).

Another special case is when one considers the so-called *isospin chemical potential* $\mu_I$ [49, 50]. Let us introduce a chemical potential variable for each flavor. In the Euclidean continuum Lagrangian this corresponds to terms

$$\sum_f \mu_f \overline{\psi}_f\, \gamma_4\, \psi_f \,, \tag{12.43}$$

which may be translated to the lattice formulation as discussed above. Focusing on two light flavors we may consider $\mu_u = \mu_I$ and $\mu_d = -\mu_I$. The term in the Lagrangian then has the form $\mu_I(\overline{u}\,\gamma_4\,u - \overline{d}\,\gamma_4\,d)$. In this case the Dirac operators are block diagonal, where the blocks are the one-flavor operators for the up and the down quark:

$$\begin{pmatrix} D(\mu_I) & 0 \\ 0 & D(-\mu_I) \end{pmatrix} = \begin{pmatrix} D(\mu_I) & 0 \\ 0 & \gamma_5 D^\dagger(\mu_I)\,\gamma_5 \end{pmatrix} \,. \tag{12.44}$$

In this step we have used (12.40). For mass-degenerate quarks the determinant of this matrix is real and positive since

$$\det[D(\mu_I)]\det[\gamma_5\, D^\dagger(\mu_I)\,\gamma_5] = \det[D(\mu_I)]\det[D^\dagger(\mu_I)] = |\det[D(\mu_I)]|^2 \,. \tag{12.45}$$

This property allows one to study nonvanishing isospin chemical potential in Monte Carlo simulations using the standard techniques discussed in Chap. 8. Isospin chemical potential is not an academic problem, since one can imagine such a situation in relativistic heavy ion collisions.

There is also a definition of the overlap Dirac operator for nonvanishing chemical potential [51, 52]. The kernel operator may be the usual Wilson Dirac operator with chemical potential $\mu$ introduced as in (12.37), which is then not $\gamma_5$-hermitian. The sign function applied in the construction of $D_{\text{ov}}$ (cf. Sect. 7.4) is a generalization to nonhermitian but diagonalizable operators, i.e., in the spectral representation (7.84) one takes the sign of the real part of the complex eigenvalues. $D_{\text{ov}}(\mu)$ is not $\gamma_5$-hermitian for $\mu \neq 0$ but inherits property (12.40). It is no longer normal and its eigenvalues are not restricted to the Ginsparg–Wilson circle. However, eigenvectors $\psi$ and $\gamma_5\psi$ have still eigenvalues related by $\lambda$ and $\lambda/(\lambda-1)$. It was demonstrated [53, 54] that this generalization reproduces correctly the thermodynamical properties of a free Fermi gas. The exact manifestation of chiral symmetry on the lattice in the presence of a nonvanishing chemical potential is still discussed, however.

### 12.2.2 The QCD phase diagram in the $(T, \mu)$ space

The situation concerning details of the phase structure, like the exact position of phase transitions and values of critical exponents, is yet to be settled. Lattice results from different kinds of Dirac operators often disagree and the numerics and statistics may not be sufficient. The phase structure discussed here therefore can only represent the present minimal consensus [55, 56].

Figure 12.5 shows the situation for two massless quarks and for two light quarks close to the physical mass. The temperature transition at $\mu = 0$ has been discussed in Sect. 12.1.2: For $N_f = 2$ it is either second order or weakly first order. For increasing quark mass the transition weakens and eventually becomes a crossover. By introducing a light third dynamical quark the phase transition becomes strong and the first-order line from the middle of the phase diagram extends toward the $\mu = 0$ region. If the mass of the third quark increases, again the situation is that of the right-hand side plot of Fig. 12.5.

There are no lattice results for zero temperature and nonvanishing real chemical potential. Straightforward perturbative treatment of QCD at high density fails as well [55]. Much of the information there comes from NJL-inspired calculations [57], Dyson–Schwinger gap equations with various interaction models like the instanton liquid model [58] and random matrix models [59, 60]. In the confinement phase one first has nuclear matter with (for two massless flavors) a small transition line extending into the phase, which separates between a resonance gas-like and a liquid-like structure. Model calculations then indicate a first-order transition between the nuclear matter region for $\mu_B$ above 1 GeV and, at higher density, a (two-flavor) color superconducting phase (2SC), where the quarks pair in a flavor singlet color anti-triplet channel [57, 58]. Adding a strange quark with a mass close to its physical value the region around the phase transition may be more complicated, e.g., allowing for an intermediate pocket of a crystalline color superconducting

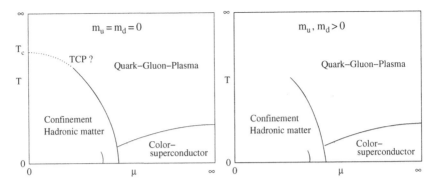

**Fig. 12.5.** The conjectured phase diagram in the $(T, \mu)$-plane for (*l.h.s.*) two massless quarks and (*r.h.s.*) two mass-degenerate light quarks. Lattice calculations so far have been useful only for small values of the chemical potential, attempting to locate the tricritical point (*l.h.s*) or second-order endpoint (*r.h.s.*) of the first-order phase transition line

phase. However, all calculations agree that for larger $\mu$ one eventually enters a color–flavor-locked (CFL) phase with a nonvanishing diquark condensate [61].

Lattice results so far have been confined to a strip close to vanishing chemical potential, typically $\mu/T < 1$. All methods have been based on analytic continuation from $\mu = 0$ or imaginary $\mu$ to nonvanishing $\mu$. Of particular importance for heavy ion collisions is whether the physical situation is close to a crossover or phase transition line and in what phase one ends up in various experiments.

## 12.3 Chemical potential: Monte Carlo techniques

The quest for the optimal simulation strategy for nonvanishing chemical potential is far from being settled. We therefore refrain from trying to cover all results available but just discuss the underlying concepts. All approaches so far have analytically continued results obtained from measurements for real determinants to the actual parameter values one is interested in. There have been two groups of such extrapolations:

- Using measurements at $\mu = 0$ and extrapolating with the help of reweighting or power series in $\mu/T$ (moment expansion).
- Using results determined for purely imaginary $\mu$ (negative $\mu^2$) and analytic continuation to real $\mu$ (positive $\mu^2$) via a power series and Padé rational expansion or reconstruction of the fugacity expansion coefficients via Fourier transformation.

Quenched simulations [62–64] first led to confusing results. It was found that the critical value of $\mu$ decreased with the pion mass: $\mu_c \propto m_\pi/2$, which would vanish in the chiral limit. On the other hand at zero temperature one expects that the transition is near $\mu_c \simeq m_p/3$, since the proton is the lightest state with nonvanishing baryon number. Stephanov [65] has pointed out that for $\mu \neq 0$ the quenched theory is not the $N_f \to 0$ limit of QCD but of a theory with $N_f$ quarks and $N_f$ conjugate quarks. This implies that for a simulation the phase of the quark determinant is of importance and dynamical fermions are necessary to obtain $\mu_c \simeq m_p/3$. Most simulations in that context are therefore done with dynamical fermions, mainly of the staggered type.

Large temperature $T \gg T_c$ corresponds to very small time extension and one expects that the system approaches effectively a 3D gauge theory. The temporal loops then play the role of an adjoint Higgs field [66, 67]. This *dimensional reduction* opens the way for Monte Carlo studies of the effective theory including chemical potential [68–71]. We restrict our discussion to methods for the full 4D system.

### 12.3.1 Reweighting

*Reweighting* is a standard tool in Monte Carlo approaches for statistical spin systems and has been successfully used to improve interpolation between Monte Carlo results at different couplings and for analytical continuation from real to complex couplings. It has been particularly useful for determining the positions of partition function zeroes for complex couplings (Lee–Yang zeroes and Fisher zeroes).

The strategy may be understood as a more general concept which we discuss here briefly for a simple bosonic system. The partition function is rewritten as a sum over the possible values of the total energy of the system:

$$Z(\kappa) \equiv \int \mathcal{D}[\Phi] \, e^{-\kappa\, S[\Phi]} = \int dE \int \mathcal{D}[\Phi] \, \delta(E - S[\Phi]) \, e^{-\kappa\, S[\Phi]}$$

$$= \int dE \, \rho(E) \, e^{-\kappa\, E} \quad \text{with} \quad \rho(E) = \int \mathcal{D}[\Phi] \, \delta(E - S[\Phi]) \, . \tag{12.46}$$

In this way the energy distribution density $\rho(E)$ may be obtained with high precision by combining Monte Carlo results from different values of the coupling – the so-called *multi-histogram technique* [72, 73]. Once the density of states $\rho(E)$ is known, $Z(\kappa)$ and its derivatives can be computed for any value of the coupling $\kappa$. This is correct, however, only if the energy distribution is known to arbitrary precision. In realistic simulations this is not the case and therefore the technically crucial point is how much the distribution generated from Monte Carlo runs at some couplings overlaps with that at some other values of the coupling where one wants to compute $Z$.

In the spirit of using the determinant as a weight factor one may use the deviation of $\det[D(\mu)]$ from $\det[D(0)]$ as a reweighting factor. From

$$\left\langle \frac{e^{-S_G[\beta',U]} \det[D(\mu',U)]}{e^{-S_G[\beta,U]} \det[D(\mu,U)]} \right\rangle_{\beta,\mu}$$

$$= \frac{1}{Z(\beta,\mu)} \int \mathcal{D}[U] \, \frac{e^{-S_G[\beta',U]} \det[D(\mu',U)]}{e^{-S_G[\beta,U]} \det[D(\mu,U)]} e^{-S_G[\beta,U]} \det[D(\mu,U)]$$

$$= \frac{1}{Z(\beta,\mu)} \int \mathcal{D}[U] \, e^{-S_G[\beta',U]} \det[D(\mu',U)] = Z(\beta',\mu')/Z(\beta,\mu),$$

$$(12.47)$$

we find that we may, at least in theory, determine $Z(\beta',\mu')$ from results evaluated at $\beta$ and $\mu$. We therefore could obtain values for

$$Z(\beta',\mu') = Z(\beta,\mu) \left\langle \frac{e^{-S_G[\beta',U]} \det[D(\mu',U)]}{e^{-S_G[\beta,U]} \det[D(\mu,U)]} \right\rangle_{\beta,\mu}. \qquad (12.48)$$

(In these equations $\beta$ denotes the inverse gauge coupling, not the inverse temperature.) Since in a Monte Carlo sampling the configurations are obtained with probabilities determined by the values of $\beta$ and $\mu$, the results crucially depend on how far one wants to extrapolate. Studying the finite temperature phase transition one then samples around ($\beta = \beta_c$, $\mu = 0$), where $\beta_c$ denotes the value of the gauge coupling for which $aN_T = 1/T_c$.

The technical issue is the overlap of the ensemble sampled at $\beta, \mu$ with the actual ensemble at the target couplings $\beta', \mu'$, and the resulting statistical reliability. The difference in the total free energy grows with the lattice volume rendering the extrapolation for larger lattice sizes more problematic. With increasing chemical potential it becomes worse and the predictive power poorer.

The first attempts have kept the gauge coupling fixed [62, 64] and extrapolated from simulation at vanishing chemical potential to a nonvanishing value for small lattices.

Fodor and Katz [74, 75] extrapolate by reweighting in both variables, the gauge coupling and the chemical potential, sampling at the transition point ($\beta = \beta_c$, $\mu = 0$), following the pseudo-critical line. This strategy appears to improve the approximation and was applied on small lattices. Analyzing the complex (Lee–Yang) zeroes of the partition function allows one to locate the endpoint of the expected first-order transition line. In a simulation with 2+1-staggered quarks near their physical mass values the point was estimated to lie at $\mu_c^E \approx 120(13)$ MeV at $T \approx 162(2)$ MeV [76], corresponding to $\mu_c^E/T \approx 0.75$.

A systematic check of the convergence properties of the reweighting approach is possible for imaginary chemical potential which can be simulated by standard methods. For a critical assessment of such a comparison, cf. [77–79].

## 12.3.2 Series expansion

The chemical potential always enters in the combination $\mu/T$. In another method of extrapolation one therefore writes the physical observables as a power series in $\mu/T$. Time-reversal invariance of the gauge configuration ensemble (compare Sect. 5.4) provides the symmetry

$$Z(\mu/T) = Z(-\mu/T) . \tag{12.49}$$

Thus $Z$ and observables that are symmetric under time reflection are even functions in $\mu/T$ and the expansion will be in even powers $(\mu/T)^2$. The pressure density (12.18) in a homogeneous system is $p = (T/V)\ln Z$. The corresponding Taylor–McLaurin series then is even in $\mu$ and reads

$$p = T^4 \sum_{n=0}^{\infty} c_{2n}(T) \left(\frac{\mu}{T}\right)^{2n} , \tag{12.50}$$

where the coefficients are derivatives of the free energy $F = -\ln(Z)/\beta$ and therefore generalized quark number susceptibilities, evaluated for vanishing $\mu$ [80–83].

The coefficient of the quadratic term in (12.50) is the quark number susceptibility at $\mu = 0$. It is a derivative of the quark number density (12.23) and a second derivative of the fermion determinant with respect to the chemical potential,

$$\chi_q = \frac{\partial n_q}{\partial \mu} = \frac{T}{V}\frac{\partial^2 \ln Z}{\partial \mu^2} . \tag{12.51}$$

Such derivatives involve traces like

$$\frac{\partial \ln \det[D]}{\partial \mu} = \mathrm{tr}\left[D^{-1}\frac{\partial D}{\partial \mu}\right] \tag{12.52}$$

and higher derivatives (cf. [83]). These are evaluated in a similar way as has been discussed in the context of the fermionic force in (8.44) and (8.47). The trace is evaluated with the help of stochastic estimators as discussed in Sect. 8.4. The coefficients in (12.50) are determined in simulations at $\mu = 0$.

In the thermodynamic limit phase transitions will be associated with singularities in some observables. The coefficients of the expansion can be used to estimate the position of the closest singularity. The quark number susceptibility exhibits a pronounced peak at the transition [84]. Finite size scaling considerations then may give hints on the type of phase transition or crossover. Several groups (in particular [82–86]) have analyzed the quark number susceptibility and other expansion coefficients in this way.

## 12.3.3 Imaginary $\mu$

In (12.42) we have noted that for imaginary chemical potential $\mu = i\eta$, $\eta \in \mathbb{R}$, the Dirac operator is $\gamma_5$-hermitian and the determinant is real. Then standard importance sampling techniques can be applied like for the case without chemical potential.

The chemical potential enters the partition function through the factors $\exp(\pm i\eta/T)$ as can be seen from (12.39). Thus thermodynamic functions are $2\pi$-periodic in $\eta/T$ and we can restrict $\eta/T$ to the interval $[0, 2\pi)$.

Let us discuss the analytic structure in the complex $z$-plane [87] where

$$z \equiv (\mu/T)^2 = -(\eta/T)^2 . \tag{12.53}$$

Dynamical fermions break the $Z_3$ symmetry of the pure gauge theory (see Sect. 12.1.2). However, the effect of a $Z_3$ transformation on the fermion fields can be compensated by a shift in the complex chemical potential, thus leading to a new symmetry. This implies a periodicity in the imaginary part of the chemical potential,

$$Z(\eta/T) = Z(\eta/T + 2\pi/3) . \tag{12.54}$$

Thus the interval $\eta/T \in [0, 2\pi)$ is split into three equivalent sectors distinguished only by the phase of the Polyakov loop. For imaginary chemical potential there are $Z_3$ transitions of first order at the boundaries of the sectors and crossover lines above a certain temperature [88, 89]. These singularities limit the domain of convergence for the series to roughly $|\mu/T| \le 1$.

One can directly simulate the theory at such negative $z$-values, i.e., imaginary chemical potential $\mu = i\eta$, and attempt to extrapolate the results to positive $z$ [88–92]. Analytic continuation may be performed in several ways. The simplest approach is a power series in $z$, which converges in a circular domain $|z - z_0| < R$, where $R$ is the distance to the closest singularity. A standard method in the theory of analytic functions is to perform a sequence of expansions around points located suitable in the convergence domain of the preceding series. Other methods involve optimal mappings or Padé expansions. The latter is a systematic method to replace the power series by that rational function which has identical expansion coefficients.

A straightforward approach is to fit the simulation results at several values of $z$ to a power series as in (12.50). The Taylor expansion has the same convergence properties as shown for the pressure in [81, 83]. It turns out that convergence is rapid for several observables studied, like the chiral condensate and screening masses [69], as well as the position of the crossover (the pseudo-critical temperature) [79, 88, 89, 92]. Padé approximants may allow extension of the extrapolation range [93] beyond the convergence circle of the power series.

Another extension to complex values of $\mu$ introduces variables $\cosh(\mu a)$ and $\sinh(\mu a)$ [94, 95], with subsequent analytic extrapolation to real chemical potential. The results for the position of the crossover are in agreement with the above-mentioned work.

### 12.3.4 Canonical partition functions

For imaginary chemical potential the fugacity expansion of the grand canonical partition function (12.22) is a Fourier sum [47, 48] and periodic in $\eta/T$ with a period of $2\pi$,

$$Z(T, \mathrm{i}\,\eta) = \sum_n \mathrm{e}^{\mathrm{i}\,n\,\eta/T} Z_n(T) \,. \tag{12.55}$$

This sum may be inverted leading to the following expression for the canonical partition functions $Z_n(T)$ at a fixed quark number $n$:

$$Z_n(T) = \frac{1}{2\pi} \int_{-\pi}^{\pi} \mathrm{d}(\eta/T) \, \mathrm{e}^{-\mathrm{i}\,n\,\eta/T} \, Z(T, \mathrm{i}\,\eta) \,. \tag{12.56}$$

Equation (12.56) thus is an exact expression for the coefficients $Z_n$ of the *fugacity expansion* (12.22).

Simulations at imaginary chemical potential can be used to determine values for $Z(T, \mathrm{i}\,\eta)/Z(T, 0)$ and thus in principle the Fourier integral (12.56) for the $Z_n$ can be evaluated numerically. A problem is the oscillations of the Fourier sum which increase with the quark number $n$, such that at higher values of $|n|$ many intermediate points of $\eta/T$ in the interval $[-\pi, \pi]$ are necessary for a reliable numerical estimate of the Fourier integral. Combinations of multi-histogram sampling and other intricate methods can be used to determine $Z_n$ and then, via (12.22), $Z(T, \mu)$ [78, 96–101]. An alternative approach that has been explored recently [102, 103] is perturbative techniques for the evaluation of the $Z_n$.

Studying the canonical partition function is particularly suitable to explore few nucleon systems at low temperature, and in principle, allows one to study the bulk properties of nuclear matter and the nuclear interactions. The metastable region is nicely exhibited and a Maxwell construction relates this signal to the grand canonical results. Nice agreement of these results from this approach with other approaches discussed earlier has been demonstrated [98]. The results so far have been on small lattices. Fixing the baryon density, though, requires to increase $n$ for increasing volumes. This may be a problem for the necessary statistics of the simulations.

# References

1. Z. Fodor: PoS **LATTICE2007**, 011 (2007)
2. F. Karsch: PoS **LATTICE2007**, 015 (2007)
3. O. Philipsen: in *Conceptual and Numerical Challenges in Femto- and Peta-Scale Physics*, edited by C. Gattringer et al., The European Physical Journal ST 152, p. 29 (Springer, Berlin, Heidelberg, New York 2007)
4. C. DeTar: PoS **LATTICE2008**, 001 (2008)
5. S. Ejiri: PoS **LATTICE2008**, 002 (2008)
6. A. M. Polyakov: Phys. Lett. B **72**, 477 (1978)
7. L. Susskind: Phys. Rev. D **20**, 2610 (1979)
8. G. 't Hooft: Nucl. Phys. B **153**, 141 (1979)
9. C. B. Lang and M. Wiltgen: Phys. Lett. B **131**, 153 (1983)
10. G. Bali: Phys. Rep. **343**, 1 (2001)

11. H. Ichie, V. Bornyakov, T. Streuer, and G. Schierholz: Nucl. Phys. A **721**, 899 (2003)
12. F. Bissey et al.: Phys. Rev. D **76**, 114512 (2007)
13. Y. Peng and R. W. Haymaker: Phys. Rev. D **47**, 5104 (1993)
14. B. Svetitsky and L. G. Yaffe: Nucl. Phys. B **210**, 423 (1982)
15. B. Svetitsky and L. G. Yaffe: Phys. Rev. D **26**, 963 (1982)
16. L. D. McLerran and B. Svetitsky: Phys. Lett. B **98**, 195 (1981)
17. J. Kuti, J. Polonyi, and K. Szlachanyi: Phys. Lett. B **98**, 199 (1981)
18. L. D. McLerran and B. Svetitsky: Phys. Rev. D **24**, 450 (1981)
19. E. Laermann and O. Philipsen: Ann. Rev. Nucl. Part. Sci. **53**, 163 (2003)
20. F. Karsch and E. Laermann: in *Quark Gluon Plasma*, edited by R. C. Hwa, p. 1 (World Scientific, Singapore 2003)
21. J. L. Cardy: *Finite-Size Scaling* (North-Holland, Amsterdam 1988)
22. J. Engels, F. Karsch, I. Montvay, and H. Satz: Phys. Lett. B **101**, 89 (1981)
23. J. Engels, F. Karsch, I. Montvay, and H. Satz: Nucl. Phys. B **205** [**FS5**], 545 (1982)
24. I. Montvay and G. Münster: *Quantum Fields on a Lattice* (Cambridge University Press, Cambridge New York 1994)
25. K. Fredenhagen and M. Marcu: Commun. Math. Phys. **92**, 81 (1983)
26. J. Bricmont and J. Fröhlich: Phys. Lett. B **122**, 73 (1983)
27. Y. Aoki, Z. Fodor, S. D. Katz, and K. K. Szabo: Phys. Lett. B **643**, 46 (2006)
28. M. Cheng et al.: Phys. Rev. D **74**, 054507 (2006)
29. F. Karsch: Lect. Notes Phys. **583**, 209 (2002)
30. Y. Aoki, Z. Fodor, S. D. Katz, and K. K. Szabo: JHEP **0601**, 089 (2006)
31. C. Bernard et al.: PoS **LAT2005**, 156 (2005)
32. R. D. Pisarski and F. Wilczek: Phys. Rev. D **29**, 338 (1984)
33. E. Shuryak: Phys. Rep. **61**, 71 (1980)
34. R. P. Feynman: *Statistical Mechanics* (Westview Press, Boulder 1998)
35. F. Karsch, E. Laermann, and A. Peikert: Phys. Lett. B **478**, 447 (2000)
36. F. Karsch: in *AIP Conf. Proc. (PANIC05)*, Vol. 842, p. 20 (Am. Inst. Phys., Melville, New York 2006)
37. P. de Forcrand et al.: Phys. Rev. D **63**, 054501 (2001)
38. I. Wetzorke et al.: Nucl. Phys. (Proc.Suppl.) **106**, 510 (2002)
39. M. Asakawa, T. Hatsuda, and Y. Nakahara: Nucl. Phys. A **715**, 863 (2003)
40. J. B. Kogut, J. F. Lagae, and D. K. Sinclair: Phys. Rev. D **58**, 054504 (1998)
41. I. Pushkina et al.: Phys. Lett. B **609**, 265 (2005)
42. S. Wissel et al.: PoS **LAT2005**, 164 (2005)
43. R. V. Gavai, S. Gupta, and R. Lacaze: PoS **LAT2006**, 135 (2006)
44. P. Hasenfratz and F. Karsch: Phys. Lett. B **125**, 308 (1983)
45. J. Kogut, H. Matsuoka, M. Stone, and H. W. Wyld: Nucl. Phys. B **225** [**FS9**], 93 (1983)
46. N. Bilic and R. V. Gavai: Z. Phys. C **23**, 77 (1984)
47. E. Dagotto, A. Moreo, R. L. Sugar, and D. Toussaint: Phys. Rev. B **41**, 811 (1990)
48. A. Hasenfratz and D. Toussaint: Nucl. Phys. B **371**, 539 (1992)
49. M. G. Alford, A. Kapustin, and F. Wilczek: Phys. Rev. D **59**, 054502 (1999)
50. D. T. Son and M. A. Stephanov: Phys. Rev. Lett. **86**, 592 (2001)
51. J. Bloch and T. Wettig: Phys. Rev. Lett. **97**, 012003 (2006)
52. J. Bloch and T. Wettig: Phys. Rev. D **76**, 114511 (2007)

53. C. Gattringer and L. Liptak: Phys. Rev. D **76**, 054502 (2007)
54. D. Banerjee, R. V. Gavai, and S. Sharma: Phys. Rev. D **78**, 014506 (2008)
55. K. Rajagopal and F. Wilczek: in *At the Frontiers of Physics / Handbook of QCD*, edited by M. Shifman, Vol. 3, p. 2061 (World Scientific, Singapore 2001)
56. M. Alford: Ann. Rev. Nucl. Part. Sci. **51**, 131 (2001)
57. M. G. Alford, K. Rajagopal, and F. Wilczek: Phys. Lett. B **422**, 247 (1998)
58. R. Rapp, T. Schäfer, E. V. Shuryak, and M. Velkovsky: Phys. Rev. Lett. **81**, 53 (1998)
59. M. A. Halasz et al.: Phys. Rev. D **58**, 096007 (1998)
60. B. Klein, D. Toublan, and J. J. M. Verbaarschot: Phys. Rev. D **68**, 014009 (2003)
61. M. G. Alford, K. Rajagopal, and F. Wilczek: Nucl. Phys. B **537**, 443 (1999)
62. I. M. Barbour et al.: Nucl. Phys. B **275**, 286 (1986)
63. J. B. Kogut, M.-P. Lombardo, and D. Sinclair: Phys. Rev. D **51**, 1282 (1995)
64. I. M. Barbour et al.: Nucl. Phys. (Proc. Suppl.) **60A**, 220 (1998)
65. M. A. Stephanov: Phys. Rev. Lett. **76**, 4472 (1996)
66. P. Ginsparg: Nucl. Phys. B **170**, 388 (1980)
67. T. Appelquist and R. D. Pisarski: Phys. Rev. D **23**, 2305 (1981)
68. K. Kajantie et al.: Phys. Rev. Lett. **79**, 3130 (1997)
69. A. Hart, M. Laine, and O. Philipsen: Nucl. Phys. B **586**, 443 (2000)
70. A. Hart, M. Laine, and O. Philipsen: Phys. Lett. B **505**, 141 (2001)
71. K. Kajantie, M. Laine, K. Rummukainen, and Y. Schröder: JHEP **04**, 036 (2003)
72. A. M. Ferrenberg and R. H. Swendsen: Phys. Rev. Lett. **63**, 1195 (1989)
73. A. M. Ferrenberg and R. H. Swendsen: Phys. Rev. Lett. **63**, 1658 (1989)
74. Z. Fodor and S. D. Katz: Phys. Lett. B **534**, 87 (2002)
75. Z. Fodor and S. D. Katz: JHEP **0203**, 014 (2002)
76. Z. Fodor and S. D. Katz: JHEP **0404**, 050 (2004)
77. S. Ejiri: Phys. Rev. D **73**, 054502 (2006)
78. P. de Forcrand and S. Kratochvila: Nucl. Phys. (Proc. Suppl.) **153**, 62 (2006)
79. P. de Forcrand and O. Philipsen: JHEP **0701**, 077 (2007)
80. S. Gottlieb et al.: Phys. Rev. D **38**, 2888 (1988)
81. S. Choe et al.: Phys. Rev. D **65**, 054501 (2002)
82. R. V. Gavai and S. Gupta: Phys. Rev. D **68**, 034506 (2003)
83. C. R. Allton et al.: Phys. Rev. D **71**, 054508 (2005)
84. C. R. Allton et al.: Phys. Rev. D **68**, 014507 (2003)
85. R. V. Gavai and S. Gupta: Phys. Rev. D **71**, 114014 (2005)
86. C. R. Allton et al.: Phys. Rev. D **66**, 074507 (2002)
87. A. Roberge and N. Weiss: Nucl. Phys. B **275**, 734 (1986)
88. P. de Forcrand and O. Philipsen: Nucl. Phys. B **642**, 290 (2002)
89. M. D'Elia and M. P. Lombardo: Phys. Rev. D **67**, 014505 (2003)
90. M. P. Lombardo: Nucl. Phys. B (Proc. Suppl.) **83**, 375 (2000)
91. P. de Forcrand and O. Philipsen: Nucl. Phys. B **673**, 170 (2003)
92. M. D'Elia and M. P. Lombardo: Phys. Rev. D **70**, 074509 (2004)
93. M. P. Lombardo: PoS **LAT2005**, 168 (2005)
94. V. Azcoiti, G. D. Carlo, A. Galante, and V. Laliena: JHEP **0412**, 010 (2004)
95. V. Azcoiti, G. D. Carlo, A. Galante, and V. Laliena: Nucl. Phys. B **723**, 77 (2005)

96. S. Kratochvila and P. de Forcrand: Nucl. Phys. B (Proc. Suppl.) **140**, 514 (2005)
97. S. Kratochvila and P. de Forcrand: PoS **LAT2005**, 167 (2005)
98. A. Alexandru, M. Faber, I. Horváth, and K. F. Liu: Phys. Rev. D **72**, 114513 (2005)
99. A. Li, A. Alexandru, and K.-F. Liu: PoS **LAT2006**, 030 (2006)
100. A. Alexandru, A. Li, and K.-F. Liu: PoS **LATTICE2007**, 167 (2007)
101. A. Li, A. Alexandru, and K.-F. Liu: PoS **LATTICE2007**, 203 (2007)
102. J. Danzer and C. Gattringer: Phys. Rev. D **78**, 114508 (2008)
103. X. Meng, A. Li, A. Alexandru, and K.-F. Liu: PoS **LATTICE2008**, 032 (2008)

# A

# Appendix

## A.1 The Lie groups SU($N$)

In this appendix we collect basic definitions and conventions for the Lie groups
SU($N$) – the special unitary groups – and the corresponding Lie algebras
su($N$). For a more detailed presentation we refer the reader to [1–3].

### A.1.1 Basic properties

The defining representation of SU($N$) is given by complex $N \times N$ matrices
which are unitary and have determinant 1. This set of matrices is closed under
matrix multiplication: Let $\Omega_1$ and $\Omega_2$ be elements of SU($N$), i.e., they obey
$\Omega_i^\dagger = \Omega_i^{-1}$ and $\det[\Omega_i] = 1$. Using standard linear algebra manipulations we
obtain

$$(\Omega_1 \Omega_2)^\dagger = \Omega_2^\dagger \Omega_1^\dagger = \Omega_2^{-1} \Omega_1^{-1} = (\Omega_1 \Omega_2)^{-1} \ ,$$
$$\det[\Omega_1 \Omega_2] = \det[\Omega_1] \det[\Omega_2] = 1 \tag{A.1}$$

and have thus established that also the product of two SU($N$) matrices is
an SU($N$) matrix. The unit matrix is also in SU($N$) and for each matrix in
SU($N$) there exists an inverse (the hermitian conjugate matrix). Thus, the set
SU($N$) forms a group. Since the group operation – the matrix multiplication –
is not commutative the groups SU($N$) are *non-abelian groups*.

### A.1.2 Lie algebra

Let us now count how many real parameters are needed to describe the matrices in SU($N$). A complex $N \times N$ matrix has $2N^2$ real parameters. The
requirement of unitarity introduces $N^2$ independent conditions which the parameters have to obey. One more parameter is used for obeying the determinant condition such that one needs a total of $N^2 - 1$ real parameters for
describing SU($N$) matrices.

Gattringer, C., Lang, C.B.: *Appendix*. Lect. Notes Phys. **788**, 327–336 (2010)
DOI 10.1007/978-3-642-01850-3_BM2    © Springer-Verlag Berlin Heidelberg 2010

A convenient way of representing $SU(N)$ matrices is to write them as exponentials of basis matrices $T_j$, the so-called *generators*. In particular, we write an element $\Omega$ of $SU(N)$ as

$$\Omega = \exp\left(i \sum_{j=1}^{N^2-1} \omega^{(j)} T_j\right) , \qquad (A.2)$$

where $\omega^{(j)}, j = 1, 2, \ldots, N^2 - 1$, are the real numbers needed to parameterize $\Omega$. We remark that the parameters $\omega^{(j)}$ can be changed continuously, making $SU(N)$ so-called *Lie groups* that are groups whose elements depend continuously on their parameters. In order to cover all of the group space, the parameters have to be varied only over finite intervals, making the groups $SU(N)$ so-called *compact* Lie groups.

The generators $T_j, j = 1, 2, \ldots, N^2 - 1$, are chosen as traceless, complex, and hermitian $N \times N$ matrices obeying the normalization condition

$$\text{tr}\,[T_j\,T_k] = \frac{1}{2}\delta_{jk} . \qquad (A.3)$$

In addition, they are related among each other by an algebra of commutation relations

$$[T_j\,,\,T_k] = i\,f_{jkl}\,T_l . \qquad (A.4)$$

The completely anti-symmetric coefficients $f_{jkl}$ are the so-called *structure constants*. Below we will give an explicit representation of the generators for the groups $SU(2)$ and $SU(3)$.

Let us verify that the representation (A.2) indeed describes elements of $SU(N)$. Using the facts that the generators are hermitian and that the $\omega^{(j)}$ are real, one finds that hermitian conjugation of the right-hand side of (A.2) simply produces an extra minus sign in the exponent (from the complex conjugation of i). Thus, it is obvious that (A.2) implies $\Omega^\dagger = \Omega^{-1}$. To show that the determinant equals 1, we use the equation

$$\det \Omega = \exp\left(\text{tr}\,[\ln \Omega]\right) = \exp\left(i \sum_{j=1}^{N^2-1} \omega^{(j)}\,\text{tr}\,T_j\right) = e^0 = 1 , \qquad (A.5)$$

where in first step we have used a formula for the determinant (see (A.54) below) and in the third step we have used the fact that the $T_j$ are traceless.

Not only the group elements but also the exponents of our representation (A.2) have an interesting structure. The linear combinations

$$\sum_{j=1}^{N^2-1} \omega^{(j)} T_j \qquad (A.6)$$

of the $T_j$ form the so-called *Lie algebra* su($N$). Their commutation properties are governed by the relations (A.4). Elements of su($N$) are also complex $N \times N$

matrices but have properties different from the elements of the group. One important difference is the fact that the unit matrix is contained in the group (all $\omega^{(j)} = 0$), while it is not an element of the algebra (all $T_j$ are traceless).

### A.1.3 Generators for SU(2) and SU(3)

The standard representation of the generators for SU(2) is given by

$$T_j = \frac{1}{2}\sigma_j , \tag{A.7}$$

with the Pauli matrices

$$\sigma_1 = \begin{bmatrix} 0 & 1 \\ 1 & 0 \end{bmatrix} , \quad \sigma_2 = \begin{bmatrix} 0 & -i \\ i & 0 \end{bmatrix} , \quad \sigma_3 = \begin{bmatrix} 1 & 0 \\ 0 & -1 \end{bmatrix} . \tag{A.8}$$

In this case the structure constants are particularly simple, given by the completely anti-symmetric tensor, i.e., $f_{jkl} = \epsilon_{jkl}$.

For SU(3) the generators are given by

$$T_j = \frac{1}{2}\lambda_j . \tag{A.9}$$

The *Gell–Mann matrices* $\lambda_j$ are $3 \times 3$ generalizations of the Pauli matrices:

$$\lambda_1 = \begin{bmatrix} 0 & 1 & 0 \\ 1 & 0 & 0 \\ 0 & 0 & 0 \end{bmatrix} , \quad \lambda_2 = \begin{bmatrix} 0 & -i & 0 \\ i & 0 & 0 \\ 0 & 0 & 0 \end{bmatrix} , \quad \lambda_3 = \begin{bmatrix} 1 & 0 & 0 \\ 0 & -1 & 0 \\ 0 & 0 & 0 \end{bmatrix} ,$$

$$\lambda_4 = \begin{bmatrix} 0 & 0 & 1 \\ 0 & 0 & 0 \\ 1 & 0 & 0 \end{bmatrix} , \quad \lambda_5 = \begin{bmatrix} 0 & 0 & -i \\ 0 & 0 & 0 \\ i & 0 & 0 \end{bmatrix} , \quad \lambda_6 = \begin{bmatrix} 0 & 0 & 0 \\ 0 & 0 & 1 \\ 0 & 1 & 0 \end{bmatrix} ,$$

$$\lambda_7 = \begin{bmatrix} 0 & 0 & 0 \\ 0 & 0 & -i \\ 0 & i & 0 \end{bmatrix} , \quad \lambda_8 = \frac{1}{\sqrt{3}} \begin{bmatrix} 1 & 0 & 0 \\ 0 & 1 & 0 \\ 0 & 0 & -2 \end{bmatrix} . \tag{A.10}$$

### A.1.4 Derivatives of group elements

Let us now show an important property of derivatives of group elements. If $\Omega(\omega)$ is an element of SU($N$) then

$$M_k(\omega) = i\frac{\partial\Omega(\omega)}{\partial\omega^{(k)}} \, \Omega(\omega)^\dagger \in su(N) , \tag{A.11}$$

i.e., the derivative times the hermitian conjugate is in the Lie algebra. In order to prove this statement we have to show the defining properties of Lie algebra elements, i.e., we must show that $M_k(\omega)$ is hermitian and traceless.

Showing the hermiticity of $M_k(\omega)$ is straightforward. By differentiating $\Omega(\omega)\Omega(\omega)^\dagger = \mathbb{1}$ with respect to $\omega^{(k)}$ one finds

$$\frac{\partial \Omega(\omega)}{\partial \omega^{(k)}}\, \Omega(\omega)^{\dagger} \;+\; \Omega(\omega)\, \frac{\partial \Omega(\omega)^{\dagger}}{\partial \omega^{(k)}} \;=\; 0\,. \tag{A.12}$$

Thus

$$M_k(\omega)^{\dagger} \;=\; \left(\mathrm{i}\,\frac{\partial \Omega(\omega)}{\partial \omega^{(k)}}\, \Omega(\omega)^{\dagger}\right)^{\dagger} \;=\; -\mathrm{i}\,\Omega(\omega)\,\frac{\partial \Omega(\omega)^{\dagger}}{\partial \omega^{(k)}} \;=\; \mathrm{i}\,\frac{\partial \Omega(\omega)}{\partial \omega^{(k)}}\, \Omega(\omega)^{\dagger} \;=\; M_k(\omega)\,, \tag{A.13}$$

where we used (A.12) in the third step.

In order to show that $M_k(\omega)$ is traceless we use the fact that the determinant of a SU($N$) matrix equals to 1 and we differentiate $\det[\Omega(\omega)]$ with respect to $\omega^{(k)}$. We obtain

$$\begin{aligned}
0 \;&=\; \frac{\partial \det[\Omega(\omega)]}{\partial \omega^{(k)}} \;=\; \frac{\partial \det[\Omega(\omega)]}{\partial \Omega(\omega)_{ab}}\, \frac{\partial \Omega(\omega)_{ab}}{\partial \omega^{(k)}} \\[1ex]
&=\; \det[\Omega(\omega)]\, \left(\Omega(\omega)^{-1}\right)_{ba}\, \frac{\partial \Omega(\omega)_{ab}}{\partial \omega^{(k)}} \;=\; \mathrm{tr}\left[\frac{\partial \Omega(\omega)}{\partial \omega^{(k)}}\, \Omega(\omega)^{\dagger}\right]\,, \quad \text{(A.14)}
\end{aligned}$$

where in the first step we applied the chain rule for derivatives. In the second step we used a standard formula for the derivative of the determinant $\det[\Omega]$ with respect to an entry $\Omega_{ab}$ of the matrix $\Omega$. In the third step we used $\det[\Omega] = 1$ and $\Omega^{-1} = \Omega^{\dagger}$. Equation (A.14) establishes that $M_k(\omega)$ is also traceless and thus we have shown $M_k(\omega) \in \mathrm{su}(N)$.

From (A.11) it follows that for the gauge transformation matrices $\Omega(x)$ with coefficients $\omega^{(k)}(x)$, depending on the space–time coordinate $x$, the combination $\mathrm{i}\,(\partial_\mu \Omega(x))\,\Omega(x)^{\dagger}$ is in the Lie algebra, since

$$\mathrm{i}\,(\partial_\mu \Omega(x))\,\Omega(x)^{\dagger} \;=\; \sum_{k}\left(\mathrm{i}\left(\frac{\partial}{\partial \omega^{(k)}(x)}\Omega\left(\omega(x)\right)\right)\Omega\left(\omega(x)\right)^{\dagger}\right)\partial_\mu \omega^{(k)}(x)\,, \tag{A.15}$$

and the right-hand side is a linear combination of $\mathrm{su}(N)$ elements with real coefficients $\partial_\mu \omega^{(k)}(x)$.

## A.2 Gamma matrices

The Euclidean gamma matrices $\gamma_\mu$, $\mu = 1, 2, 3, 4$ can be constructed from the Minkowski gamma matrices $\gamma_\mu^M$, $\mu = 0, 1, 2, 3$. The latter obey

$$\{\gamma_\mu^M, \gamma_\nu^M\} \;=\; 2\,g_{\mu\nu}\,\mathbb{1}\,, \tag{A.16}$$

with the metric tensor given by $g_{\mu\nu} = \mathrm{diag}(1, -1, -1, -1)$ and $\mathbb{1}$ is the $4 \times 4$ unit matrix. Thus when we define the Euclidean matrices $\gamma_\mu$ by setting

$$\gamma_1 = -\mathrm{i}\gamma_1^M\,,\ \gamma_2 = -\mathrm{i}\gamma_2^M\,,\ \gamma_3 = -\mathrm{i}\gamma_3^M\,,\ \gamma_4 = \gamma_0^M\,, \tag{A.17}$$

we obtain the Euclidean anti-commutation relations

$$\{\gamma_\mu, \gamma_\nu\} = 2\delta_{\mu\nu}\, \mathbb{1}\, . \tag{A.18}$$

In addition to the matrices $\gamma_\mu$, $\mu = 1, 2, 3, 4$ we define the matrix $\gamma_5$ as the product

$$\gamma_5 = \gamma_1\gamma_2\gamma_3\gamma_4\, . \tag{A.19}$$

The matrix $\gamma_5$ anti-commutes with all other gamma matrices $\gamma_\mu$, $\mu = 1, 2, 3, 4$ and obeys $\gamma_5^2 = \mathbb{1}$.

An explicit representation of the Euclidean gamma matrices can be obtained from a representation of the Minkowski gamma matrices (see, e.g., [4]). Here we give the so-called chiral representation where $\gamma_5$ (the chirality operator) is diagonal:

$$\gamma_{1,2,3} = \begin{bmatrix} 0 & -i\sigma_{1,2,3} \\ i\sigma_{1,2,3} & 0 \end{bmatrix}, \ \gamma_4 = \begin{bmatrix} 0 & \mathbb{1}_2 \\ \mathbb{1}_2 & 0 \end{bmatrix}, \ \gamma_5 = \begin{bmatrix} \mathbb{1}_2 & 0 \\ 0 & -\mathbb{1}_2 \end{bmatrix}, \tag{A.20}$$

where the $\sigma_j$ are the Pauli matrices (A.8) and $\mathbb{1}_2$ is the $2 \times 2$ unit matrix. More explicitly the Euclidean gamma matrices read

$$\gamma_1 = \begin{bmatrix} 0 & 0 & 0 & -i \\ 0 & 0 & -i & 0 \\ 0 & i & 0 & 0 \\ i & 0 & 0 & 0 \end{bmatrix}, \ \gamma_2 = \begin{bmatrix} 0 & 0 & 0 & -1 \\ 0 & 0 & 1 & 0 \\ 0 & 1 & 0 & 0 \\ -1 & 0 & 0 & 0 \end{bmatrix}, \ \gamma_3 = \begin{bmatrix} 0 & 0 & -i & 0 \\ 0 & 0 & 0 & i \\ i & 0 & 0 & 0 \\ 0 & -i & 0 & 0 \end{bmatrix},$$

$$\gamma_4 = \begin{bmatrix} 0 & 0 & 1 & 0 \\ 0 & 0 & 0 & 1 \\ 1 & 0 & 0 & 0 \\ 0 & 1 & 0 & 0 \end{bmatrix}, \ \gamma_5 = \begin{bmatrix} 1 & 0 & 0 & 0 \\ 0 & 1 & 0 & 0 \\ 0 & 0 & -1 & 0 \\ 0 & 0 & 0 & -1 \end{bmatrix}. \tag{A.21}$$

In addition to the anti-commutation relation (A.18) the gamma matrices obey (here $\mu = 1, \ldots, 5$)

$$\gamma_\mu = \gamma_\mu^\dagger = \gamma_\mu^{-1}\, . \tag{A.22}$$

When we discuss charge conjugation, we need the charge conjugation matrix $C$ defined through the relations ($\mu = 1, \ldots, 4$)

$$C\gamma_\mu C^{-1} = -\gamma_\mu^T\, . \tag{A.23}$$

Using the explicit form (A.21) it is easy to see that in the chiral representation (A.20) the charge conjugation matrix is given by

$$C = i\gamma_2\gamma_4\, . \tag{A.24}$$

It obeys

$$C = C^{-1} = C^\dagger = -C^T\, . \tag{A.25}$$

We finally quote a simple formula for the inverse of linear combinations of gamma matrices ($a, b_\mu \in \mathbb{R}$):

$$\left(a\mathbb{1} + i\sum_{\mu=1}^{4}\gamma_\mu b_\mu\right)^{-1} = \frac{a\mathbb{1} - i\sum_{\mu=1}^{4}\gamma_\mu b_\mu}{a^2 + \sum_{\mu=1}^{4} b_\mu^2}\, . \tag{A.26}$$

This formula can be verified by multiplying both sides with $a\mathbb{1} + i\sum_\mu \gamma_\mu b_\mu$.

## A.3 Fourier transformation on the lattice

The goal of this appendix is to discuss the Fourier transform $\widetilde{f}(p)$ of functions $f(n)$ defined on the lattice $\Lambda$. The lattice is given by

$$\Lambda = \{n = (n_1, n_2, n_3, n_4) \mid n_\mu = 0, 1, \ldots N_\mu - 1\}, \qquad (A.27)$$

and in most of our applications we have $N_1 = N_2 = N_3 = N$, $N_4 = N_T$. For the total number of lattice points we introduce the abbreviation

$$|\Lambda| = N_1 N_2 N_3 N_4. \qquad (A.28)$$

We impose toroidal boundary conditions

$$f(n + \hat{\mu} N_\mu) = e^{i 2\pi \theta_\mu} f(n) \qquad (A.29)$$

for each of the directions $\mu$. Here $\hat{\mu}$ denotes the unit vector in $\mu$-direction. Directions with periodic boundary conditions have $\theta_\mu = 0$, anti-periodic boundary conditions correspond to $\theta_\mu = 1/2$.

The momentum space $\widetilde{\Lambda}$, which corresponds to the lattice $\Lambda$ with the boundary conditions (A.29), is defined as

$$\widetilde{\Lambda} = \left\{ p = (p_1, p_2, p_3, p_4) \mid p_\mu = \frac{2\pi}{a N_\mu}(k_\mu + \theta_\mu), k_\mu = -\frac{N_\mu}{2} + 1, \ldots, \frac{N_\mu}{2} \right\}. \qquad (A.30)$$

The boundary phases $\theta_\mu$ have to be included in the definition of the momenta $p_\mu$ such that the plane waves

$$\exp(i\, p \cdot na) \quad \text{with} \quad p \cdot n = \sum_{\mu=1}^{4} p_\mu n_\mu \qquad (A.31)$$

also obey the boundary conditions (A.29).

The basic formula, underlying Fourier transformation on the lattice, is (here $l$ is an integer with $0 \le l \le N - 1$)

$$\frac{1}{N} \sum_{j=-N/2+1}^{N/2} \exp\left(i\frac{2\pi}{N} lj\right) = \frac{1}{N} \sum_{j=0}^{N-1} \exp\left(i\frac{2\pi}{N} lj\right) = \delta_{l0}. \qquad (A.32)$$

For $l = 0$ this formula is trivial. For $l \neq 0$ (A.32) follows from applying the well-known algebraic identity

$$\sum_{j=0}^{N-1} q^j = \frac{1 - q^N}{1 - q} \qquad \text{to} \qquad q = \exp\left(i\frac{2\pi}{N} l\right). \qquad (A.33)$$

We can combine four of the 1D sums in (A.32) to obtain the following identities:

$$\frac{1}{|\Lambda|} \sum_{p \in \tilde{\Lambda}} \exp\left(\mathrm{i}\, p \cdot (n - n')a\right) = \delta(n - n') = \delta_{n_1 n_1'}\, \delta_{n_2 n_2'}\, \delta_{n_3 n_3'}\, \delta_{n_4 n_4'} \,, \quad (A.34)$$

$$\frac{1}{|\Lambda|} \sum_{n \in \Lambda} \exp\left(\mathrm{i}(p - p') \cdot na\right) = \delta(p - p') \equiv \delta_{k_1 k_1'}\, \delta_{k_2 k_2'}\, \delta_{k_3 k_3'}\, \delta_{k_4 k_4'} \,. \quad (A.35)$$

We stress that the right-hand side of (A.35) is a product of four Kronecker deltas for the integers $k_\mu$ which label the momentum components $p_\mu$ (compare (A.30)).

If we now define the Fourier transform

$$\tilde{f}(p) = \frac{1}{\sqrt{|\Lambda|}} \sum_{n \in \Lambda} f(n)\, \exp\left(-\mathrm{i}\, p \cdot na\right) \,, \quad (A.36)$$

we find for the inverse transformation

$$f(n) = \frac{1}{\sqrt{|\Lambda|}} \sum_{p \in \tilde{\Lambda}} \tilde{f}(p)\, \exp\left(\mathrm{i}\, p \cdot na\right) \,. \quad (A.37)$$

The last equation follows immediately from inserting (A.36) in (A.37) and using (A.34).

## A.4 Wilson's formulation of lattice QCD

In this appendix we collect the defining formulas for Wilson's formulation of QCD on the lattice. The dynamical variables are the group-valued link variables $U_\mu(n)$ and the Grassmann-valued fermion fields $\psi^{(f)}(n)_{\alpha\, c}$, $\overline{\psi}^{(f)}(n)_{\alpha\, c}$. They live on the links, respectively the sites of our lattice (A.27). Vacuum expectation values are calculated according to

$$\langle O \rangle = \frac{1}{Z} \int \mathcal{D}\left[\psi, \overline{\psi}\right] \mathcal{D}[U]\, \mathrm{e}^{-S_F[\psi, \overline{\psi}, U] - S_G[U]}\, O[\psi, \overline{\psi}, U] \,, \quad (A.38)$$

where the partition function is given by

$$Z = \int \mathcal{D}\left[\psi, \overline{\psi}\right] \mathcal{D}[U]\, \mathrm{e}^{-S_F[\psi, \overline{\psi}, U] - S_G[U]} \,. \quad (A.39)$$

The measures over fermion and gauge fields are products over the measures for the individual field variables:

$$\mathcal{D}\left[\psi, \overline{\psi}\right] = \prod_{n \in \Lambda} \prod_{f, \alpha, c} \mathrm{d}\psi^{(f)}(n)_{\alpha\, c}\, \mathrm{d}\overline{\psi}^{(f)}(n)_{\alpha\, c} \,, \quad \mathcal{D}[U] = \prod_{n \in \Lambda} \prod_{\mu=1}^{4} \mathrm{d}U_\mu(n) \,.$$

$$(A.40)$$

For the individual link variables $U_\mu(n)$ one uses the Haar measure discussed in Sect. 3.1. For the fermions the rules for Grassmann integration from Sect. 5.1 apply. The gauge field action for gauge group SU($N$) is given by

$$S_G[U] = \frac{\beta}{N} \sum_{n\in\Lambda} \sum_{\mu<\nu} \text{Re tr}\left[\mathbb{1} - U_{\mu\nu}(n)\right], \tag{A.41}$$

where the plaquettes are defined as

$$U_{\mu\nu}(n) = U_\mu(n)\,U_\nu(n+\hat\mu)\,U_{-\mu}(n+\hat\mu+\hat\nu)\,U_{-\nu}(n+\hat\nu)$$
$$= U_\mu(n)\,U_\nu(n+\hat\mu)\,U_\mu(n+\hat\nu)^\dagger\,U_\nu(n)^\dagger\,. \tag{A.42}$$

The fermion action is a sum over $N_f$ flavors:

$$S_F[\psi,\overline\psi,U] = \sum_{f=1}^{N_f} a^4 \sum_{n,m\in\Lambda} \overline\psi^{(f)}(n)\,D^{(f)}(n|m)\,\psi^{(f)}(m) \tag{A.43}$$

and the lattice Dirac operator is given by

$$D^{(f)}(n|m)_{\substack{\alpha\,\beta \\ a\,b}} = \left(m^{(f)} + \frac{4}{a}\right)\delta_{\alpha\beta}\,\delta_{ab}\,\delta_{n,m} - \frac{1}{2a}\sum_{\mu=\pm1}^{\pm4}(\mathbb{1} - \gamma_\mu)_{\alpha\beta}\,U_\mu(n)_{ab}\,\delta_{n+\hat\mu,m}\,. \tag{A.44}$$

In (A.42) and in the last equation we use the conventions

$$\gamma_{-\mu} = -\gamma_\mu\,, \quad U_{-\mu}(n) = U_\mu(n-\hat\mu)^\dagger\,, \quad \mu = 1,2,3,4\,. \tag{A.45}$$

We remark that Wilson's Dirac operator (A.44) is $\gamma_5$-hermitian, i.e., it obeys

$$\gamma_5\,D\,\gamma_5 = D^\dagger\,. \tag{A.46}$$

## A.5 A few formulas for matrix algebra

In quantum mechanics one usually deals with hermitian or unitary matrices, while in lattice QCD often more general matrices occur. In this appendix we list a few results for general complex matrices together with short remarks concerning their proof (for a more detailed account see, e.g., [5]).

The basic result for general complex matrices is that they are unitarily equivalent to upper triangular matrices: Let $M$ be a complex-valued $N \times N$ matrix. Then there exits a unitary matrix $U$ and an upper triangular matrix $T$, such that

$$U^\dagger M U = T\,. \tag{A.47}$$

This result can be proven by induction in $N$. The elements $t_j$ on the diagonal of $T$ are the roots of the characteristic polynomial of $M$ since

$$P(\lambda) = \det[M - \lambda \mathbb{1}] = \det[T - \lambda \mathbb{1}] = \prod_{j=1}^{N} (t_j - \lambda) . \qquad (A.48)$$

An important consequence of this result is a unique classification of matrices that can be diagonalized with a unitary transformation. A complex matrix $M$ is called *normal* if it commutes with its hermitian conjugate, i.e., $[M, M^\dagger] = 0$. It is obvious that hermitian or unitary matrices are normal. The announced result is: If and only if $M$ is normal, then there exists a unitary matrix $U$ such that

$$U^\dagger M U = D , \qquad (A.49)$$

where $D$ is diagonal. It is straightforward to see that a matrix $M$ which is unitarily equivalent to a diagonal matrix is normal. To prove the other direction we first note that the normality of $M$ implies the normality of the triangular matrix $T$ corresponding to $M$. When evaluating explicitly the two sides of the normality condition, $T^\dagger T = T T^\dagger$, for the upper triangular matrix $T$, one concludes that $T$ must be diagonal and the statement is proven. Equations (A.49) and (A.48) imply that a normal matrix has a complete set of orthonormal eigenvectors, the columns of $U$.

The existence of a complete orthonormal set of eigenvectors $v^{(j)}$ with eigenvalues $\lambda^{(j)}$ can be used to represent the matrix $M$ in the form

$$M = \sum_{j=1}^{N} \lambda^{(j)} v^{(j)} v^{(j)\dagger} , \qquad (A.50)$$

the so-called *spectral representation*. On the right-hand side of this equation matrix/vector notation was used to write the dyadic product $v^{(j)} v^{(j)\dagger}$. The spectral representation of the matrix can be used to define a function of $M$ in terms of the function for the eigenvalues, if this exists. This gives rise to the *spectral theorem*

$$f(M) = \sum_{j=1}^{N} f\left(\lambda^{(j)}\right) v^{(j)} v^{(j)\dagger} . \qquad (A.51)$$

We finally discuss a formula for the expansion of the determinant:

$$\det[\mathbb{1} - M] = \exp\left(\operatorname{tr}[\ln(\mathbb{1} - M)]\right) . \qquad (A.52)$$

In this equation $M$ is a complex matrix and the logarithm (where it exists) is defined through its series expansion. The proof of (A.52) applies (A.47):

$$\det[\mathbb{1} - M] = \det[\mathbb{1} - T] = \prod_{j=1}^{N} (1 - t_j) = \exp\left(\sum_{j=1}^{N} \ln(1 - t_j)\right) \qquad (A.53)$$

$$= \exp\left(-\sum_{j=1}^{N} \sum_{n=1}^{\infty} \frac{1}{n} (t_j)^n\right) = \exp\left(-\sum_{n=1}^{\infty} \frac{1}{n} \operatorname{tr}[T^n]\right) = \exp\left(\operatorname{tr}[\ln(\mathbb{1} - M)]\right) .$$

In the fifth step we have used the fact that when evaluating powers of a triangular matrix the diagonal elements do not mix with other entries of the matrix. In the last step we used $\mathrm{tr}[T^n] = \mathrm{tr}[M^n]$ which follows from (A.47).

Since a matrix $A$ may always be written as $A = \mathbb{1} - M$, the result (A.52) is often stated as

$$\det[A] = \exp\left(\mathrm{tr}[\ln A]\right) . \tag{A.54}$$

# References

1. H. Georgi: *Lie Algebras in Particle Physics* (Benjamin/Cummings, Reading, Massachusetts 1982)
2. H. F. Jones: *Groups, Representations and Physics* (Hilger, Bristol 1990)
3. M. Hamermesh: *Group Theory and Its Application to Physical Problems* (Addison-Wesley, Reading, Massachusetts 1964)
4. M. E. Peskin and D. V. Schroeder: *An Introduction to Quantum Field Theory* (Addison-Wesley, Reading, Massachusetts 1995)
5. P. Lancaster: *Theory of Matrices* (Academic Press, New York 1969)

# Index